Lecture Notes in Physics

360

Y. Mellier B. Fort G. Soucail (Eds.)

Gravitational Lensing

Proceedings of a Workshop
Held in Toulouse, France
September 13–15, 1989

Springer-Verlag Berlin Heidelberg GmbH

Editors
Yannick Mellier
Bernard Fort
Geneviève Soucail
Observatoire Midi Pyrénées
14, avenue Belin, F-31400 Toulouse, France

ISBN 978-3-662-13783-3 ISBN 978-3-540-47087-8 (eBook)
DOI 10.1007/978-3-540-47087-8

Originally published by Springer-Verlag Berlin Heidelberg New York in 1990
Softcover reprint of the hardcover 1st edition 1990

2158/3140-543210 – Printed on acid-free paper

Preface

During the conference on gravitational lenses organized in honor of Bernard Burke in June 1989, the amount of new results and ideas presented was such that most of the participants realized that it would be necessary to meet again very soon. The possibility of organizing something in France was reinforced a few months later by some discussions with Israel Kovner during his visit to Toulouse. Finally we decided to set the date in September, since it is supposed to be the most pleasant month in southern France, with guaranteed nice and warm weather! Moreover, it was proposed to dedicate this meeting to one of the pioneers of the gravitational lens studies, namely Sjur Refsdal, who had the possibility of seeing that most of his earlier predictions have now been realized, 10 years after the discovery of the first multiple-imaged QSO.

We were happy to see how most of the participants answered the invitations so quickly and with enthusiasm. In the end 70 people attended the workshop, coming from all around the world. 34 papers were presented: all of them are included in this volume. They show the present status of the theoretical aspects of gravitational lensing as well as many observations of the phenomena related to it: multiple QSOs, surveys of QSOs and QSO–galaxy associations. The microlensing effect led to a fascinating debate, since some possible events have been detected recently, as did the time-delay measurement on the double-QSO. Finally 10 papers are related to the rings and arcs. It appeared that a half-day session alone was too short for the subject, which should now be considered a full sub-topic of gravitational lensing in view of all the spectacular new observations obtained at optical and radio wavelengths. We all got the feeling that the observations of gravitational distortion of the huge background galaxy population by the large scale overdensities of the universe, like the clusters of galaxies, will be very essential for the future.

In order to include some entertainment in the discussions, an excursion to the medieval city of Carcassonne was proposed, as well as a conference on some special points of history which influenced all of the South of France in the 13th century. The idea was to present something other than the French Revolution, evoked enough during the year of Bicentennial commemoration!

The workshop was held in the conference room of the Laboratoire d'Automatique et d'Analyse des Systèmes. We are very grateful to its director Mr. Coste for his welcome, and his secretary Mrs. Ippolito for all the facilities they offered during the workshop. The meeting was sponsored by the Centre National de la Recherche Scientifique, the Groupe de Recherche Cosmologie and Thomson. The financial organization was also greatly simplified with the help of the Banque Courtois. Finally, we thank very much the municipality of Toulouse for their reception in the beautiful rooms in the Capitole.

Last but not least, we would like to express our gratitude to some key people who participated in the organization of this workshop with efficency, namely Mireille Cailloux, Barbara Davoust, Blanche Lours and Sylvie Marc.

Observatoire Midi–Pyrénées
April 3, 1990

Bernard Fort
Yannick Mellier
Geneviève Soucail

CONTENTS

List of participants

Bartelmann M.	MPA, Garching, FRG
Bender R.	Heidelberg, FRG
Blandford R.D.	CIT, Pasadena, USA
Borgeest U.	Hamburg, FRG
Burke B.F.	MIT, Cambridge, USA
Corrigan, R.	Inst. of Astronomy, Cambridge, UK
Deguchi S.	Nobeyama Radio Obs., Nagano, Japan
Ellis R.	University of Durham, UK
Erdl H.	MPA, Garching, FRG
Falco E.	CfA, Cambridge, USA
Faraoni V.	SISSA, Trieste, Italy
Fort B.	Observatoire de Toulouse, France
Freyberg M.	MPA, Garching, FRG
Grieger B.	Hamburg, FRG
Grossman S.	Steward Obs., Tucson, USA
Hammer F.	Observatoire de Meudon, France
Harding M.E.	Inst. of Astronomy, Cambridge, UK
Heflin M.	MIT, Cambridge, USA
Hewett P.	Inst. of Astronomy, Cambridge, UK
Hewitt J.	Princeton, USA
Hindmarsh M.	University of Newcastle, UK
Kassiola A.	Observatoire de Toulouse, France
Kellogg M.E.	CfA, Cambridge, USA
Kochanek C.S.	CIT, Pasadena, USA
Kovner I.	Weizmann Inst., Israel
Kunze W.D.	Hamburg, FRG
Le Borgne J.F.	Observatoire de Toulouse, France
Le Fèvre O.	CFHT, Hawaii, USA
Lelièvre G.	Observatoire de Paris, France
Mc Breen B.	University College, Dublin, Ireland
Magain P.	Université de Liège, Belgium
Marchandon S.	Observatoire de Meudon, France
Mathez G.	Observatoire de Toulouse, France
Mellier Y.	Observatoire de Toulouse, France
Metcalfe L.	ESTEC, Noorwijk, Netherlands
Meylan G.	STScI, Baltimore, USA
Narasimha D.	Tata Inst., Bombay, India
Nemiroff R.	Naval Research Lab., Washington, USA
Nieto J.L.	Observatoire de Toulouse, France
Nityananda R.	Raman Inst., Bangalore, India
Nottale L.	Observatoire de Meudon, France
Picat J.P.	Observatoire de Toulouse, France
Pello, R.	Universidad de Barcelona, Spain
Petrosian V.	Stanford, USA

Reboul H.	Université de Montpellier, France
Refsdal S.	Hamburg, FRG
Sanahuja, B.	Universidad de Barcelona, Spain
Saust A.B.	Copenhagen University Obs., Denmark
Schild R.	CfA, Cambridge, USA
Schneider P.	MPA, Garching, FRG
Schramm J.	Hamburg, FRG
Schramm T.	Hamburg, FRG
Sebag J.	Obs. de Paris, France
Smette A.	ESO, Garching, FRG
Soucail G.	Observatoire de Toulouse, France
Stabell R.	Inst. of Theoretical Astrophysics, Oslo, Norway
Surdej J.	Université de Liège, Belgium
Swings J.P.	Université de Liège, Belgium
Thomas P.	University of Toronto, Canada
Tyson J.A.	AT& T Bell Labs., USA
Vanderiest C.	Observatoire de Meudon, France
Vauclair G.	Observatoire de Toulouse, France
Véron P.	Observatoire de Haute-Provence, France
Vilenkin A.	Tufts University, Medford, USA
Von Linde J.	Hamburg, FRG
Wambganss J.	MPA, Garching, FRG
Webster R.	University of Toronto, Canada
Witt H.J.	Hamburg, FRG
Wlérick G.	Observatoire de Meudon, France
Yee H.K.	Université de Toronto, Canada

FINAL PROGRAM

Wednesday, September 13

8h30	Registration / Poster setup	
9h15	Welcome address	M. Blanc

CHAIRMAN: R. Blandford

9h30	*Introductory talk*	R. Nityananda
10h15	*Determination of cosmological parameters with gravitational lenses*	S. Refsdal
10h30	Coffee Break	
11h	*Theoretical aspects of gravitational lensing (45 min)*	I. Kovner
12h	*When is lensing probable?*	R. Nemiroff
12h15	Lunch	

CHAIRMAN: P. Schneider

14h	*Gravitational redshift effects (30 min)*	L. Nottale
14h30	*What the present observed lens systems tell about the lensing agent?*	D. Narasimha
14h45	*Gravitational lens models with elliptical geometry*	T. Schramm
15h	Coffee Break	
15h30	*Observational aspects of gravitational lensing (40 min)*	J. Surdej
16h10	*A search for gravitationally lensed highly luminous quasars:*	
	1. observational strategy	J.P. Swings
16h20	*2. new results*	P. Magain
16h30	*An optical imaging search for gravitational lenses and quasar pairs*	G. Meylan
16h45	End of the session	
17h30	Reception at the Town Hall	

Thursday, September 14

CHAIRMAN: L. Nottale

8h45	*Quasar/galaxy associations (45 min)*	R. Webster
9h45	*Lensing effects on distant 3C sources*	O. LeFèvre
10h	*Gravitational lensing of quasar absorption line systems*	P. Thomas
10h15	Coffee Break	
10h45	*Radio gravitational lens surveys (45 min)*	B. Burke
11h45	*The theory and practise of radio ring lenses (30 min)*	C.S. Kochanek
12h15	*First epoch VLBI observation of the gravitational lens system: 2016+112*	M.B. Heflin
12h30	Lunch	
14h00	Visit to Carcassonne	
18h30	Conference on Cathare History	
19h30	Conference Dinner in St Félix-de-Lauragais	
22h	Return to Toulouse	

Friday, September 15

CHAIRMAN: S. Refsdal

8h45	*Gravitational microlensing (45 min)*	P. Schneider
9h45	*Observations of Microlensing in 2237+0305*	R.T. Corriga
10h	*Gravitational microlensing as a clue to quasar structure*	B. Grieger
10h15	Coffee Break	
10h30	*Spectrophotometry of 0957+561 and the "microlensing effect"*	C. Vanderriest
10h45	*The time delay of 0957+561 A,B: from 10 years of optical monitoring*	R. Schild
11h	*Statistical investigations of amplification near caustics*	H.J. Witt
11h15	Poster display (**CHAIRMAN: R. Ellis**)	
12h15	Lunch	

CHAIRMAN: J. Surdej

13h45	*Rich clusters of galaxies: a new class of gravitational lenses (30 min)*	B. Fort
14h15	*The background galaxy population: mapping dark matter in galaxy clusters (30 min)*	T. Tyson
14h45	*Redshift distribution of very faint galaxies: implications for lensing*	R. Ellis
15h	*Searching for cosmic strings*	M. Hindmars
15h15	*X-ray observations of gravitational lenses*	E.M. Kellog
15h30	Coffee Break	
16h	*Mass distribution in clusters based on gravitational lens models of luminous extended arcs*	V. Petrosian
16h15	*Modeling of the arcs in A370 and A2390*	Y. Mellier
16h30	*Gravitationally lensed images in Abell 370*	S. Grossman
16h45	*New views through the gravitational telescope*	F. Hammer
17h	*Spatial correlation in gravitationally lensed images*	S. Deguchi
17h15	*Concluding remarks*	R. Blandford
17h45	End of the meeting	

I

Gravitational Optics

PICTURES OF GRAVITATIONAL LENSING

Rajaram Nityananda
Raman Research Institute, Bangalore, India

Gravitational lenses seem to offer everything one could ask from an exciting branch of astronomy. New observations are pouring in. Theoretical interpretation, while by no means easy, is not impossible and some of the oldest and most important questions in cosmology are being addressed, if not yet answered. We will hear much about all this in our meeting, but I would like to touch upon something different which is almost an unexpected bonus. In modeling lenses, we are led to rather beautiful and unexpectedly rich questions in classical geometrical optics - a field we might otherwise tend to take for granted. Appropriately, one might say that the story began here in Toulouse more than three hundred years ago when Fermat invented the principle of least time. Not far away, in Paris, the Academy honoured Huyghens who used wavefronts so brilliantly to understand the whole range of optical phenomena. It seems very difficult to understand how more than a hundred years after this brilliant start, Malus could doubt that a family of rays reflected off two general curved mirrors was normal to a surface.

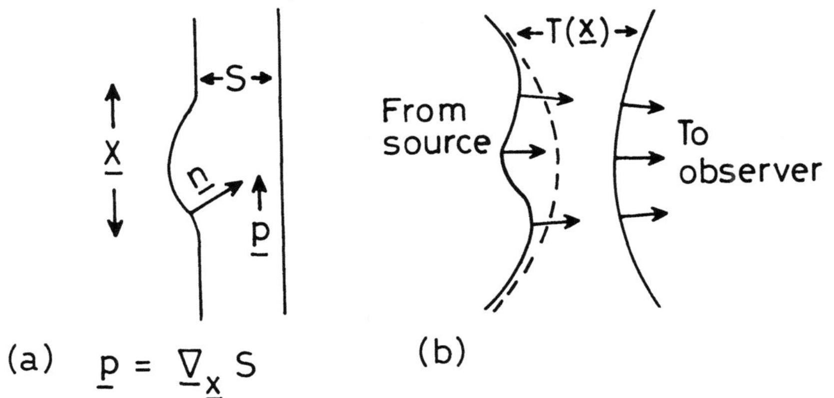

Fig. 1. (a) A wavefront showing the time delay S with respect to a reference plane, and the momentum $\underline{p}$ (transverse deflection of the unit normal $\underline{n}$).
(b) A geometrical construction based on wavefronts for images in lensing. Note the dashed line which is the situation without gravitational deflection and the associated time delay.

All this was cleared up by Hamilton who introduced two powerful ideas - phase space and the action S as a function of coordinates. As sketched in Fig. 1a, these have a very simple meaning in our context. The function S(x,y) or S($\underline{x}$) gives the shape of the wavefront or equivalently, the time delay measured from a reference plane. The transverse momentum $\underline{p}$ is, apart from a factor, the tilt of the wavefront with respect to the same reference, and is the transverse gradient of S.

Hamilton's ideas took root in mechanics but, strangely, are still relatively neglected in the teaching of optics, their original source. As an illustration, time delays in lensing were confusing many of us a decade or more after Refsdal's clear original (1964) discussion based on wavefronts. The three sketches in Fig. 2 bring out the different ways one can look at lensing in a simple two dimensional case. Fig. 2a is the familiar ray picture. The incident rays (not shown in the figure) come in along z and encounter a lens which produces deflections along $\pm$x. The ray at x = 0 is undeviated while adjacent ones focus at the tip of the caustic. Rays at larger values of x meet their neighbours at larger z further back along the caustic. A number of qualitative features are already evident. The crowding of the rays illustrates the increase in the total intensity for observers within and near the caustic. The region outside near the tip is where observers would see highly magnified single images. The merger, intensification, and disappearance of a pair of images as the observer crosses the caustic is also seen.

Fig. 2b shows the wavefronts normal to the given family of rays. As the first crossing of adjacent rays occurs at the tip of the caustic, the corresponding pieces of wavefront must cross each other, while still facing the positive z direction. The result is a surface with multiple sheets attached to each other at cusps (ridges when we include the third dimension). As such a wavefront passes the observer, the gaps between the sheets give the time delays.

The third picture of gravitational lensing (Fig. 2c) is in phase space. Each curve is a plot of p versus x (i.e. wavefront tilt versus position) for a given wavefront. The incident plane wavefront is just the x axis (p = 0 everywhere). After passing the

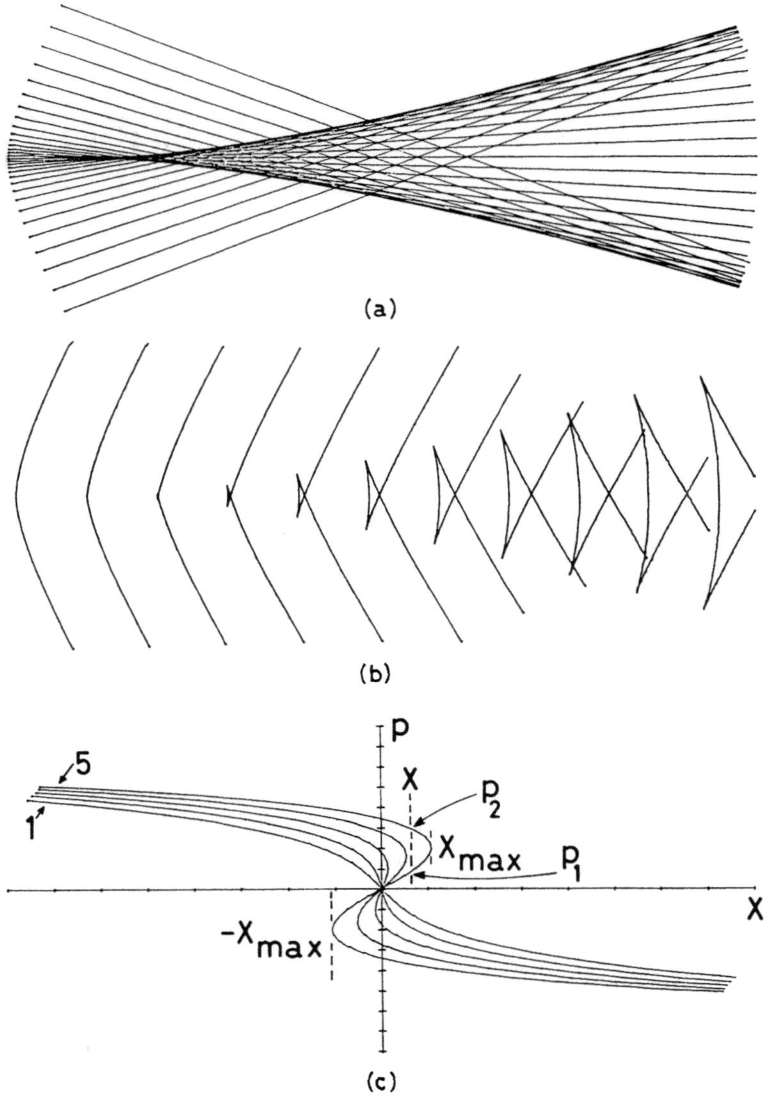

Fig. 2. Different ways of picturing multiple images, magnifications and time delays based on (a) rays (b) wavefronts (c) phase space. See text for details.

lens, rays at negative x are deflected upwards (positive p) and those at positive x deflected downwards. The labelled curve 1 Fig. 2c is just basically this graph of deflection versus impact parameter. The evolution of the wavefront is represented on this diagram by two very simple rules. As z increases, the rate of change of x is given by p, while p itself is conserved in free propagation without further deflections. This means that each point on the graph moves to its right (or left) at a rate depending on its distance above (or below) the x axis. The family of curves shown in Fig. 2c were drawn in this way. Clearly, at large enough z the curve folds over while remaining perfectly smooth. There is now a range of x for which we have three values of p, correspondingly to those observers who lie within the caustic at a given value of z. Near x_{max} (the upper limit of this range) it is clear from the curve that $x_{max} - x$ is proportional to $(p_2 - p_1)^2$ where p_1 and p_2 are the two values of p at x. This tells us that the angular separation $(p_1 - p_2)$ of the two images vanishes as $(x_{max} - x)^{1/2}$ as the observer nears the caustic. As Kayser and Refsdal (1983) pointed out, we can get the time delay between the two sheets by $S_1 - S_2 = \int(p_1 - p_2)dx$ since p is the gradient of S. This scales as $(x_{max} - x)^{3/2}$. The magnification is obtained by populating the wavefront with equally spaced rays to start with. The distribution of points (each representing a ray) along the curve is nonsingular at x_m. The density of rays in real space x becomes singular because of the projection faction $dp/dx \propto (x_{max} - x)^{-1/2}$ which occurs when we transform from p to x. The magnification M of either image thus scales as $(x_{max} - x)^{-1/2}$. Since observers are presumably distributed along x with uniform probability, we have $dx \propto M^{-3}dM$ giving us the probability of high amplifications near this caustic. Thus the simple phase space picture of Fig. 2c leads to the correct global picture as well as scaling laws for multiple imaging and magnification.

All we now need to do is to add one more spatial dimension to the picture but this is quite nontrivial! Arnold's (1984) book on catastrophe theory illustrates what wavefronts and caustics look like in three dimensions. One has to keep reminding oneself that the rich variety of forms comes out of projecting a smooth surface in a four dimensional phase space down onto real space. Such a wavefront has a tremendous amount of information packed into it -

the image geometry, magnifications and time delays for a whole family of observers located on a plane of constant z. Unfortunately, it is a singular surface and hence perhaps not easy to visualise in general. I have made some attempt (1988) to sketch a five sheeted wavefront for the quadrupole lens. The corresponding surface in phase space has the advantage of being smooth and hence amenable to imagination, Taylor expansion etc. However, it lives in four dimensions - one more than we are used to.

More modestly restricting oneself to one observer at a time, a simple geometrical construction shows that we can picture all the interesting properties of the image in terms of a single smooth function of x and y - the time delay. This is illustrated in Fig. 1b. A diverging spherical wave front has just passed a lens, becoming dimpled by the gravitational time delay. This is being compared to a reference spherical wave centred on the observer. For an image, a ray from the source has to reach the observer. This means that it has to belong to the family normal to the first surface as well as the second i.e. the perpendiculars to the two surfaces have to match. If we plot the clearance between the two surfaces - the time delay function T(x,y) - the extrema give the locations of images. (This would not have surprised Fermat!) The bumps on the wavefront are all important so one could call this approach gravitational lens phrenology. Some elegant uses of this viewpoint are (i) Schneider's (1984) general amplification theorem (ii) the classification of possible image geometries on the basis of the saddle points of this surface (Blandford and Narayan 1986, see also Goodman et. al. 1987 in which this method is used to understand the various catastrophes).

I would now like to suggest that some insight can be gained by gravitational lens palmistry - the interpretation of various lines drawn on the wavefront to predict its future. Again, we look just after it has passed the lens. At any given point, the normal to the tangent plane gives the direction of the ray. The behaviour of a small bundle of neighbouring rays near this point can be understood by the quadratic terms in the shape of the wavefront referred to this tangent plane. In general, we have two principal directions of curvature, each determining a focal length in one plane and resulting in two focal lines of infinite intensity (in geometrical optics). As our point moves over the wavefront, these lines sweep out the caustic surface. For an observer plane at a fixed distance z, the points on the wavefront which "focus" at z will lie on one of two curves $R_1(x,y) = z$ and $R_2(x,y) = z$ where R_1 and R_2 are the two principal radii of curvature. These are just what are called critical curves in the lens literature since a pair of images would be seen to merge on these curves for an observer at the corresponding point on the caustic. The systematic study of the (extrinsic) curvatures on a surface goes back to Darboux and the use in optics has been reviewed by Berry and Upstill (1980). This article is in fact an excellent gateway to "catastrophe optics". What is given below is just an application of these general ideas to gravitational lenses which I hope will whet the reader's appetite for the original sources.

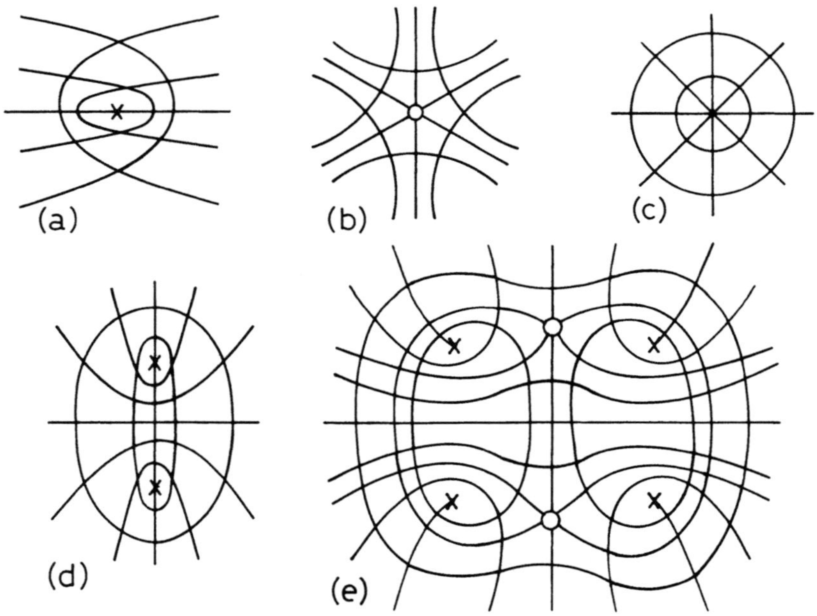

Fig. 3. The principal directions of curvature for a wavefront which has just passed a lens. (a) Index +1/2 (b) Index −1/2 (c) A circularly symmetric lens with a +1 umbilic point at the centre (d) Breakup of +1 into two +1/2's when a quadrupole perturbation is included. (e) Two well separated circular mass distributions producing four +1/2's (crosses) and two −1/2's (circles).

It is interesting to draw the map of directions of principal curvature. These form as orthogonal grid which is well defined except at those locations where $R_1 = R_2$. These special points on the wavefront are called the umbilics. To understand the behaviour of the curvature grid near these points, one needs to look at the cubic terms in the shape of the wavefront since the quadratic terms by themselves give equal curvatures. As one travels around an umbilic point, the two principal directions rotate by a half integer times 2π, as illustrated in Fig. 3a and b. In some ways, this reminds us of William Burke's (1981) analysis of gravitational lens images as zeros of a vector field drawn on the lens plane (the vector represents the amount by which a ray passing at that point misses the observer). In that case, the index (multiple of 2π) was ± 1. In ours, it is $\pm 1/2$ since we are dealing with a second rank tensor (wavefront curvature) or more geometrically, because an ellipse or hyperbola looks the same when flipped through 180°. Fig. 3c shows the curvature grid for a circularly symmetric lens where the two principal directions are clearly radial and tangential. It looks as if we have an index of +1 here but this is just one more instance of how one can be misled by circularly symmetric lenses. For example let us place two overlapping circularly symmetric lenses with their centres slightly displaced along x to break the symmetry. A little thought and/or calculation shows that there are now two umbilic points separated along y as in Fig. 3d. The argument goes as follows. At the origin, the focusing along the y direction is clealy stronger because the mass distribution (and the potential) are elongated along x. As we move up along y, well away from the mass distribution, we have defocusing along y (since it is now a radial direction) but focusing along x. Clearly, at some intermediate value of y, the focusing in the two directions must be equal and we have an umbilic point. There is similarly one point along -y. Since the grid far away must still look like a circularly symmetric lens, (the higher multipoles die off faster) the total index is still +1 and our two umbilics have index +1/2. What we have here is of course just the quadrupole lens.

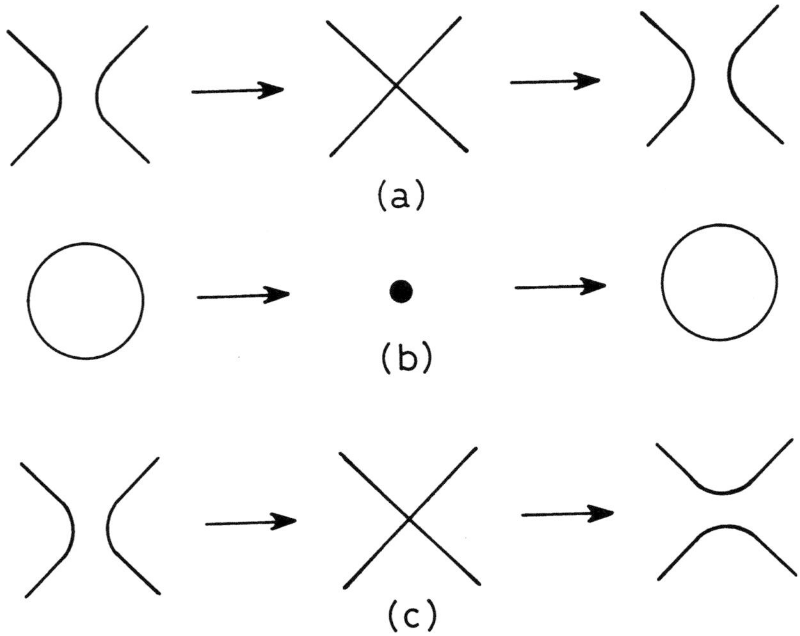

Fig. 4. Behaviour of critical lines (of constant principal curvature radii R_1 and R_2) near an umbilic point where $R_1 = R_2$. (a) hyperbolic umbilic (b) elliptic umbilic (c) not an umbilic but a place where R_1 or R_2 has a saddle point as a function of x and y.

The relevance of the umbilics is that they are the only points where critical curves of the R_1 and R_2 families can meet. As we increase z, the critical curves can change roles only by meeting at umbilic points. This is illustrated in Fig. 4 which shows two possibilities - called for obvious reasons hyperbolic and elliptic. Interestingly, the index +1/2 umbilics are all hyperbolic, but the -1/2 index can give either a hyperbolic or elliptic case and the hyperbolic +1/2's come in two varieties (Berry and Upstill 1980).

Our two mass lens can show more umbilics. If we move the two masses further apart, the pattern of umbilics looks like Fig. 3e. This rather complicated looking pattern can actually be understood rather simply. In the limit that the masses are far apart, the tidal force of each weakly perturbs the circular symmetry near the centre of the other, giving rise to a pair of +1/2's near each as happened in Fig. 3d. Since 1/2+1/2+1/2+1/2 = 2, we need two -1/2's in the pattern (at least!) to make the total +1 at infinity. The location and nature of these can again be understood physically. At a point equidistant from the two masses where the two radius vectors are at right angles, the shears cancel. Thus we have an umbilic which calculation confirms to be what topology indicated - of index -1/2.

The situation is particularly clear when the umbilic lies in a region with zero surface density, as when we have two point masses. The cubic terms in the potential, and hence the shape of the wavefront, satisfy the two dimensional Laplace equation. They must look like $r^3\cos 3\theta$ or $x^3 - 3xy^2$ is a suitable coordinate system. As the threefold symmetry suggests, this has an index of -1/2 (Fig. 3b) and calculation of the critical curves shows that it is an elliptic umbilic. For the two point mass case, we can think of these two elliptic umbilics as places where the wavefront is locally flat. The radial defocusing caused by one mass is compensated by the tangential focusing of the other at these two points, which can therefore give images as intense as when there was no lens. After the fact, we can understand how Robert Nemiroff (1988) made an ingenious (but fine tuned) lens model producing images of comparable intensity and arbitrary separation with two arbitrary small masses.

Perhaps my fondness for these curvature maps with umbilics of index $\pm 1/2$ is partly because I met them many years ago as well known patterns of alignment in liquid crystals! In any case, I would like to suggest that specifying the umbilics and their indices gives a fair measure of how complex a given model is, telling us more about the caustics than just the number of masses or parameters used to build up a lens. In a real sense, the quadrupole lens in the simplest and the two well separated masses come next.

Even with such sketchy coverage, nothing has been said about thick lenses, microlensing, or arcs to mention just three areas where the optics as well as the astronomy is fascinating. I will shamelessly borrow my last remark from the last talk I gave on lenses (Nityananda 1988) a year ago. It is remarkable how many basic ideas in the field originate from the work of Sjur Refsdal of the Hamburg Observatory. It is therefore pleasant and no surprise to learn that these proceedings are being brought out to mark his many contirbutions and wish him many more.

REFERENCES

Arnol'd, V.I. 1984 Catastrophe Theory, Springer.

Berry, M.V. and Upstill, C. 1980 Progress in Optics, Vol. 18 E. Wolf ed. North Holland.

Blandford, R. and Narayan, R. 1986 Astrophys. J. 310, 568.

Burke, W.L. 1981 Astrophys. J. Lett. 244, L1.

Goodman, J. et.al. 1987. Mon. Not. Roy. Astron. Soc. 229, 73.

Kayser, R. and Refsdal, S. 1983 Astron. and Astrophys. 128, 156.

Nityananda, R. 1988 Raman Centenary Conference, Bangalore, Proceedings ed. G. Srinivasan and S. Ramaseshan, Indian Academy of Sciences, Bangalore 1990.

Refsdal, S. 1964 Mon. Not. Roy. Astron. Soc. 128, 295.

Schneider, P. 1984 Astron. and Astrophys. 140, 119.

DETERMINATION OF COSMOLOGICAL PARAMETERS BY THE GRAVITATIONAL LENS EFFECT

S. Refsdal

Hamburger Sternwarte, Gojenbergsweg 112,
D–2050 Hamburg 80

The title of my talk is not very precise, since most gravitational lens work aims at determining cosmological parameters, or in any case quantities which are of interest for cosmology.

This makes it easier for me to switch the main part of my talk from the determination of H_0 and the mas M_G of lens galaxies to a proposal for using parallel observations from the earth and a remote space observatory (for instance Cassini) to get more information from microlensing phenomena. In view of the recent observation of QSO 2237+030 by Corrigan *et al.* (1989, in this volume) such a project seems very promising.

But first a few comments on the determination of H_0 and M_G. The key to a possible independent determination of H_0 and M_G is the time delay Δt which one can hope to observe as a shift Δt in the light curves of gravitationally lensed, multiple quasars. Only for the "oldest" system QSO 0957+561 A, B, regular monitoring has been carried out. Florentin Nielsen [1] gave the first value of $\Delta t \sim 1.55$ years, and data by Schild (in this volume) and Vanderriest *et al.* [2] now favour a Δt value of ~ 1.15 years. Caution is however in place, since microlensing may cause independent variations in the light curves, and a large part of the correlations seems to be based on one event, see also Falco's contribution in this volume.

As a personal comment I would like to add that the statistical evaluation of the data is far from being trivial due to the gap of 4–5 months in the observations every year. With time delay of N years ($N = 1, 2...$) one get much more points to correlate than with $\Delta t = (N + 1/2)$ years, only with an unreliable interpolation this interpolation can be "overcome". This may lead to a preference for choosing time delays of $\sim N$ years (or of $N + 1/2$ years).

Apart from the uncertainty in Δt, which probably will be overcome within a few years, it is the "unknown" mass ans mass distribution in the surrounding cluster that causes the largest uncertainty in the determination of H_0. Here I can only refer to papers of Borgeest and myself [3] and by Falco *et al.* [4] and references therein. The present best guess, based on $\Delta t \sim 1.15$ years, gives values in good agreement with those obtained with conventional methods, with a slight preference for H_0–values around 80, but even 50 cannot be ruled out.

For the determination of M_G, Borgeest has shown that the uncertainties due to the cluster is much less, and he finds the mass inside an angular radius of 3.1" to be:

$$M_G/M_\odot = 7.510^{11} \Delta t_{years} \mp 20\% \tag{1}$$

In order to get a comparable accuracy by the determination of H_0 one need additional information, for instance on the value of $M_G.H_0$ by applying the virial theorem on the galaxy, or information on the cluster mass, again by applying the virial theorem. There are plans for observing the velocity dispersion in the galaxy as well as in the cluster, but the observations are very difficult and the theoretical interpretation is not simple.

A very promising method of determining the cluster mass and mass distribution is to observe the distorsion of background galaxies caused by gravitational lensing as reported by Tyson in this volume. If Tyson's method could be applied on the cluster lying in front of QSO 0957+561 a big step towards a more accurate value of H_0 would have been made.

I now turn to my second topic where another kind of time delay, δt, plays a key role. It is well known that microlensing due to stars can cause distant compact sources (QSOs) to vary [5] [6] due to transverse motion. Particularly during a so called high amplification event (HAE), when the observer crosses a caustic, large and rapid variations may occur. Such an event seems to have been observed now for the first time by Corrigan *et al.* (this volume). They have found that one of the four images in QSO 2237+030 has become $0.6mag$ brighter in less than one year for thereafter getting fainter again. The other images have remainded constant. Theoretical calculations show that the time delay δt between the different images are about one day or less, therefore intrinsic variations can be ruled out and microlensing seems to be the only reasonable explanation for the observed phenomena. The pronounced peak in the light curve even suggests a HAE. Assuming that the caustic sweeps through the solar system with a velocity of about $V \sim 600km/s$, one would from a remote observatory close to Saturn (distance $\sim 10A.U.$ see a timeshifted lightcurve [7] with $\delta t \sim 10A.U./2V \sim 15days$, where the factor 2 roughly takes care of geometrical projection effects. For the event mentioned above the most rapid variation must have been at least $|\dot{m}| \sim 0.002/day$, probably more. This correspond to a magnitude difference between here and Saturn (Cassini) of about $\delta m \sim 0.03$. A determination of δt would therefore have been possible if a suitable space observatory had been available. One could then have obtained a good estimate of V and the size of the QSO [7]. With an accurate lightcurve (from the earth) one could also have determined its one dimensional luminosity profile [8].

The unknown direction of the caustics causes an uncertainty in the determination of V which can be overcome from observations from one additional space observatory [7], here CRAF offers an interesting possibility with a distance from the earth of up $7A.U.$

Since Cassini and CRAF are mainly supposed to observe the planetary system one would have to carry out monitoring from the earth, and only make parallel observations from space when rapid variations show up. In principle such a programm could include many quasars which for some reasons are expected to show rapid variations, not only QSO 2237+030. An observed time shift $\delta t \neq 0$ would then be a proof of microlensing, whereas $\delta t = 0$ probably would mean intrinsic variations (could also mean microlensing with very large V or a special orientation of the caustic)

The largest uncertainty concerning the suitability of Cassini and CRAF for this project is the pointing stability. On a preproposal Briefing in Noordwijk on November 9, 1989, this question, which is crucial for how faint QSOs can be observed, could not be finally settled.

Let me finally point out some other interesting possibilities with observations from remote space observatories:

1.) A general survey of a large number of QSOs with parallel observations from the earth could give very important information on the general importance of micro-lensing and therefore also information on the mass distribution in the universe. A baseline larger than $10A.U.$ is however desired.

2.) Regular observations would usually be possible all the year round, so that continuous lightcurves could be obtained. This is of great importance for instance by the determination of the time delay Δt as indicated above.

References

[1] Florentin–Nielsen, R.: (1984), *Astron. Astrophys.* **138**, L19

[2] Vanderriest, C., Schneider, J., Herpe, G., Chervreton, M., Moles, M., and Wlerick, G.: (1989), *Astron. Astrophys.* **215**, 1

[3] Borgeest, U. and Refsdal, S.: (1984) *Astron. Astrophys.* **141**, 318

[4] Falco, E.E., Gorenstein, M.V. and Shapiro, I.I.: (1989) *Astrophys. J.* **327**, 693

[5] Chang, K. and Refsdal, S.: (1979) *Nature* **282**, 561

[6] Kayser, R., Refsdal, S. and Stabell, R.: (1986) *Astron. Astrophys.* **156**, 36

[7] Grieger, B., Kayser, R. and Refsdal, S.: (1986) *Nature* **324**, 126

[8] Grieger, B., Kayser, R. and Refsdal, S.: (1988) *Astron. Astrophys.* **194**, 54

COSMIC GRAVITATIONAL DIAGNOSTICS

Israel Kovner
Institute of Astronomy, Cambridge, UK
and Physics Department,* Weizmann Institute, Rehovot, Israel

ABSTRACT

I review theoretical problems posed by recent observations. Many of the results and questions presented are new, to my knowledge.

Is there too much QSO-galaxy association to be explained by gravitational lensing ? We can reexamine observations and theories (minding, however, Roger Blandford's warning to ponder the fate of Albigensians). In particular, we may wonder how much the background QSO counts are affected by lensing. Microlensing on $\Delta\theta \lesssim 10^{-3}$ $''$ scales, or marginal lenses on larger scales might conceivably affect the brightest QSO counts.

A population of marginal lenses with tips of caustics at $z_{tip} \gtrsim 3$ would magnify QSOs at $z_s \sim 1-2$ by 2-3 magnitudes, making no multiple images. However, the clumpiness of the universe produced by (at least) the known populations of galaxies would perturb the marginal lenses, bring some caustics to $z_{tip} < 2$, and cause some fraction of magnified QSOs to be triple images of comparable fluxes. Thus, a theory involving marginal lenses must account for the absence of marginal triplets discovered yet.

We must also mind the clumpiness when we try to determine the Hubble constant. Cosmological populations of compact objects can be strongly constrained from the clumpiness they would introduce. The cell approximation, based on an analog of the optical theorem (flux conservation) can simplify the treatment of practical questions.

There is a topological similarity between optical effects which produce marginal lenses and the effects which produce giant arcs in clusters of galaxies (dense lenses): In both cases, sources are situated near cusps of the caustics of our past light cone. In particular, a marginal lens perturbed by a smaller scale dense lens (e.g. cluster by a galaxy) is similar to the lens responsible for the Koo's arc, broken in three by a clump of galaxies.

Arcs, rings, and the background made up by Tyson's galaxies belong to the most interesting tools of cosmic diagnostics to date. Besides probing galaxies and clusters, they can also probe the puzzling "dark" lens double QSO candidates. The microwave background anisotropies, when discovered, can join the club and probe the cosmography at high redshifts.

I. INTRODUCTION

Probably the first proposal for cosmic diagnostics via gravitational lenses was made by Sjur Refsdal: Given sufficient information about the lens, the Hubble constant can be obtained from the arrival time differences between multiple images of a variable source.[Rl] Since the discovery of the "Double Quasar",[WCW] multiply

---

* Incumbent of the Charles H. Revson Foundation Career Developement Chair

imaged QSOs gave evidence that cores of the lensing galaxies are denser than $\Sigma_{core} \gtrsim 10^4 - 10^5 M_\odot pc^{-2}$ (from the bounds on the third image in the Double Quasar[Gn] and from the multiplicity of the Huchra's object Q2237+0305[YS]). Giant luminous arcs in clusters of galaxies[SFMP,LP,Pi] may have been the first in a series of important discoveries, as the QSO-galaxy association[Fn,WHHW], and microlensing[IC] soon followed.

The French Revolution of April 1986 was started by Soucail, Fort, Mellier, and Picat who submitted a *Letter* to *Astronomy and Astrophysics*[SFMP] with a report on the A370 arc (stating that gravitational lensing was unlikely !). Lynds and Petrosian[LP] announced beautiful photographs of the A370 and 2244-02 arcs. Paczyński[Pi] published these photographs and suggested gravitational lensing and the redshift test. These arcs may have opened the epoch when we can deduce astrophysical information heretofore inaccessible, from the effects of gravitational lensing[Kr4,KBLN].

The space-time limitations of the workshop allow me to address only a small fraction of interesting questions, even after filtering with strong biases. The biases are to my personal curiosity, to avoidance of overlaps with other reviews of this workshop, to astrophysical diagnostics, and to problematic questions prompted by recent observations. Thus I apologize to the proponents of all the interesting ideas omitted, and to those who may find the degree of briefness disagreeable. The list of references is by far incomplete, however there are other reviews.[GLR] I address QSO-galaxy association in §§ II-IV, lens effects on background QSO counts in §§ IV-VI, marginal lenses in §§ V, VI, and IX, inhomogeneity of the universe in § VI and VII, dark double QSO lens candidates in § VIII, arcs and arclets in §§ VIII and IX, topological similarity between marginal lenses and cusps in dense lenses in § X, and the backgrounds formed by Tyson's galaxies and by cosmic microwave radiation in § XI.

II. QSO-GALAXY ASSOCIATION

Arp[Ap] claimed frequent association between galaxies and QSOs with discrepant redshifts, to support noncosmological redshifts. Canizares[Cs1] suggested that magnification by gravitational lensing can produce QSO-galaxy association. Let $n_o(f_s)$ be the average number of QSOs of flux $f > f_s$, per unit solid angle, in absence of lensing effects ("intrinsic" integral counts). It is easy to see that magnification by factor A causes enhancement of counts by factor

$$\mathcal{E}(f_s, A) = n_o\,(f_s/A) \;\, / \;\, A n_o\,(f_s)\,. \tag{1}$$

Given A, $\mathcal{E} > 1$ and $\mathcal{E} < 1$ are possible, depending on f_s.[VO] Also, given f_s large enough to have $\mathcal{E} > 1$ possible, there is a maximal possible enhancement.[Kr5] This is illustrated in Fig. 1: $\log \mathcal{E}$ is given by the difference between the solid curve, $\log n_o(f)$, and a dashed line, $(f_s/f) n_o(f_s)$, at the abscissa separated by $-2.5 \log A$ from $-2.5 \log f_s$. Let us consider the slopes of integral and differential counts,

$$\nu_i(f) = -0.4\; d\log n_o / d\log f, \qquad \text{or} \qquad \nu_d(f) = -0.4\; d\log\left(\nu_i n_o\right) / d\log f, \tag{2}$$

respectively. Reports on large QSO samples usually give ν_d,[BSP] however the parameter which directly determines possible signs of $\log \mathcal{E}$ is $\nu_i(f_s)$ (for a power law, $\nu_d = \nu_i$): For $n_o(f) \propto 1/f$ (dashed lines, $\nu_d = \nu_i = 0.4$, the "neutral slope") magnification has no effect, $\mathcal{E} = 1$. For sample thresholds too faint, $\nu_i(f_s) < 0.4$, only reduction of counts, $\mathcal{E} < 1$, is possible. For a sufficiently high flux threshold with $\nu_i(f_s) > 0.4$, enhancement $\mathcal{E} > 1$ is possible for a limited range of magnifications, $1 < A < A_1$, and it has a maximum, $\mathcal{E}_{max}$. In real lenses $A \neq const$, however it is possible to find bounds on $\mathcal{E}$ subject to various constraints.[Kr5]

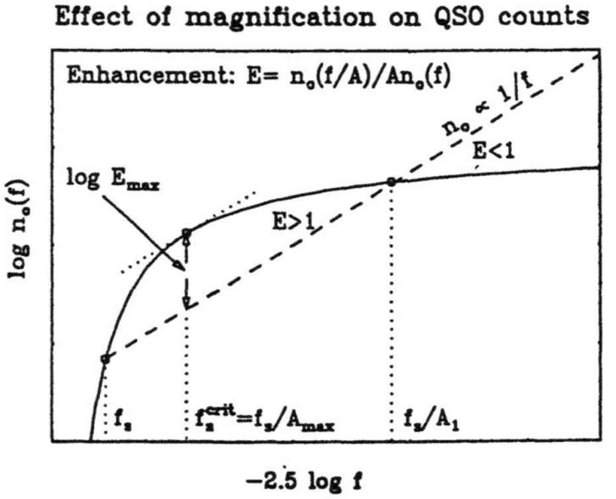

Figure 1: *A typical shape of integral QSO number counts* (solid). *Magnifications A_{max} and A_1 produce maximal enhancement, $\mathcal{E} = \mathcal{E}_{max}$, and no enhancement, $\mathcal{E} = 1$, respectively. For $A > A_1$ the number counts are reduced, $\mathcal{E} < 1$. If the sample threshold flux is $f_s > f_s^{crit}$, any $A > 1$ can only reduce number counts (although demagnification, $A < 1$, can give $\mathcal{E} > 1$).*

We would like to compare the observed association[Fn,WHHW] with "test" predictions from various lensing models. To do so, we must use the "background" QSO counts[BSP] and thus restrict the analysis to the BSP redshift interval, $0.5 < z < 2.2$. We also have to take into account that $WHHW$ were uncertain in detecting QSO-galaxy pairs separated by $< 3''$. The enhancement calculated this way ($\mathcal{E}_{whhw} = 6$) is much above the conventional predictions for galaxies ($\mathcal{E} \sim 1.5 - 2$). To predict $\mathcal{E} \geq \mathcal{E}_{WHHW}$, it would be necessary to assume large observational errors in $\gtrsim 2$ parameters of QSO counts, as shown in Fig. 2 (see also Nn and $Sr2$). Given conventional general assumptions, even artificial lenses constructed to maximize $\mathcal{E}$ under $WHHW$ conditions[Kr5] produce $\mathcal{E}_{max} = 4 < \mathcal{E}_{whhw}$.

A part of the $WHHW$ association might have been contributed by statistical and observational uncertainties (in particular in the definition of the sample of galaxies). There is however a clear motivation to reexamine the theoretical views on QSO-galaxy association. Clustering, microlensing, or unconventional mass distributions are not very relevant on their own, for $\mathcal{E}_{whhw}$, since the "maximal lens bounds" on $\mathcal{E}$[Kr5] constrain any lenses, given two essential hypotheses: (1) that the BSP counts, $dN_{QSOs} = n_{bsp}(f)df$, are little affected by lensing, and (2) that the dependence of the magnification probability distribution, dP_A, on the intrinsic QSO flux can be neglected. Before going into details of these hypotheses, let us note that the large (could be effective) galactic velocity dispersions suggested by $WHHW$ and by Nn would produce secondary images, which could not be missed but were not observed (Fig. 2a, a similar argument was used by GG for Q1548+115a,b).

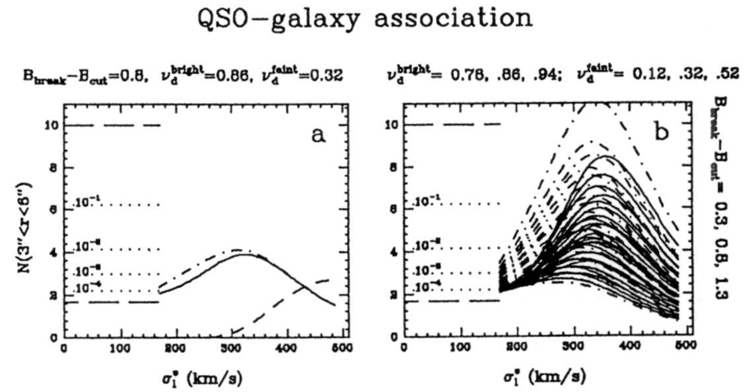

Figure 2: *Predictions for QSO-galaxy association, for the $WHHW$ conditions, versus the 1-d velocity dispersion of an effective L^*-galaxy (a single morphological type), given conventional assumptions*^{Kr5,Kr7}*, for singular isothermal galaxies: the number of "primary" images of QSOs within a ring about a galaxy* (solid); *the number of "secondary" images in the same ring* (dash, with primaries outside and more magnified, not found by $WHHW$); *and the number of primaries with roughly maximal possible microlensing included* (dot dash). *The* long-dash *levels indicate the number of $WHHW$ pairs (10) which are examined against BSP counts, and the prediction (1.7) for a Poissonian distribution. Numbers near* dotted *levels indicate a probability to find* $\geq$ *10 pairs, given an expectation equal to the level.* **a:** *Predictions for sample parameters taken from $WHHW$ and BSP.* **b:** *Predictions for 27 combinations of assumed observational errors in 3 parameters: the position of the break in the slope of QSO counts with respect to the $WHHW$ sample threshold, $B_{break} - B_{cut}$, and the slopes of the differential counts at the bright and faint ends, ν_d^{bright} and ν_d^{faint}.*

III. THIN SKY APPROXIMATION

The well defined QSO samples are still not very large, thus number counts $F(f) = -dn(f)/df$ can be more useful for some test predictions, than the still uncertain redshift-dependent luminosity function $dN_{QSOs} = \Phi(L,z)dLdz$. In general, the probability distribution for magnification by lensing depends on the source redshift, $dP_A = p(A, z_s)dA$. Thus, as f/L depends on z_s, there is some "macro-"correlation between dP_A and f. There can also be a "micro-"correlation, if microlensing is important: The finite size of the relevant emission region can determine the cutoff of $p(A,z)$ at large A ^{Cs2,Sr1} (for point sources in geometrical optics, $p(A) \to const \cdot A^{-3}$ as $A \to \infty$); this size is correlated with L, and thus with f. However, when the lens galaxy is given as an input (usually at $z_g \lesssim 0.3$, for relevant samples), the differences between dP_A for QSOs at $z_s \sim 1-2$ are not large. Therefore, for many practical questions of statistics,^{VO} we can neglect the dependence of dP_A on f and apply it directly to the number counts. However, the error introduced has yet to be investigated.

This error can be addressed in a simple model for macrolensing, for a separable combination of density and luminosity QSO evolution, $\Phi(L,z) = \psi(z)\phi[L/L(z)]$: Then it is possible to define a flux-independent "pseudo-probability" distribution, $dP_A^* = p^*(A)dA$. It acts on $F(f)$ just as a probability distribution does, i.e. $F(f) \rightarrow \bar{A}^{-1}\int F(f/A)p^*(A)d\ln A$. However there is an important difference: $p(A) \geq 0\ \forall A$, whereas $p^*(A) < 0$ is possible for some A. Preliminary results, for $p^*(A)$ calculated on basis of the BSP model for QSO evolution and the singular isothermal sphere approximation for galaxies, do not indicate a large error in neglecting the dependence of dP_A on f.[Kr7]

IV. *IS QSO BACKGROUND AFFECTED BY LENSES ?*

This question is almost as old as our awareness of the QSOs[BB]. Weighty arguments in favor of a negative answer are provided by the *"magnification-multiplicity"* conjecture: QSO counts cannot be much affected by lensing unless a considerable fraction of QSOs is multiply imaged. I shall explain below why it is still an open question, but first let us see whether a large fraction of QSOs may be multiply imaged on some angular scale. This is clearly excluded on scales $1'' \lesssim \Delta\theta \lesssim 10'$, from the gravitational lens surveys of $\sim 10^{3.5}$ QSOs. It is also excluded on angular scales $10^{-3}\,'' \lesssim \Delta\theta \lesssim 1''$ from the known samples of compact radio sources, of which a large fraction is identified with bright QSOs at $z > 0.5$.[Kr6]

It is conceivable that lenses with $\Delta\theta \gg 1'$ could produce multiple images with time delays larger than QSO lifetimes. However, the Rees-Sciama effect[RS] by such lenses would affect the brightness of the microwave background (not counting the Sunyaev-Zel'dovich effect in lenses with sufficient gas content[OV]). Within an order of magnitude, the typical MBR temperature deviation and the typical arrival time difference are related as[Kr7] $\Delta t \sim H_\circ^{-1}\,(\delta T/T)^2_{MBR}$. Therefore, the present day limits $(\delta T/T)_{MBR} < 10^{-4}$ on the $\Delta\theta > 10'$ scale rule out systematic strong lens effects over the sky with time delays $\gg 10$ years (though rare unusual cases may not yet be ruled out). As $\gtrsim 10^3$ QSOs are known for $\gtrsim 10$ years, typical QSO lifetimes smaller than $\sim 10^3\ y$ are unlikely.

We cannot directly test for multiple imaging at $\Delta\theta < 10^{-3}\,''$. However (micro)lenses acting on scale $\Delta\theta < 10^{-4.5}\,''$ cannot magnify much the emission line regions of QSOs. This provides a test based on comparison of the QSO distribution in equivalent widths, EW, and in fluxes:

Prediction $\mathcal{E} \gtrsim \mathcal{E}_{WHHW}$ would be possible if the bright QSO counts slope ($\nu_d^{bsp} = 0.86$, i.e. $F \propto f^{-3.2}$) were entirely due to lens effects on intrinsically much steeper counts[Kr5] (for point sources, asymptotically at $f \rightarrow \infty$, lenses produce $F \propto f^{-3}$ and $\nu_d = \nu_i = 0.8$; however larger ν_d can occur when the size of the source is important[Sr1]). Then, if this lensing occurs on scales $\Delta\theta \lesssim 10^{-4.5}\,''$, this slope would only describe $n(f)$ for the continuum. Now, if the intrinsic distribution of continuum/line ratios, $dN_{EW^{-1}} = F_{c/l}\left(EW^{-1}\right)dEW^{-1}$, is also intrinsically much steeper (at small EW) than dN_{QSOs} observed (at large f), then $\nu_d^{EW} \equiv -0.4 - 0.4d\ln F_{c/l}\left(EW^{-1}\right)/d\ln EW^{-1} \approx \nu_d^{bsp}$. Preliminary examination of a large sample of quasars with measure EW collected by H. Netzer suggests that $\nu_d^{EW} > \nu_d^{bsp}$ (against much microlensing), although a full investigation of the possible selection effects is required (in particular, the difference $\nu_d^{EW} - \nu_d^{bsp}$ depends on the intrinsic slopes and on the degree of importance of the lensing).

The same principle can be used for a test based on comparison of slopes of number counts at different wavelengths. For instance, X-ray emission regions in QSOs are believed to be smaller than the continuum emitting regions. Thus the X-ray flux can be magnified at least as much as the continuum, by lensing. Therefore, the possibility that the observed slopes may be determined by lensing can be tested by comparing the slopes β_X and β_{opt}, and can be ruled out if $\beta_X > \beta_{opt}$ is established significantly (with the same caution as regards $\nu_d^{EW} - \nu_d^{bsp}$, above).

V. MAGNIFICATION-MULTIPLICITY CONJECTURE

If we follow the observer's past light cone, distorted by some lenses, the null-geodesics can eventually start crossing each other (e.g. Fig. 3 of BKKN). The past-caustics are their envelopes and separate the sets of events which can be connected with the observer by different numbers of null-geodesics. If we draw a bundle of lines along these null-geodesics, with a certain density per solid angle near the observer, the density of these lines increases near the caustics (and becomes infinite on the caustics). This means that sources positioned in regions of high line density will have one or more very magnified images. As we usually assume that the source and lens positions are random and uncorrelated with each other, we think that in an ensemble of very magnified cases some are bound to be produced by sources "inside" the caustics and be thus multiimage cases.

However, Paczyński and Wambsganss[PW] found a counter-example in numerical simulations of an extreme density bias: lenses composed of very "weak" elements at random positions, such that only very rare superpositions of many elements could produce much magnification. Density bias makes *"marginal lenses"* (almost or barely able to split images[Kr3]) statistically important. *PW* found regimes in which very magnified single image cases were much more frequent than multiimage cases.

In objection to *PW*, considering $A \rightarrow \infty$, it seems impossible to magnify much QSOs at redshifts $z_s \sim 1-2$ without some of the QSOs being multiply imaged. However, considerable effects on QSO counts may be caused by just $2-3$ *mag* of magnification (e.g. for *WHHW* sampling conditions and conventional assumptions $A_{max} \approx 10$[Kr4]). An example of constant magnification contours for a marginal lens, in a plane which contains the line of sight (redshift direction) is shown in Fig. 3.

Since most known QSOs are at $z_s \lesssim 2$, a population of marginal lenses with tips of the caustics at $z_{tip} \gtrsim 3$ may possibly produce *"singlet dominance":* an "affected" QSO background, dominated at the bright end by single, magnified QSO images. Can it be possible in our universe ?

VI. MARGINAL LENSES

Clusters of galaxies, individual galaxies, conceivable dark halos, or objects superposed along the line of sight can possibly be marginal: The appearance of arcs in clusters 0500-24[Gd] and A2319[FMP] suggests that the clusters of galaxies can be marginal (e.g. Fig. 6 of *Kr*4 for a typical such arc, and see *WGSW* for another interpretation of the 0500-24 arc). The probable presence of foreground galaxies and the absence of secondary images of BL Lac objects AO0235+164 and PKS0537-441 suggests low density cores[NS], at least in some galaxies. The Lyα clouds suggest a numerous population of minihalos[Rs] which can have low density cores.

Let us assume there is a $z_{tip} \gtrsim 3$ population of marginal lenses and try to guess (for lack of a proper investigation) whether the singlet dominance is then possible.

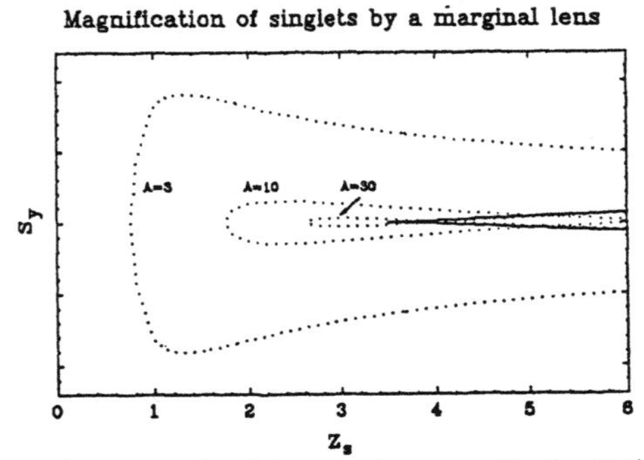

Figure 3: *Redshift slice of a marginal lens: the caustic* (solid) *and the constant magnification contours* (dot). *A source inside the cusp has 3 images.*

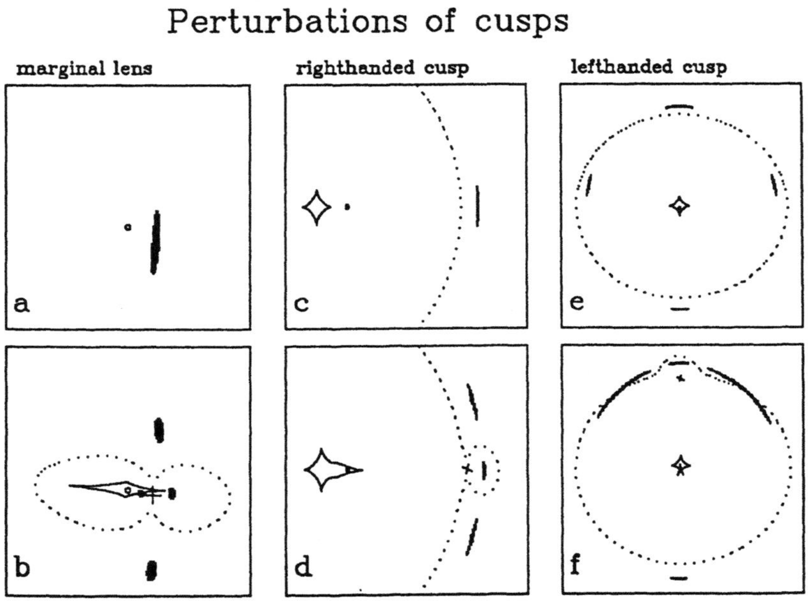

Figure 4: *Perturbations of cusps and marginal lenses:* **a,c,e:** *Large scale components, marginal*[Kr3] *(***a** *) and dense elliptical*[Kr2] *(***c,e** *).* **b,d,f:** *Small scale dense components at positions marked by crosses. Only the tangential caustics* (solid) *and critical lines* (dot) *are shown. The dark pixels are images of an extended source shown as a small circle. A "right-handed" cusp has the same topology (or rather topography of the arrival time surface) as a marginal lens, whereas the perturbation of a "left-handed" cusp produces a butterfly in the tangential caustic.*[Kr7] *The images formed by a composite marginal lens (***b** *) are a quadruplet. For dense perturbations, quadruplets are roughly as frequent as triplets (cf Figs. 5d,e of Kr1).*

There are two pieces of evidence of particular importance:

(1) ***Singular lens statistics:*** even numbers of images are known in all but one lens case[GLR] (Q2016+112, which is also special in that it has the highest redshift, $z = 3.3$, and seems to have at least two galaxies involved). The cores of responsible galaxies must be dense enough to demagnify the central image below the detection threshold.[NSC]

(2) ***Minimal clumpiness from galaxies:*** if only the known populations of galaxies are taken into account, they make the universe slightly "clumpy", so that a few percent of L^*-galaxies approximated by singular isothermal spheres (SIS) cannot be regarded as isolated lenses.[Kr6]

Suppose there are two populations of galaxies: One has very dense cores and is responsible for all the multiply lensed QSOs we know. Another has low density cores and makes up the marginal lenses. Marginal lenses are more susceptible to perturbations than dense lenses, so that just the minimal clumpiness would shift a notable fraction of the caustic tips to lower redshifts.

If galaxies are not usually marginal, but there are other marginal objects, these objects must affect a much larger fraction of the sky, than galaxies, to produce the singlet dominance. Thus some $z_{tip} \sim 1 - 2$ must be produced by clumpiness in any case. "Canonical" marginal lenses (3rd kind of $Kr3$) produce triple images of comparable magnitudes (their magnifications satisfy equation (3) below[KP,Kr7]). If a dense lens is superposed on a marginal lens of a larger angular scale (a composite marginal lens, e.g. a galaxy in a marginal cluster, 4th kind of $Kr3$), double images are sometimes possible, however triple or quadruple images of comparable brightness are as or more probable (e.g. Figs. 5d,e of $Kr1$ give a fair representation). An example of a marginal cluster, alone and perturbed by a galaxy is shown in Figs. 4a,b.

Therefore, our best guess would be that the marginal lenses cannot entirely violate the magnification-multiplicity conjecture. The multiple/single ratio can be much smaller than for dense lenses, but it is unlikely that many of the known cases are marginal, because of the singular lens statistics.

VII. CLUMPY UNIVERSE

For an arbitrary line of sight there is one $L \geq L^*$ galaxy within $0'.5 - 1'$ from it (the galaxy can be a strong lens, within a few arcseconds from itself, for a source at $z_s \gtrsim 2$; and of course smaller galaxies are more numerous). This must be taken into account in a variety of issues: — perturbations of marginal lenses (above); — QSO-galaxy association (a galaxy cannot be regarded an isolated lens beyond $10'' - 30''$ from its center, and, in addition, clustering of galaxies must contribute to the association when galaxies are sought near QSOs); — Hubble constant and constraints on the cosmological density of compact objects (below).

If we want to measure H_o by Refsdal's method, we would like to know how much and how commonly neighbor galaxies perturb the lens we think we see. Perturbations which can be linearized, can be treated in the "nonlinear telescope" approximation[Kr2], and their effects can be deduced and taken into account in the analyses.[FBK] In particular, a homogeneous distribution of dark mass in front of the lens galaxy (essentially just a modification of the angular diameter distances) can be detected if some mass parameter, say, the velocity dispersion of the galaxy can be measured independently. However, there may be considerable probabilities for nonlinear effects[GFS,KA,Kr6], which have not yet been well studied.

Zel'dovich[Zh] pointed out that clumpiness of the mass in the universe can be regarded as a modification of angular diameter distances: a beam of rays passing between clumps is (relatively) defocused by some effective negative mass equal to the mass concentrated in the clumps. The use of the modified (Dyer-Roeder[DR]) distances requires caution.[Wg,Kr6] It is possible to incorporate the effective negative mass directly into the lens equation in the "cell approximation": select relevant clumps (say galaxies, or hypothetical compact objects), devide the sky into cells most of which contain one clump, and require explicit conservation of brightness by the lens equation, on the average, within every cell.[Kr6] It is essentially an explicit implementation of an analog of the optical theorem (conservation of flux through a closed surface encompassing 4π of the observer's sky).

This approach provides means to calculate various lensing probabilities, it is applicable to any lenses (i.e. not necessarily gravitational), and it is considerably simpler than the explicit calculations of metrics and geodesics.[FS] Correct probabilities for lensing[Nf,Kr6,FS] allow us to derive low bounds on cosmological densities of point mass lenses: $\Omega_l \lesssim 0.01$ for $M > 10^{12} M_\odot$ from lens surveys, $\Omega_l \lesssim 0.1$ for $M > 10^6 M_\odot$ from compact radio sources, and possibly much lower constraints for $M > 10^{13} M_\odot$ from bounds on arclets among Tyson's galaxies[Kr7] (see Fig. 6 of *PG* !).

VIII. MULTIPLE QSOs

There are 4 pairs of similar QSOs split by $5'' - 7''$: Q0957+561 at $z = 1.4$ split by $6''$, Q2345+007 at $z = 2.2$ split by $7''$, Q1120+019 at $z = 1.5$ split by $7''$, and Q1429-008 at $z = 2.1$ split by $5''$.[MQ]. Q0957+561 is a clear lens case, however the other cases are still in question, in particular lensing galaxies are not yet established. In contrast, there are 11 cases of $2'' - 4''$ separations: 5 quadruplets, 1 triplet, 3 doublets, and 2 radio rings.[GLR] The two groups differ in typical multiplicities and separations.

The large doublets may either be produced by unusual "dark" lenses or be physically distinct objects. In the former case, deep imaging of the "dark" doublets can reveal "arclets" of $\sim$ a few arcseconds size formed by a background galaxy.[Kr7]. In addition, if a population of these "dark" lenses exists and is fairly represented by the dark doublets, there should be several arclets in the published very deep images of Tyson's galaxies[Tn1]. Tyson[Tn2] reported a similar arclet in a cluster of galaxies albeit with a clearly visible lensing galaxy. It may be possible to establish an excess of double QSOs, above what can be produced by lensing, by considering cases with extreme differences between magnitudes of the images[Kr7]: e.g. $\lesssim 10\%$ of lens cases are expected to be about as extreme as Q1120+019 ($I_1/I_2 \approx 75$).

A "side" result of searches for gravitational lenses would be evidence for an excess of small separations between physically distinct QSO pairs (above the expectations[PB] from just the spatial correlation). This in turn would give evidence (from the lack of pairs with larger separations) for stimulating or triggering QSO activity by interactions between galaxies. In fact, the discovery of clearly distinct pairs[DDQ] is already suggestive.

IX. ARCS

The discovery of the giant luminous arcs proved that some clusters of galaxies have $\Sigma_{core} \gtrsim 10^{9.5} M_{\odot} kpc^{-2}$. The giant cases can be used individually to constrain the shapes, the depths, and the centers of the responsible gravitational wells:
The very thin and even, "pretty" shapes of the arcs suggest Σ_{core} well above the minimum necessary for multiple imaging, as nearly marginal lenses tend to produce "ugly" arcs[Kr4] (this may be a problem for the GN model, in which some fine tuning seems to be implied). The possibility that the counter-arc is absent in A370 suggests a naked (or almost naked) tangential cusp in the lens caustics,[Kr4] and thus a constraint from below on the elongation of the potential. The distribution of color and brightness along an arc can sometimes indicate the number of images (of a part of the source galaxy) it contains[Kr4,Het] and possibly perturbations by small scale inhomogeneities of the potential[BPL]. Checking the reality of extra images in the 2244-02 arc[Het] can add constraints to the potential. However, it may be impossible at present to discriminate between different (reasonable) radial profiles of the mass distribution, as, say, the image configurations obtained with de Vaucouleurs profiles[Het] seem to be as easily produced by the simpler pseudo-isothermal profiles[Kr4].

Arcs in Cl0500-24[Gd] and A2319[FMP] suggest marginality of the lenses. In particular, the very straight arc in A2319 and several small elongated images of faint galaxies suggest a very elongated lens, may be a pancake-like supercluster-size dark mass with our line of sight in the pancake plane[Kr3]. Even if this is true, it may still be puzzling why the cluster A2319 does not introduce at least some curvature in the arc. An alternative[FMP] possibility of a heavy dark halo in a position which is not marked by any feature in the distribution of observed galaxies is at least as unusual in itself as is the arc it may explain.

X. TOPOLOGICAL SIMILARITIES

As interesting is the "Koo's arc" in Cl0024+16, discovered independently by Turner and colleagues[TK]: The large arc probably has a counter-arc, and a perturbation by a clump of galaxies apparently breaks the large arc into 3 segments. Most of the known large arcs are images of galaxies near cusps of the caustics of dense lenses: They occur in nearly level valleys of arrival time surfaces,[KP,Kr4] with topographies (in the simplest cases) similar to the topographies of arrival time surfaces for marginal lenses. Thus, the Koo's arc is topologically similar to a perturbation of a marginal lens by a dense lens (of a smaller deflection scale),[Kr3,Kr7] Fig. 4. The segments of a triple arc formed by a "canonical" marginal lens[Kr3] or by an extended source inside a "canonical" cusp[KP] must obey relations

$$l_1 + l_3 = l_2, \qquad \varphi_1 l_1 + \varphi_3 l_3 = \varphi_2 l_2, \tag{3}$$

where l_i and φ_i, $i = 1,2,3$ are the lengths and the positions of centers along the arc, of the segments numbered in sequence[Kr7] (as one can easily find from the lens equations of KP). Usual models for smooth lenses would violate these relations by $(a\ few) \cdot 10\%$,[KP], whereas a small scale dense perturbation can produce a much larger violation. Koo's arc strongly violates equation (3), suggesting constraints on the perturbing clump of galaxies.

Another example of topological similarities is the family of ellipse-quadrupole model lens potentials[Kr2,Kr4]). Toplogical similarities, in general, have not yet been much appreciated, however they can be useful.

XI. BACKGROUNDS: TYSON's GALAXIES AND MICROWAVE ANISOTROPIES

The large number of small, elongated images of background galaxies[Tn2] can provide means to analyze cluster potentials, although such means have yet to be developed beyond the first steps made so far.[Wr1,Tn2] If all the galaxies were intrinsically circular discs of the same size and at the same redshift, and were certainly identifiable among the cluster galaxies, their distortions would directly measure the surface mass density of the lens: The magnification matrix is $\mathcal{A} = [\delta_{ij} - \partial^2\Phi/\partial x_i \partial x_j]^{-1}$, where Φ is the lens potential, and (x_1, x_2) are Cartesian coordinates in a small area on the sky, measured in radians. The surface mass density is $\Sigma = \nabla^2\Phi \cdot (c^2/8\pi G)(d_{os}/d_{ol}d_{ls})$, where d_{ol}, d_{ls}, and d_{os} are observer-lens, lens-source, and observer-source angular diameter distances. In reality, we need statistical methods, to take into account the distribution of galaxies in shapes, sizes, luminosities, and redshifts.

The slope of number counts of Tyson's galaxies in some colors (R, I, J) is very close to the "neutral" $\nu_i = 0.4$ (*cf* § II), thus number counts of galaxies are little affected by magnification (I disagree with *GN* in this point). However, ν_i is larger at shorter wavelengths,[Tn2] thus "galaxy-cluster association" is possible. In addition, saturation has been claimed at the faintest observed magnitudes (as in the rightmost part of Fig. 1), thus the faintest counts can have "galaxy-cluster anti-association".

When discovered, MBR anisotropies may compete with Tyson's background of galaxies, in their diagnostical value.[BS,SNC] The known bounds on these anisotropies have been already used to constrain $\lambda \gtrsim 10\ Mpc$ gravitational waves[Lr]. MBR anisotropies will test the existence of a high redshift antipode: The angular diameter distance to the recombination shell can be deduced from the Taylor expansion of the MBR brightness autocorrelation function[DK]. MBR anisotropies, if and when detectable on the relevant angular scales, may also test the existence of marginal lenses.

I am very grateful to $\sim 10^{1.5}$ colleagues for enlightening discussions and comments. Parts of this work were supported by SERC/RTA grant GR/E 51480, and by the funds granted by the Charles H. Revson foundation. The statements made and views expressed are solely the responsibility of the author.

REFERENCES

Ap : Arp, H. 1987, *Quasars, redshifts, and controversies,* (Interstellar Media, Berkeley).

BB : Barnothy, J.M. 1965, *A. J.*, **70,** 666; + Barnothy, J.M., and Barnothy, M.F. 1972, *Ap. J.*, **174,** 477.

BPL : Bergmann, A.G., Petrosian, V., and Lynds, R. 1989, *Ap. J.*, submitted.

BS : Blanchard, A., and Schneider, J. 1987, *Astr. Ap.*, **184,** 1.

BSP : Boyle, B.J., Shanks, T., and Peterson, B.A. 1988, *M. N. R. A. S.*, **235,** 935.

Cs1 : Canizares, C.R. 1981, *Nature*, **291,** 620.

Cs2 : Canizares, C.R. 1982, *Ap. J.*, **263,** 508.

DDQ : Distinct double QSOs: Djorgovski, S., Perley, R., Meylan, G., and McCarthy, P. 1988, *Ap. J.*, **321,** L17; + Crampton, D. *et al.* 1989, *Ap. J.*, **330,** 184.

DK : Durrer, R. and Kovner, I., *Ap. J.*, in press, preprint WIS-89/69/Nov.-Ph

DR : Dyer, C.C., and Roeder, R.C., 1973, *Ap. J.*, **180,** L31.

FBK : Falco, E.E., Gorenstein, M.V., and Shapiro, I.I. 1985, *Ap. J.*, **289,** L1; + Borgeest, U. 1986, *Ap. J.*, **309,** 467; + Kovner 1987, *Ap. J.*, **318,** L1 (Erratum in 1987, *Ap. J.*, **323,** L155).

FMP : contributions by Fort, B., Mellier, Y., and Pello-Descayre, R. at this workshop.

Fn : Fugmann, W. 1988, *Astr. Ap.*, **204,** 73.

FS : Futamase, T., and Sasaki, M. 1989, *Phys. Rev. D*, **40,** 2502.

Gd : Giraud, E. 1988, *Ap. J.*, **334,** L69.

GFS : Gorenstein, M.V., Falco, E.E., and Shapiro, I.I. 1988, *Ap. J.*, **327,** 693.

GLR : Gravitational lens reviews: Peacock, J.A. 1983, in *24 Liége Colloquium, Quasars and Gravitational lensing"*, p.86; + Gott, J.R. 1987, in *IAU Symposium 117, Dark Matter in the Universe,* ed. J.Kormendy and G.R. Knapp, (Dortrecht: Reidel), p. 219; + Canizares, C.R. 1987, in *IAU Symposium 124, Observational Cosmology,* ed. A. Hewitt, G. Burbidge, and Li Zhi Fang (Dortrecht: Reidel), p. 729; + Blandford, R.D., and Kochanek, C.S. 1987, in *Dark Matter in the Universe,* Proceedings of the 13th Jerusalem Winter School for Theoretical Physics, ed. J. Bahcall, T. Piran, and S. Weinberg (World Scientific, Singapore); + Blandford, R.D., Kochanek, C.S., Kovner, I., and Narayan, R. 1989, *Science,* **245,** 824; + Turner, E.L. 1989, in *The 14th Texas Symposium on Relativistic Astrophysics* , in press (Princeton preprint POP-302); + reviews in *Gravitational Lenses* , eds. J.M. Moran, J.N. Hewitt, K.Y. Lo (Springer-Verlag: Berlin), p. 159; + reviews of this workshop.

GG : Gott, J.R., and Gunn, J.E. 1974, *Ap. J.*, **190,** L105.

Gn : Gorenstein, M.V. *et al.* 1988a, in *IAU Symposium 129, The Impact of VLBI on Astrophysics and Geophysics,* ed. M.J. Reid and J.M. Morgan (Dordrecht: Reidel), p.201.

GN : Grossmann, S.A., and Narayan, R. 1989, *Ap. J.*, **344,** 637.

Het : Hammer, F. *et al.* 1989, *Astr. Ap.*, **208,** L7; + Hammer, F., and Rigaut, F. 1989, *Astr. Ap.*, in press.

IC : Irwin, M.J. *et al.* 1989, *Ap. J. (Letters)*, in press; + Corrigan, R. this workshop.

KA : Kochanek, C.S., and Apostolakis, J. 1988, *M. N. R. A. S.*, **235,** 1073.

KBLN : Kochanek, C.S., Blandford, R.D., Lawrence, C.R., and Narayan, R. 1989, *M. N. R. A. S.*, **238,** 43.

KP : Kovner, I., and Paczyński, B. 1988, *Ap. J.*, **335,** L9.

*Kr*1 : Kovner, I. 1987, *Ap. J.*, **312,** 22.

*Kr*2 : Kovner, I. 1987, *Ap. J.*, **316,** 52.

*Kr*3 : Kovner, I. 1987, *Nature,* **325,** 507; + 1987, *Ap. J.*, **321,** 686.

*Kr*4 : Kovner, I. 1989, *Ap. J.*, **337,** 621; + 1988, in *The Post-Recombination Universe,* ed. N. Kaiser and A.N. Lasenby (Kluwer Academic Publishers), p. 315.

*Kr*5 : Kovner, I. 1989, *Ap. J.*, **341,** L1; + submitted, preprint WIS-89/13/March-Ph.

$Kr6$: Kovner, I., submitted, preprint WIS-89/20/April-Ph.

$Kr7$: Kovner, I., submitted and in preparation.

LP : Lynds, R., and Petrosian, V. 1986, *Bull. Am. Astr. Soc.* **18,** 1014; + 1989, *Ap. J.*, **336,** 1.

Lr : Linder, E.V. 1988, *Ap. J.*, **326,** 517.

MQ : Multiple QSOs - recent references: Gorenstein, M.V. *et al.* 1988, *Ap. J.*, **334,** 42, for Q0957+561; Nieto, J.-L. *et al.* 1988, *Ap. J.*, **325,** 644, for Q2345+007; Meylan, G., and Djorgovski, S. 1989, *Ap. J. (Letters)*, **338,** L1, for Q1120+019; Hewett, P.C. *et al.* 1989, *Ap. J. (Letters)*, in press, for Q1429-008. This list may be complemented by Q1635+267 at $z = 2$ split by $4''$, recently revisited by Turner, E.L. *et al.* 1988, *A. J.*, **96,** 1682; and by a case in preparation by Meylan and colleagues.

Nf : Nemiroff, R.J. 1989, *Ap. J.*, **341,** 579.

Nn : Narayan, R. 1989, *Ap. J.*, **339,** L53; + Hogan, C.J., Narayan, R., and White, S.D.M. 1989, *Nature*, **339,** 106.

NS : Narayan, R., and Schneider, P. 1989, MPA preprint 469.

NSC : Narasimha, D., Subramanian, K., and Chitre., S.M. 1985, *Nature*, **321,** 45.

OV : Ostriker, J.P., and Vishniak, E.T. 1986, *Nature*, **322,** 804.

PB : Phinney, E.S., and Blandford, R.D. 1986, *Nature*, **321,** 569.

Pi : Paczyński, B. 1987, *Nature*, **325,** 572.

PG : Press, W.H., and Gunn, J.E. 1973, *Ap. J.*, **185,** 397.

PW : Paczyński, B., and Wambsganss, J. 1989, *Ap. J.*, **337,** 581.

Rl : Refsdal, S. 1964, *M. N. R. A. S.*, **128,** 295;

Rs : Rees, M.J. 1986, *M. N. R. A. S.*, **218,** 25p.

RS : Rees, M.J., and Sciama, D.W. 1968, *Nature*, **217,** 511. + *ibid* 307; + this workshop, and references therein.

SNC : Saslaw, W.C., Narasimha, D., and Chitre, S.M. 1985, *Ap. J.*, **292,** 348.

$SFMP$: Soucail, G., Fort, B., Mellier, Y., and Picat, J.P. 1987, *Astr. Ap.*, **172,** L114.

$Sr1$: Schneider, P. 1987, *Ap. J.*, **316,** L7; + this workshop.

$Sr2$: Schneider, P. 1989, *Astr. Ap.*, **221,** 221.

TK : Turner, E.L. *et al.* in preparation; + Kellog, this workshop.

$Tn1$: Tyson, J.A. 1988, *A. J.*, **96,** 1; + Tyson, J.A., and Seitzer, P. 1988, *Ap. J.*, **335,** 552.

$Tn2$: Tyson, J.A. 1989, this workshop, + private communication.

VO : Vietri, M. and Ostriker, J.P. 1983, *Ap. J.*, **267,** 488; + Ostriker, J.P., and Vietri, M. 1986, *Ap. J.*, **300,** 68.

WCW : Walsh, D., Carswell, R.F., and Weymann, R.J. 1979, *Nature*, **279,** 381.

Wg : Weinberg, S., 1976, *Ap. J.*, **208,** L1.

$WGSW$: Wambsganns, J., Giraud, E., Schneider, P., and Weiss, A. 1989, *Ap. J.*, **337,** L73.

$WHHW$: Webster, R.L., Hewett, P.C., Harding, M.E., and Wegner, G.A. 1988, *Nature*, **336,** 358. + Webster, R.L., this workshop

Wr : Webster, R.L. 1985, *M. N. R. A. S.*, **213,** 871.

YS : Yee, H.K.C. 1988, *A. J.*, **95,** 1331; + Schneider, D.P. *et al.* 1988, *A. J.*, **95,** 1619.

Zh : Zel'dovich, Ya.B. 1964, *Astr. Zh.*, **41,** 19 (*Soviet Astr.*, **8,** 13).

Gravitational redshifts and lensing by large scale structures

Laurent Nottale
C.N.R.S. - D.A.E.C. Observatoire de Meudon.
92195 Meudon Cedex, France

Abstract. Various effects of perturbation of the distance-redshift relation by density inhomogeneities are considered: (1) Gravitational redshifts intrinsic to clusters of galaxies; (2) Redshifts and time delays of background sources; (3) Gravitational magnifications of light by very large scale structures. Observational evidence concerning some of these effects is also briefly discussed.

1. INTRODUCTION.

The consideration of lensing effects from the most general view point leads to asking questions about every gravitational optics effect due to foreground matter on background sources. While most of these effects are consequences of the elementary phenomenon of light rays deviation by gravitational fields, and as such mainly concerns light fluxes (luminosity and/or morphology), one should not forget the possible consequences of the second basic general relativistic prediction, i.e. time dilation and frequency shifts. The Einstein gravitational redshift effect has been already observationally demonstrated for the solar limb, white dwarfs and laboratory experiments[1], but is usually considered to be negligible in the extragalactic domain.

However we believe that such a conclusion should be revised, owing to several facts: (i) while the opacity of stars allows to measure only the frequency shift relative to the difference of gravitational potential between their edge and the infinity, the centers of objects like clusters of galaxies are observable; (ii) the discovery of gravitational arcs behind clusters[2,3] has definitively demonstrated the ability of the richest of them to reach critical densities to lensing, as previously suggested[4]; (iii) the possibility of existence of large amounts of dark matter.

The redshift effects play also an important role concerning the theoretical problem of energy conservation and its relation to the magnitude-redshift relation. It was shown by Weinberg[5] that, *provided one may define Robertson-Walker coordinates* for distant sources, the total energy in a clumpy universe (Friedmannian in the mean) should equal that of the completely uniform model, so that the Mattig[6] relation remains valid in the mean for uniformly distributed sources. A question which still appears unclear is what "Robertson-Walker coordinates" means. In fact it corresponds precisely to neglecting the spectral shift effects, so that a clear quantification of them is needed to judge the validity of this theorem.

2. GRAVITATIONAL REDSHIFTS.

Two different spectral shift effects are to be studied. The first one is the generalization of the Einstein effect, (i.e. a consequence of a difference in gravitational potential between source and observer): in its application to the cosmological domain, one should account for the fact that extragalactic objects are embedded in a cosmological background of non zero density. The second one is a more original effect, (it is due to the crossing by light of a density inhomogeneity involved into differential expansion with respect to the cosmological background), since it is expected to occur even when the source and the observer lie in similar gravitational potentials.

2.1. The Einstein redshift from the centres of clusters.

2.1.1. Solution in the "two-step" vacuole model.

One first possible way to deal with this problem is to work into an exact solution of Einstein field equations including as well a Friedmann cosmological background as a central inhomogeneity. Thanks to the general possibility to match Schwarzschild and Friedmann solutions (i.e. one may match their systems of coordinates in such a way that the metric potentials and their first derivatives remain continuous on the boundary), several inhomogeneous solutions of the gravitation equations may be built, all of them being characterized by the presence of a vacuole[7,8]: one digs a comoving spherical hole into a Friedmann Universe, and places back in the center the extracted mass. Seen from the exterior of the vacuole, there is no additional mass with respect to the original uniform cosmological model, which ensures its stability. Now there are several ways to describe the central region. Point masses may be used[9], but also any extended solution which is known to match with an exterior Schwarzschild solution: e.g. another Friedmann solution of density larger than the cosmological one[4,10-16], or any kind of interior Schwarzschild model[17].

Let us consider the "two-step" vacuole model, made of two Friedmann models separated by a Schwarzschild vacuole. Let r_i and r_e be respectively the internal and external radii of the vacuole. The "null apparent mass condition" writes in terms of the cosmological density ρ_e (at the time of passage across the external boundary of a light ray which would reach the observer at epoch t_o) and of the central density $\bar{\rho}_i$ (at the time of passage of the same photon across the internal boundary) :

$$\rho_e \cdot r_e^3 = \bar{\rho}_i \cdot r_i^3$$

By explicitely matching the three coordinates systems and solving the equations of propagation of a light ray from the center of the inhomogeneity to the observer (assumed to lie in the background Friedmann model), one may compare the redshift of this center in the inhomogeneous model to the value it would have had into the equivalent totally uniform model. This has been done to lowest order in Ref. 13 and to a higher order in Ref. 14. One finds:

$$\delta z_c = 3/2 \,.\, G/c^2 \,.\, M\,(1/r_i - 1/r_e)$$

which may be also written into the form:

$$\delta z_c = 3/2 \,.\, G/c^2 \,.\, M/r_i \,.\, [1-(\rho_e/\bar{\rho}_c)^{1/3}],$$

where the index "c" stands for the cluster center and where, z_c being the cluster cosmological redshift, $\rho_e \approx \rho_o(1+z_c)^3$.

In other words, the intrinsic redshift at cluster center is given for this simple density profile by the standard Einstein redshift, corrected by a term which makes it vanish when the inhomogeneity merges into the cosmos. Though rather logical, such a result can not be considered as *a priori* trivial, since one should remember that there is no mass additional to the background cosmological model and that, as a consequence of spherical symmetry, the inhomogeneity is undetectable by any purely *gravitational* experiment. However, now that we have been able to quantify the relative contributions of the inhomogeneity and of the universe, an estimation of the effect by using a more realistic density profile is needed.

2.1.2. More realistic density profile.

The static spherically symmetric interior Schwarzschild solution of constant density may be generalized to any density profile[1]. The metric potentials may be expressed directly in terms of redshift z. Let $M(r)$ be the total mass included into a sphere of radius r; setting:

$$G.M(r)/c^2 = m(r) \quad \text{and} \quad \zeta = (1+z)^{-1},$$

the redshift is solution of the differential equation:

$$\frac{\mathrm{d}}{\mathrm{d}r}\{[1-2m(r)/r]^{1/2}.\ 1/r\ .\ \mathrm{d}\zeta/\mathrm{d}r\} = \zeta\ .\ [1-2m(r)/r]^{-1/2}\ .\ \frac{\mathrm{d}}{\mathrm{d}r}[m(r)/r^3]$$

Let us now specialize the model and choose a density profile $\rho(r)=a/r^2$ in order to describe a typical cluster of galaxies. The equation becomes:

$$r^2\zeta'' - r\zeta' + [\frac{8\pi Ga/c^2}{1-8\pi Ga/c^2}].\zeta = 0$$

The solution[17] confirms that using the unrealistic profile of the vacuole model leads to an underestimation of the actual effect. One gets to lowest order for the redshift of light coming from the central regions of the cluster ($r<r_o$):

$$z(r) = [3/2 - \frac{r^2}{2r_o^2} + \ln(R/r_o)].\ \frac{m}{R}$$

where r_o stands for a cut-off radius inside which the density is assumed to have reached a maximum value $\rho_o = a/r_o^2$, and R stands for the cluster radius. The redshift of a photon coming from a point of the cluster lying at $r > r_o$ is given by the function:

$$z(r) = [1 + \ln(R/r)].\,\frac{m}{R}$$

This would finally imply a center-edge effect:

$$\delta z \approx \frac{G}{c^2}\,\frac{M}{R}\;[1/2 + \ln(R/r_o)]$$

Though expectedly small for most clusters, such an effect could eventually become observable for the richest of them, particularly those with a strongly peaked density profile in their central regions. For example, a cluster of 3 Mpc radius, $5.10^{15}\ M_\odot$ total mass and 30 kpc internal cut-off radius would show a redshift difference between its central region and its limb of $\delta z \approx 5m/r \approx 100$ km/s, once translated in terms of radial velocity. The observational difficulty to verify such a prediction comes from the high velocity dispersions in clusters. A large number of redshifts could help to overcome it by statistics. A possible method could be to compare the redshift of the brightest cluster galaxy, assumed to trace the potential well of the cluster, to the average redshift of cluster halo galaxies. If one wants to set the effect with a 3σ error, one needs to reach a statistical error on the mean of ≈ 30 km/s, i.e. ≈ 1000 redshifts are needed if one starts from clusters with a velocity dispersion of ≈ 1000 km/s. We are now not too far from being able to work out such a test, for example by adding data coming from different clusters, after having properly set their different scales. The method proposed hereabove would be preferable to the study of the average redshift as a function of angle from center, since the integration on the line of sight strongly decreases the effect. Indeed one expects an average redshift on the central line of sight:

$$z_S(0) = \frac{5}{3}\,\frac{m}{R}$$

Another method would be to consider again the proposal of Nottale[18], consisting in statistically comparing the redshifts of clusters into pairs of different richness. Rood and Struble[19] concluded in 1982 that the data was still too poor to reach a conclusion (the statistical error was not smaller than 150 km/s), but the situation may have changed by now thanks to the fast increase of the number of redshifts measured by multiple object spectroscopy.

The interest of such a test would be, additionally to a new verification of the Einstein general relativity theory in the extragalactic domain, to yield a new and independant measure of the total mass of clusters of galaxies, and as a consequence of the presence of dark matter.

2.2. Redshift effect on background sources.

The redshift and time delay resulting from the crossing of a cluster by the light coming from a background source constitute a fainter effect, harder to set theoretically. It was first studied by Rees and Sciama[20] in the frame of the vacuole model. However their calculation was only approximate, in that they did not explicitely work out the matching of metrics throughout the boundaries. This was done by Dyer[10], who get a result exactly twice the Rees and Sciama value. This factor of two is reminiscent of a number of similar factors from "classical" to "relativistic" calculations. In particular it can be compared to the same correction applying in general relativity to the "Newtonian" estimation of the light rays deviation by masses ($2m/r \rightarrow 4m/r$), owing to the fact that the redshift effect can be considered as the equivalent for temporal coordinate of what is the light deviation for spatial coordinates. However Dyer's conclusion remained that this gravitational redshift was inobservable, since masses larger than 10^{19} $M_\odot$ would be needed.

This conclusion was revised by a generalization of this calculation, made in terms of immediately observable quantities, to clusters in a phase of fast contraction[14]. In order to establish the effect, one should compare two light rays, one travelling in the totally uniform Friedmann model and the other crossing an inhomogeneity of null apparent mass added to that model. It is assumed that both photons reach the observer at the same date. Two phenomena contribute to the effect:

(i) A spectral shift coming from a differential expansion in the central high density Friedmann model compared to the outer Universe. Calling $\boldsymbol{H}_c$ the value of the Hubble constant at the centre of the cluster at the time when the photon passes, it is given by[10,14]:

$$\delta z_1 = 4GM/c^2 \,.(H_o/c - \boldsymbol{H}_c/c)$$

The dynamical nature of this effect is confirmed by its vanishing when $H_o = \boldsymbol{H}_c$. This is most of the time a reshift, since the excess mass is expected to slow down the expansion, so that $H_o > \boldsymbol{H}_c$. A contraction would correspond to $\boldsymbol{H}_c$ negative.

(ii) A time delay, coming from the fact that the light velocity is different in the inhomogeneity of what it is in the cosmos. It will result into an additional blueshift term:

$$\delta z_2 = -4GMH_o/c^3 \,.\, \ln(r_e/r_i)$$

where it is recalled that $r_e/r_i = (\bar{\rho}_i/\rho_e)^{1/3}$. The total spectral shift is finally given by the sum of these two terms. Such an effect is clearly too small to be observed on a discrete source. However an interesting possibility is its consequence on the cosmic microwave background radiation (MBR). It would result into a temperature decrement in the direction of a cluster of galaxies relative to the average temperature given by:

$$\delta T/T = -4GMH_o/c^3 \,.\, [1 - \ln(\bar{\rho}_i/\rho_e)^{1/3} - H_c/H_o]$$

The effect remains small if $|H_c/H_o| \approx 1$. But if there exist clusters involved at some epoch of their evolution into a phase of rapid collapse, large negative values of H_c could be expected. For example to collective infall velocities of 1000 km/s at a distance of 1 Mpc from center would correspond a ratio $H_c/H_o \approx -20$ for $H_o = 50$ km/s.Mpc. Such a cluster, if its *total* mass was $M \approx 5.10^{16}$ $M_\odot$, would yield a relative MBR temperature decrease of $\approx 4.10^{-5}$. This could contribute, in parallel with the Sunyaev-Zeldovich[21] effect, in a non negligible way to the MBR decrement of $\approx 10^{-4}$ now confirmed in some clusters as 0016+16 or A2218. But the gravitational effect is a pure redshift (as is the variation of MBR temperature with cosmological time), which means that it corresponds to an identical temperature decrease in the high and low energy parts of the spectrum. On the contrary the Sunyaev-Zeldovich effect is an inverse Compton effect of MBR photons on the electrons of the intra-cluster gas, and as such corresponds to a temperature decrease in the Raleigh-Jeans part of the spectrum but to an increase in the Wien one. Hence it is predicted that if both effects contribute they would be expected to cancel in spatial observations of the Wien part of the blackbody spectrum.

This result is confirmed by a direct calculation madé in the single frame of the two-step vacuole model. Assuming that the interior and exterior Friedmann models are issued from the same Big-Bang singularity allows to compute the value of the inner Hubble constant[14]. One finds:

$$H_c = -4[\pi G \rho_c/3]^{1/2}$$

which finally yields a temperature decrement:

$$\delta T/T = -4\sqrt{2}\,(m/r_i)^{3/2}$$

where $m = GM/c^2$. Values of $\delta T/T$ larger than 10^{-5} can be obtained for clusters whose total mass is greater than 10^{16} $M_\odot$. The problem of the setting of the effect in the frame of more realistic density profiles and velocity fields remains open. If, as is the case for the generalized Einstein redshift (see hereabove), such an account increases the theoretical expectation, this would increase also the possibility that the gravitational effect is indeed at work in the direction of the richest clusters of the Universe.

3. GRAVITATIONAL MAGNIFICATIONS BY LARGE SCALE STRUCTURES.

3.1. Propagation of a light beam.

The equations of propagation of a light beam have been written in the general relativistic frame by Sachs[22]. These optical scalar equations (OSE) relate the three possible deformations of the beam, i.e. expansion, rotation and shear to the geometry of the Universe, described by the Riemann and Ricci tensors. They were written by Kantowski[9] in terms of the light beam cross sectional area A. These methods have been extensively used since the first suggestion by Zeldovich[23] that the clumpy distribution of matter in the Universe may perturb the distance-redshift relation. First applied to the effects of emptiness[23-27,5], which finally revealed to be small, their application to density excesses[28,11-17,4], met the gravitational lensing theory which had been developed in parallel in the frame of the linearized Einstein theory[29]. Exact solutions to the propagation of light rays in the two-step vacuole model, which is itself an exact solution of Einstein equations, have been given by Nottale[11-14], Nottale and Hammer[4] and Hammer[15,16]. Though heavier than the Einstein linearized methods when applied to the computation of shear effects, the OSE method remains better adapted to the account of the fact that inhomogeneities are embedded in a background universe of non vanishing density. In particular solutions to the problem of lensing by large scale structures may be obtained in this frame.

When looking at large scale magnification effects, only the matter terms can be considered in the OSE. This is because the shear effects by individual masses average themselves to yield in the mean the matter effect which is expected from a smoothing out of the same mass distribution[5,17,30-33] .

It has been shown[4,17,34] that the OSE may be linearized in terms of new variables. One defines an "optical distance", which is the following function of redshift:

$$\Delta(z) = q_o^{-2} \left[\frac{q_o z + 1 - q_o}{\sqrt{1+2q_o z}} + q_o - 1 \right]$$

and a reduced light beam diameter:

$$\lambda = \frac{(1+z)^2}{\sqrt{1+2q_o z}} \sqrt{A}$$

In terms of these new variables the light beam propagation becomes linear in a Friedmann model of present density ρ_o. Indeed the shearless OSE writes in that case:

$$d^2\lambda/d^2\Delta = 0$$

Another advantage is that the optical distance is a monotonic function of redshift (which is not the case for the angular diameter distance when $q_o>0$) and that the usual distance term in gravitational lensing formula is now linearized (see Fig. 1 in Ref. 17):

$$D_d.D_{ds}/D_s = \frac{c}{H_o} \frac{1+2q_o z_d}{(1+z_d)^3} \Delta_d \, (1 - \Delta_d/\Delta_s)$$

Now the effect of a perturbation to the average uniform model becomes a perturbation to linearity. We adopt in the following the Dyer-Roeder method, in which we admit that one may neglect the metrics perturbations, while the optics perturbations are assumed to be well described by a Ricci tensor in which the actual density is input. The perturbed ESO writes[17] :

$$d^2\lambda/d^2\Delta + k.\omega^2(\Delta).\lambda + \xi^2 \cdot \lambda^{-3} = 0$$

where $k=\pm 1$, ξ is the usual shear term, unaffected by the change of variable, and where the generalized driving term is given by:

$$k.\omega^2(z) = 3\, q_o \frac{(1+2q_oz)^2}{(1+z)^3}\left[\frac{\rho(z)}{\rho_o(1+z)^3} - 1\right]$$

The effects of density fluctuations on a light beam are now immediately understandable. If it propagates in a medium of average density larger than the local cosmological density, $k>0$ so that one gets for the shearless beam a "sinus" behaviour, i.e. a convergence (then a magnification) with respect to the unperturbed case; when $\rho(z)<\rho_o(1+z)^3$, one gets a sinh divergence, i.e. attenuation. We consider in the following an application of these methods to the obtention of new solutions to the problem of gravitational lensing by various structures.

3.2. Extended lenses.

Consider a line of sight starting from a source at distance Δ_S, then travelling into a uniform Friedmann model up to a distance Δ_C+R, then into an extended lens characterized by a constant value of ω^2 up to an optical distance Δ_C-R from the observer, and then finally travelling to the observer (at $\Delta=0$) into the initial uniform model. The solution of the OSE in this case has been given in Ref. 17. A third order expansion of this result in terms of the dimensionless radius R yields a magnification M given by:

$$M^{-1/2} = 1 - 2\omega^2 R\Delta_c\,[1-\Delta_c/\Delta_S] + 2R^3/3\,[2\omega^4\Delta_c\,(1-\Delta_c/\Delta_S) - \omega^2/\Delta]$$

The remarkable result here is that the second order term vanishes, while the first order one is the usual κ of the thin lens shearless formula $M=(1-\kappa)^{-2}$. The scale of R being that of redshifts, this means that the thin lens magnification remains correct even for large scale structures. For example it could be applied to an edge-on galaxy filament: even 300 Mpc long, this would correspond to $R\approx 0.05<<1$.

3.3. Multiple lensing along the line of sight.

The hereabove methods have been used[34] to obtain an explicit formula for the (one image) matter magnification due to several lenses distributed at any place between the source and the observer. Let κ_{ij} be the usual κ term (see hereabove) corresponding to a lens at redshift z_i and a source at redshift z_j. The magnification by n lenses ($z_1, z_2, ...z_n$) on a source (z_s) may be written as:

$$M^{-1/2} = 1 - \Sigma_i\,\kappa_{is} + \Sigma\Sigma_{i<j}\,\kappa_{ij}\kappa_{js} - \Sigma\Sigma\Sigma_{i<j<k}\,\kappa_{ij}\kappa_{jk}\kappa_{ks}\ ....$$

This result was confirmed[35,36], then was used to study the lensing effect of foreground matter on absorption line QSOs[17,37]. A highly significant correlation was found between the prediction of the hereabove formula, applied to the observed absorption redshifts (z_i) and emission redshift (z_s) of quasars and their "absolute" luminosity. It was concluded that a strong selection effect by lensing was at work among these sources and that about half of the apparent evolution was a consequence of it. Recent work on the relation of lensing and (absorption-line) QSOs may be found in the present volume[38,39].

3.4. Very large scale density fluctuations.

Consider now a line of sight along which the density is slowly varying, in such a way that the condition $d\omega/d\Delta << \omega^2$ is always verified. The similarity of the OSE to the one-dimensional Schrödinger equation allows to use the BKW approximation in this case, and so to get an explicit new solution whatever the density profile from source to observer:

$$\mathrm{Amp} = \frac{\omega(0)\,\omega(\Delta)\,\Delta^2}{\sin^2 \int_0^\Delta \omega(\Delta)\, d\Delta}$$

This result opens the possibility to have large fluctuations of luminosity of distant objects compared to the Mattig prediction, even in the absence of a well definite lens. For example, if due to statistical fluctuations, there existed a line of sight for which $\omega^2/3q_o \approx 1$ (this corresponding to excess densities by a factor 2 to 6), a light beam with vertex at the observer would be expected to reconverge at $z=4$ if $q_o=1/2$, i.e. a quasi infinite amplification for a point source lying there.

4. Conclusion and prospect.

The theoretical consideration of the various possible effects of large scale structures (richest clusters of galaxies, superclusters, filaments, very large scale random density fluctuations) on the light of sources situated behind them leads to the conclusion that their influence on distant cosmological observations may have been up to now underestimated[4,40]. The difficulty with these kinds of effects, compared to standard gravitational lensing, is that they reduce to luminosity magnifications and/or spectral shifts, without any clear morphological signature such as image multiplication, deformation or variability.

Concerning the gravitational temperature decrement of the MBR, a possible way to reach its observationability would be to look in the direction of *several superposed rich clusters*. This is not an unlikely situation: if one estimates to 10000 the total number of rich Coma/A370-like clusters on the whole sky, the probability that two clusters be superposed by less than $\theta=2'$ is $P=\pi.\sigma.\theta^2 \approx 9.10^{-4}$ so that at least 9 cases are expected.

Concerning magnification effects, their quantification in the absence of image multiplication passes by statistical methods. It was proposed in 1975 by Karoji and Nottale[41] to study the lensing effect of rich clusters on standard candles by statistically comparing the Hubble diagram of those which lie behind clusters to the remaining sample. This method was later applied with success to Brightest Cluster Galaxies[42], absorption-line QSOs[17], and distant radiogalaxies[43], and should be generalized once complete catalogs of distant clusters will become available.

References.

1. Weinberg, S., 1972, *Gravitation and Cosmology,* [Wiley and Sons]

2. Soucail, G., Fort, B., Mellier, Y., Picat, J.P., 1987, *Astron. Astrophys.*, **172**, L14

3. Lynds , R., Petrosian, V., 1986, *Bull. Astron. Am. Soc.*, **18**, 1014
4. Nottale, L., Hammer, F., 1984, *Astron. Astrophys.*, **141**, 144
5. Weinberg, S., 1976, *Astrophys. J. Lett.*, **208**, L1
6. Mattig, W., 1958, *Astron. Nachrichten*, **284**, 109
7. Einstein, A., Strauss, E.G., 1945, *Rev. Mod. Phys.* **17**, 120
8. Schucking, E., 1954, *Zs. J. Phys.* **137**, 595
9. Kantowski, R., 1969, *Astrophys. J.*, **155**, 89
10. Dyer, C.C., 1976, *Mon. Not. Roy. Astr. Soc.*, **175**, 429
11. Nottale, L., 1982, *Astron. Astrophys.*, **110**, 9
12. Nottale, L., 1982, *Astron. Astrophys.*, **114**, 261
13. Nottale, L., 1983, *Astron. Astrophys.*, **118**, 85
14. Nottale, L., 1984, *Monthly Notices Roy. Astron. Soc.*, **206**, 713
15. Hammer, F., 1985, *Astron. Astrophys.*, **144**, 408
16. Hammer, F., 1985, *Astron. Astrophys.*, **152**, 262
17. Nottale, L., 1988, *Ann. Phys. Fr* .**13**, 223
18. Nottale, L., 1976, *Astrophys. J. Lett.*, **208**, L103
19. Rood, H.J., Struble, M.F., 1982, *Astrophys. J. Lett.*, **252**, L7
20. Rees, M.J. & Sciama, D.W., 1968, *Nature*, **217**, 511
21. Sunyaev, R.A., Zeldovich, Ya.B., 1972, *Comm. Astr. Space Sci.*, **4**, 173
22. Sachs, P.K., 1961, *Proc. Roy. Soc. London*, **A264**, 309
23. Zeldovich,Ya. B., 1964, *Soviet. Astr.-Astron.J.*, **8**, 13
24. Gunn, J.E., 1967, *Astrophys. J.*, **150**, 737
25. Dyer, C.C., Roeder, R.C., 1972, *Astrophys. J. Lett.*, **174**, L115
26. Dyer, C.C., Roeder, R.C., 1973, *Astrophys. J. Lett.*, **180**, L31
27. Dyer, C.C., Roeder, R.C., 1974, *Astrophys. J.* , **189**, 167
28. Nottale, L., 1980, Thèse de Doctorat ès Sciences, Université Paris VI
29. Bourassa, R.R., Kantowski, R., 1975, *Astrophys. J.*, **195**, 13
30. Young, P., 1981, *Astrophys. J.*, **244**, 756
31. Dyer, C.C., Roeder, R.C., 1981, *G.R.G.* , **12**, 1157
32. Peacock, J.A., 1986, *Mon. Not. Roy. Astr. Soc.*, **223**, 113
33. Marchandon, S., 1989, this volume
34. Nottale, L., Chauvineau, B., 1986, *Astron. Astrophys.*, **162**, 1
35. Kovner, I., 1987, *Astrophys. J.* , **321**, 686
36. Schneider, P., Weiβ, A., 1987, *Astron. Astrophys.*, **171**, 49
37. Nottale, L., 1987, *Ann. Phys. Fr.*, **12**, 241
38. Thomas, P.A., Webster, R.L., 1989, this volume
39. Surdej, J., 1989, this volume
40. Turner, E.L., Ostriker, J.P., Gott III, J.R., 1984, *Astrophys. J.*, **284**, 1
41. Karoji, H., Nottale, L., 1976, *Nature*, **259**, 31
42. Hammer, F., Nottale, L., 1986, *Astron. Astrophys.*, **167**, 1
43. Hammer, F., Nottale, L., Le Fèvre, O., 1986, *Astron. Astrophys. Lett.*, **169**, L1

WHAT THE PRESENT OBSERVED LENS SYSTEMS TELL ABOUT THE LENSING AGENT

D. Narasimha and Sunita Nair
Theoretical Astrophysics Group
Tata Institute of Fundamental Research
Homi Bhabha Road, Bombay 400 005
India

1. Introduction:

The lens systems consisting of multiple images or source with extended structures provide valuable information regarding the mass and length scale of the lens. In the simple case of a single point object lensing a background point source we can determine the mass of the system if the distance to the source and the lens are known, using the relation,

$$M = r_1 r_2 c^2 (\frac{D_l}{4G})(\frac{D_s}{D_{ls}})$$

where M is the mass of the point lens, r,observed angular separation between the lens and the image, and D_l, D_s and D_{ls} are respectively the angular diameter distances from the the lens to the observer, source to the observer and the source to the lens. Moreover, from the intensity ratio between the two images we can test our assumption that the lens has size much less than the angular distance from the images. Consequently, in principle a point lens can also be used to determine the distance to the lens or to the source or the distance ratio D_{ls}/D_s if the other parameters are known. However, even when only two images are present it is possible to deduce that the lens is extended and it consists of either multiple components or higher moments because the lens and the images will not in general appear collinear. In such systems we can only get a lower limit on the mass of the lens, and in some cases possibly a limit on the size of the lens also.

If the images are extended the lens model becomes more definite for the following reasons:

1. The ratio of the linear magnification between the images in two different directions can be determined from observations, and consequently the elements of the transformation matrix between the images, not just its determinant, can be computed. In addition, if the centre of the lens is observable, we can get an idea of the orientation of the axes of a simple elliptical lens. (In view of the discussions in this workshop regarding the definition of this term: by elliptical we mean a lens whose projected mass distribution perpendicular to the line of sight has elliptical symmetry)

2. While the observed continuum flux ratio between the images is apt to be influenced by minilensing, the parameters for a structure extending over a tenth of an arcsecond or more will not feel the effect unless the minilens is massive or is situated near a caustic.

3. If we can assume that there exists a point (or a small region) of the image at which both the linear and tangential magnification are large, the projected surface mass density of the lens at that point is very nearly equal to the critical mass density given by,

$$\Sigma_c = \frac{c^2}{4\pi G}\left(\frac{D_{ls}}{D_l D_s}\right)$$

However, a more useful practical estimate can be obtained when the image is an arc having size comparable to the radius of the cone of inversion of the lens. In this case, if the projected mass distribution of the lens is nearly circularly symmetric, the average surface mass density within the circle enclosed by the arc is approximately equal to the critical density. Since the radius of curvature of the observed arcs is known reasonably accurately, we can estimate the enclosed mass of the lens (modulo the Hubble constant, H_0). However, as the lens becomes more asymmetric, the estimation becomes poorer. Nevertheless, in this manner we could get an idea of the large scale variation of the density in a cluster producing multiple arcs.

At present lens configurations consisting of point images of separation less than an arcsecond to about 7 arcseconds and arcs having linear size of the order of half an arcminute have been observed. Only five or six of these have sufficient observational results to test the lens models; for the other systems, the models should be treated as mere indicative numbers. In the next section we present such typical numbers for some of the lenses; these numbers are based on model-independent calculations, observations and explicit modelling. We have taken the Hubble constant H_0 to be 50 km/sec/Mpc and q_0=1/2. However, apart from scaling, the numbers are independent of these cosmological parameters. In the following section we discuss some implications of these models on the large scale distribution of matter in the Universe. The model-dependency of the result is discussed in the last section.

2. Mass and Length Scales as Inferred from Lensing:

The observed lenses can be broadly divided into three categories:

1. Systems in which the mass associated with the luminous matter alone can explain the lensing action.

2. Configurations which require lensing by massive unseen component whose length-scale is large compared to its observed optical counterpart.

3. Lensing due to clusters of galaxies, where the observed velocity dispersion and size appear to be in confirmity with the scales inferred from the images.

In Table 1 we have presented the lens parameters for six systems of small angular separation. The exact model is not important, but for these systems *a halo of a less than approximately 10 kpc core radius* and mass a few times the mass of the luminous component is unacceptable. A halo of much larger size is possible, but we cannot get information regarding a nearly constant surface mass distribution from the separation and intensity ratio or the time delay of the images. But we might state that the lensing action in these cases is essentially by the luminous matter.

Table 1. Gravitational Lenses Producing Images of Small Separation

SYSTEM	**Max. Sep.** (arcsec)	**Mass of Lens** ($10^{11} M_\odot$)	**Core Radius** (kpc)	**Comments**
+PG1115+080	2.3	6	2.3	–
+MG1131+046	2.1	10	2.5	Assumed: $z_l = 0.5$ $z_s = 1.5$
+UM673	2.2	5	2.0	–
+H1413+117	0.8	9	3.0	Assumed: $z_l = 1.5$
+MG1654+1346	2	8	2.7	Radio Ring
+1830–211	1	1.4	1.0	Assumed: $z_s = 0.2$

There are at least four lens candidates which compel us to invoke dark matter having mass greater than $10^{12} M_\odot$ as the lensing object. The case for massive halo appears to be strong for the object MG2016+112, as indicated by the broad emission lines, radio continuum and the newly observed VLB structures (Cf. Heflin *et al.*,1989). But for other systems we can only infer a typical scale. However, the main conclusion from an analysis of these systems is that *the mass and length scales inferred from lensing are consistent with the corresponding numbers obtained from the X-ray observations of massive isolated ellipticals.* (cf. Fabian and Thomas, 1986; Quinn and Hernquist, 1986)

Table 2. Systems requiring large Unseen Mass as Lensing Agent

SYSTEM	max. separation	LUMINOUS PART mass ($10^{11} M_\odot$)	LUMINOUS PART core radius (kpc)	UNSEEN COMPONENT mass ($10^{13} M_\odot$)	UNSEEN COMPONENT core radius (kpc)
MG2016+112	3.4	4	2.2	1	30
0023+171	~ 5	1.5	3	0.8	16
1635+267	3.8	5	1.5	2	30
2345+007	7.1	5	1.5	1.5	15

The large number of giant arcs and rings being observed in the recent years *support and compel* us to invoke mass distribution in the lensing clusters of galaxies at length scale of a few hundred kiloparsecs.(cf Soucail *et al.*, 1987; Mellier *et al.*, 1988). As seen from Table 3, the virial mass estimated from the velocity dispersion and observed number counts of galaxies are consistent with the lens parameters. But the size obtained from the X-ray data is generally larger than the value inferred from lensing, though the position of the centre of mass appears to agree.

Table 3. Gravitational Lensing
of Extended Sources by Clusters

OBSERVED LENS PARAMETERS System	σ_{los} (km/s)	Core radius (kpc)	$\Sigma(0)$ ($g.cm^{-2}$)	Σ_{av} ($g.cm^{-2}$)	Σ_c ($g.cm^{-2}$)
A370	1350	100	1.2	0.92	0.73
A963	1100	130	0.7	0.6	0.65

3. Argument for Massive Compact Dark Objects:

While the numbers presented in the previous section are apt to vary slightly according to the density profile used for the models, we do not expect the general inferences to be sensitive to the details of the models. However, in the case of 2345+007 no lens of magnitude lower than 23 could be detected in spite of deep imaging. Further, the images are not much brighter than normal unlensed quasars at the same redshift. This is suggestive of the lensing action of a more centrally condensed object. Such a mass distribution with core radius 5kpc and velocity dispersion of the order of 350km/sec can explain the observed properties of the system quite well. Possibly unobserved lenses in many other systems might belong to this class. But if the object is a galaxy where star formation did not occur, it should be observable through its redshifted 21cm radio emission (cf Narasimha and Chitre, 1989). Probably we could differentiate between the lensing due to massive centrally condensed dark objects and halos associated with giant galaxies through their action on extended sources. While an extended massive halo will produce miniarcs of linear size 2" to 5" due to lensing of background galaxies, compact candidates are likely to produce a slightly distorted image and its faint counterpart at a few arcseconds away. Deep survey should be able to rule out massive dark halos and dark compact objects from the failure to observe either kind of images of distant galaxies.

4. Limitations of Lens as a Probe of the Structure of Universe:

It should be noted that any information regarding the images in a two-dimensional surface will not enable us reconstruct the three-dimensional matter distribution of the lens uniquely. Even the simple parameters obtained from lensing give a somewhat biased view of Universe. As an example, we are more likely to detect the lensing action of objects of high ellipticity in any incomplete sample. We encounter a more serious problem while using the lens statistics because the *relation between the mass, size and symmetry* of the lens influence the image properties. Consider a circularly symmetric projected mass distribution varying as $(r_c^2 + r^2)^{-1}$, r_c being the core radius. From Figure 1. we get an idea of when multiple image formation is possible. Let the velocity dispersion of a class of lensing objects were to vary as r_c^n. If $n > 2$ we would observe lens systems with large separation while, if $n < 2$ more systems of small separation are expected. Consequently, if lenses of only one class were present they could have been studied from the distribution function of the images. However, when more than one class of objects are present and a specified lens consists of multiple components, the separation between the images depends on which component dominates. The net result is that, the expected *distribution function of the image separation or intensity ratio are very much dependent on the number of classes of lensing objects as well as the relation between mass and size of the lens.*

Summary

Limits on the mass and size of the lens can be obtained from the observed properties of the images. For point lens the system is overdetermined. However, the structure of an extended lens is not uniquely determined from the images. The present observed lenses are suggestive of large scale matter distribution at three scales: galaxies of core radius 1-3 kpc and mass $10^{11} - 10^{12} M_{\odot}$, massive dark halos of core radius 10-30 kpc and mass $10^{13} M_{\odot}$ and compact clusters of core radius $100 - 200$ kpc. But inferences drawn from lens statistics are likely to be biased. We are likely to observe galaxies of higher ellipticity or compact clusters more often in a finite Universe using instruments having limitations. At a specific redshift of the lens, a specific class of lenses will have projected surface mass density closer to the critical value and consequently, their influence will be more pronounced.

Acknowledgement: D.N. is grateful to CNRS, France for a Research Associateship and to the Observatoire Midi-Pyrenees for hospitality.

References:

Fabian, A. C. and Thomas, P. A., 1986 in *"Structure and Dynamics of Elliptical Galaxies"*, IAU Symp. no. 127, *ed.* de Zeeuw, T. p155.

Heflin, M. *et al.*, 1989 in this conference.

Mellier, Y. *et al.*, 1988, *Astron. Astrophys.* **199**, 13.

Narasimha, D. and Chitre, S. M., 1989, *Astron. J.* **97**, 327.

Quinn, P. J. and Hernquist, L., 1987 in it "Structure and Dynamics of Elliptical Galaxies", IAU Symp. no. 127, *ed.* de Zeeuw, T. p249.

Soucail, G. *et al.*, 1987,*Astron. Astrophys. Letters* **172**, L14.

Fig. 1. Formation of Multiple Images by Gravitational Lenses

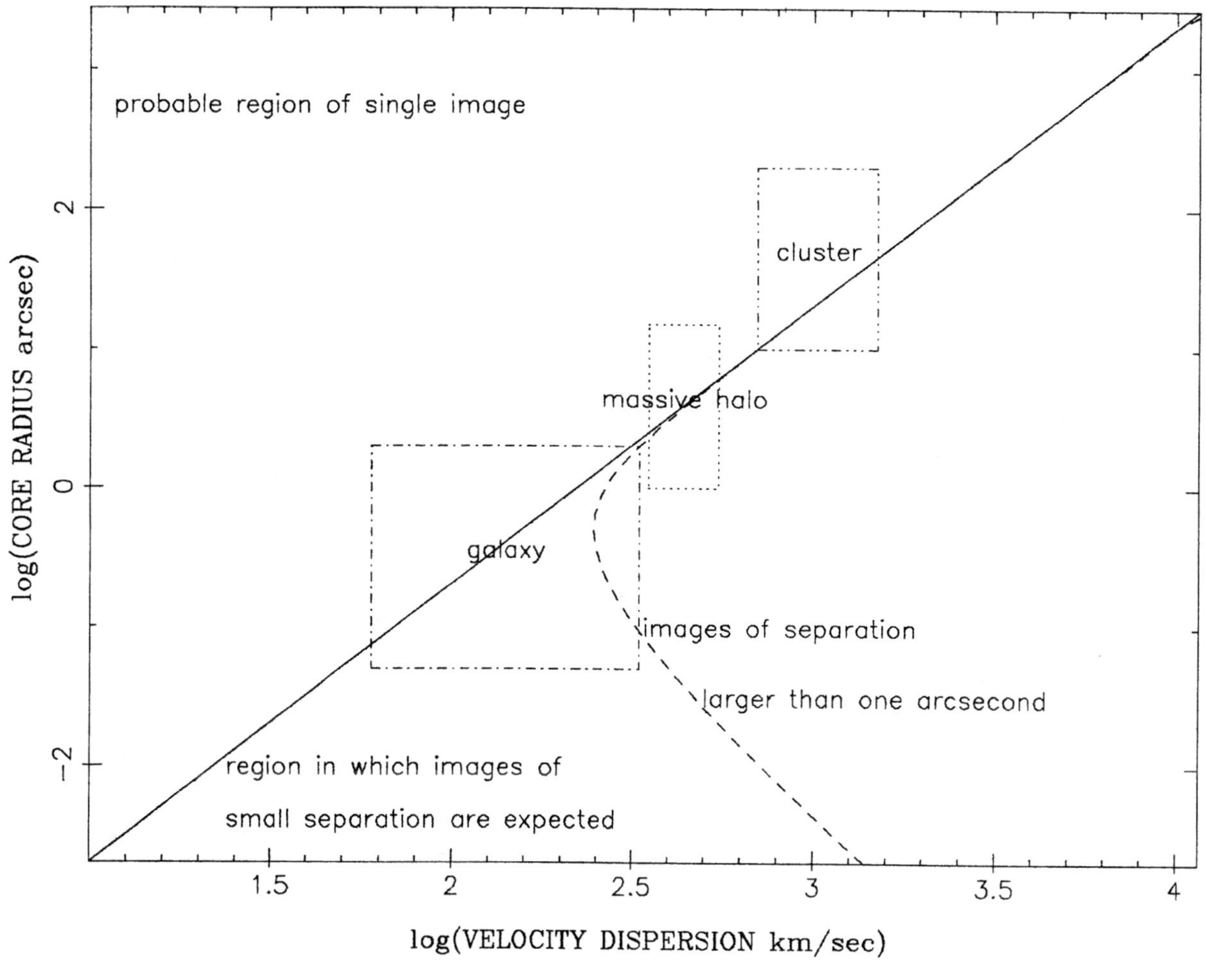

Gravitational Lens Models with Elliptical Geometry

T. Schramm
Hamburger Sternwarte, Gojenbergsweg 112, D–2050 Hamburg 80

1 The problem

The observation of multiple images of distant sources, probably lensed by intervening galaxies or clusters of galaxies, makes it desirable to have some models to describe the lensing features of these galaxies. What we need exactly, is the gradient of the twodimensional gravitational potential produced by the projected matter distribution of the galaxy. Unfortunately galaxies seem to hide most of their material. Thus our ideas of the *real* mass distribution in galaxies come on one hand from theoretical models of stellar systems and on the other hand from the hope that the dark matter might share the symmetry of the luminous matter. If we try to find the potentials of matter distributions we have to remember that we deal with potential–density *pairs* which do not share their symmetries in general. Elliptical potentials usually have non-elliptical matter distribution and vice versa. For complicated morphologies of galaxies there is no chance to find simple expressions describing their potentials and we are stuck to the piecemeal summation of the contributions of each mass element. Fortunately, for spherical and elliptical systems the situation is not so bad. There are two main approaches to describe galaxies with elliptical geometry. First potentials with enforced elliptical symmetry and second potentials generated by elliptical mass distributions.

2 Potentials with enforced elliptical symmetry

If we define a set of ellipses in angular coordinates (x, y) on the sky

$$(1-\epsilon)x^2 + (1+\epsilon)y^2 = m^2 \tag{1}$$

we find a potential with elliptical symmetry

$$\Phi = \Phi(m). \tag{2}$$

By applying Poisson's equation we see that the corresponding density σ is not only a function of m. The isodensities are no ellipses. Even worse, we find for particular shapes of the potential (e.g. the logarithmic, elliptical potential, Binney and Tremaine, 1987, p. 45) regions of negative density. Most authors (e.g. Blandford et al., 1988; Kochaneck, this volume) prefer the potential of the generalized isothermal sphere. Using Kochaneck's notation it reads

$$\Phi = b(s^2 + m^2)^{1/2} \tag{3}$$

where b denotes the deflection scale and s the core size of the potential. The density remains positive and for small ϵ the isodensities look quite elliptical with an eccentricity of about three times the eccentricity of the potential. For higher eccentricities the isodensities turn out to be dumb-bell shaped. As one can see from the spherical case, the density varies approximately with r^{-1} for large radii. This can be a good model inside a galaxy. However, since the total mass is growing approximately linear with radius the regions outside of a galaxy where the total mass stays constant are not well represented. We note that the potential Φ is generated by the infinitely extended density distribution with diverging total mass. This is no problem for the spherical case, because at a location r the outer regions do not contribute to the gradient of the potential at r. This nice feature of spherical density distributions can however only be extended to the very special density distributions with homoeoidal elliptical symmetry (see next section). To test the influence of the outer parts of the mass distribution on the deflection at the inner regions, we cut out the central part of the distribution along an isodensity σ_c. Using a Monte Carlo integration method (Press et al., 1986, p.221) we computed the deflection angle produced by the interior mass inside σ_c. By subtracting the deflection angle α_t due to the total potential (Eq. (3)) we found the deflection due to the outer infinite regions α_{ext} at a location inside the isodensity. First preliminary results show that for small eccentricities α_{ext} can be about one percent of α_t (for parameters $\sigma_c = \sigma(2,0), b = 0.97, s = 0.07, \epsilon = 0.14$). For higher eccentricities ($\sigma_c = \sigma(2,0), b = 1.0, s = 0.1, \epsilon = 0.5$) we found values for α_{ext} up to fifteen percent of α_t.

3 Potentials of elliptical density distributions

Potentials with enforced elliptical symmetry can lead to unphysical deflector models. An alternative approach is to start with the density distribution of elliptical symmetry. But first we have to clarify the meaning of "elliptical". We speak of a *homoeoidal* (elliptical) symmetry, if the isodensities are similar and similar situated, concentric ellipses, given by an equation like Eq. (1). A homoeoid then is a ring of constant surface density bounded by two homoeoidal ellipses.If not mentioned otherwise, we will only treat infinitesimal thin rings. The other important elliptical symmetry is called *confocal*. Here the iso-lines are confocal and similar situated, concentric ellipses. Now we ask for the potential and the deflection field generated by a homoeoidal surface mass distribution. The 3D-case was treated by many authors. The results can be found in Chandrasekhar (1969) and Routh (1892). Their results for homoeoidal 3D-shells can be transformed to the potentials and force-fields for infinitely long homoeoidal, elliptical cylinders, simply by taking the limiting case of one infinitely long semi-axis aligned with the direction of sight. We achieve a cylindrical potential independent of the third coordinate, which is equivalent to the required 2D-potential. Properly translated all theorems about 3D-homoeoids stay true:

1. The deflection angle inside a homoeoid of constant surface mass density is zero.
2. This is even true for thick homoeoids if the curves of equal surface density are homoeoidal ellipses.
3. The equipotential curves exterior to a homoeoid are ellipses, confocal to the homoeoid.
4. All confocal homoeoids of equal mass produce at a location r exterior to them the same deflection, which only depends on the mass and the parameters of the confocal ellipse through r.

These theorems could be rather helpful for the modelling of gravitational lenses since they give some geometrical insights in the behavior of homoeoidal mass distributions. A broad discussion of this approach to the potential for homoeoidal mass distributions can be found in Schramm (1990) where we give the complete algorithm to compute the deflection angle.

In 1973 Bourassa et al. introduced a formulation for the deflection angle of a projected spheroidal mass distribution using a mixed real- and complex-valued formalism which is quite cumbersome to apply. Their results are equivalent to ours, noting that every triaxial, homoeoidal 3D mass distribution shows in projection homoeoidal 2D symmetry (Stark, 1977).

A very elegant reevaluation of our results can be done by using complex calculus and the theory of conformal mappings, which is also applyable for other symmetries. A short introduction is given in Schramm (1990, Appendix B).

References

Blandford, R., Kochanek, C.S., Kovner, I., Narayan, R.: 1988, *Caltech Preprint*, **GRP-177**

Bourassa, R.R.,Kantowski, R., and Norton, T.D.: 1973, *Astrophys. J.*,**185**,747.

Binney J.J. and Tremaine, S.: 1987, *Galactic Dynamics.* Princeton University Press, Princeton, New Jersey.

Chandrasekhar, S.: 1969, *Ellipsoidal Figures Of Equilibrium.* Yale University Press, New Haven.

Press, W.H. et al.: 1986, *Numerical Recipes.* Cambridge University Press, Cambridge.

Routh, E.J.: 1892, *A Treatise On Analytical Statics.* Volume 2, Cambridge University Press, Cambridge.

Schramm, T.: 1990, *Astron. Astrophys*, in press.

Stark, A.A.: 1977, *Astrophys. J.*, **213**, 368.

Amplification near a gravitational lens macro-caustic

W.D. Kunze
Hamburger Sternwarte, Gojenbergsweg 112, D–2050 Hamburg 80

The aim of our work is to investigate what happens when a light source (quasar) is crossing the macro (anti-) caustic of a lensing galaxy.

What is the maximum brightness amplification that can occur?

How large is the influence of stars - situated near the image(s) of the source - acting as individual micro-lenses?

Which parameters of the galaxy (macro-lens) and which parameters of the stars (micro-lenses) are important for the amplification behaviour of the source ?

What are typical time-scales in which strong brightness-changes will appear?

It is well known that a large change in apparent brightness will occur when a source crosses the macro caustic of a lensing galaxy. Two new images which are both strongly amplified appear (or disappear).

We worked with the King-modell used by Young et al. (1980, 1981) by modeling the QSO 0957 + 561 and considered a source crossing the caustic corresponding to the *inner* critical curve of the galaxy. With this smooth galaxy we get light curves as shown in Fig.1. The maximum amplification is given by:

$$A_{max}(R_S) \approx 220\sqrt{\frac{10^{-2}\mathrm{pc}}{R_S}} \qquad (1)$$

where R_S is the radius of the source.

But this description is much to simple because it treats the galaxy as a smeared out mass distribution neglecting the individual stars which act, of course, also as gravitational lenses. One then gets a large number of bright micro images and it is not simple to estimate the total amplification, so numerical calculations are needed. We did inverse ray shooting.

In and around the shooting area we replaced a fraction of the continously distributed mass with randomly distributed stars, at first with all masses m_i equal. We calculated different cases with $0.1 < m_i < 1.0 M_\odot$. We computed up to 80,000 lensing stars.

It is seen that the stars have a strong influence on the lightcurve. The maximum amplification is very much reduced. The strong high amplification event in a rather short time scale is smeared out due to the scattering by the stars to a longer lasting event with a smaller amplification.

We find that the effect of micro-lensing is correlated to the mass of the stars. The effect gets more pronounced with increasing number of stars (larger fraction of the total mass in stars, i.e. larger σ_S , in units of σ_0), see Fig. 2.

Furthermore, with the same fraction of mass in stars (i.e. same σ_S) the effect increases with increasing star mass and with decreasing source size, see Fig. 3.

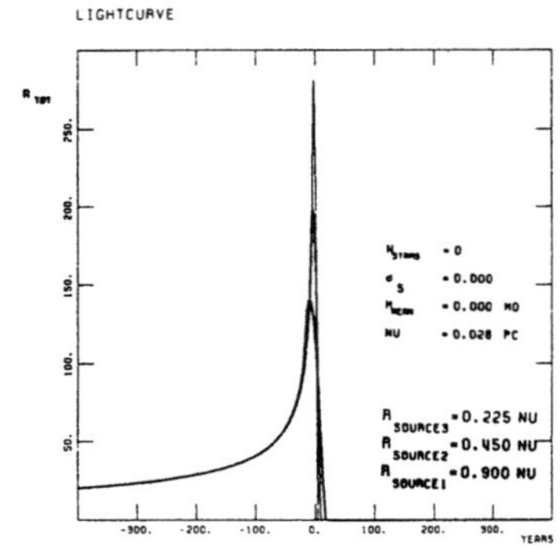

Fig.1 : Lightcurve due to lensing of the smooth galaxy.

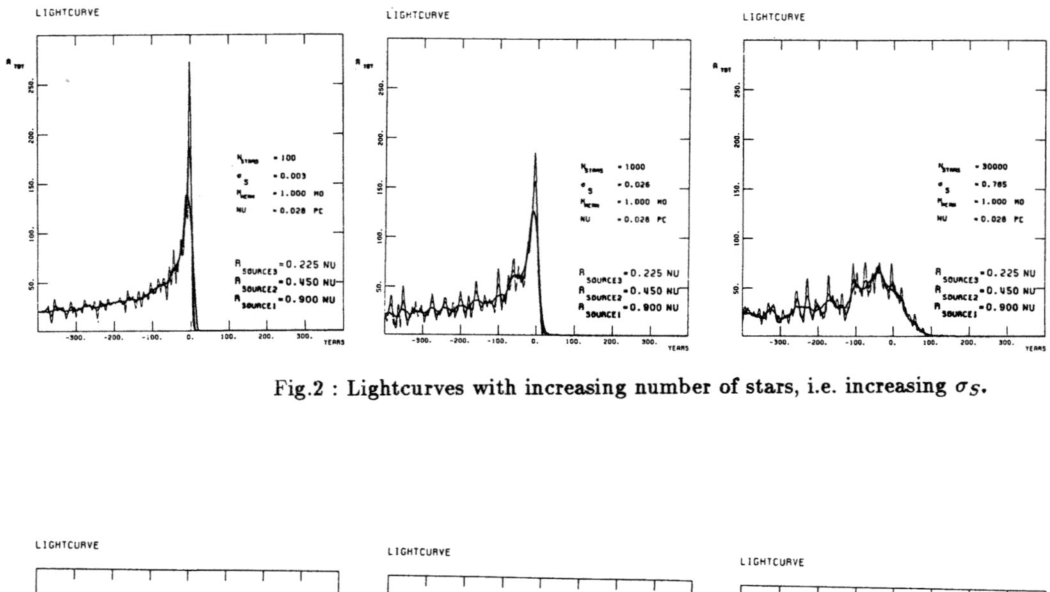

Fig.2 : Lightcurves with increasing number of stars, i.e. increasing σ_S.

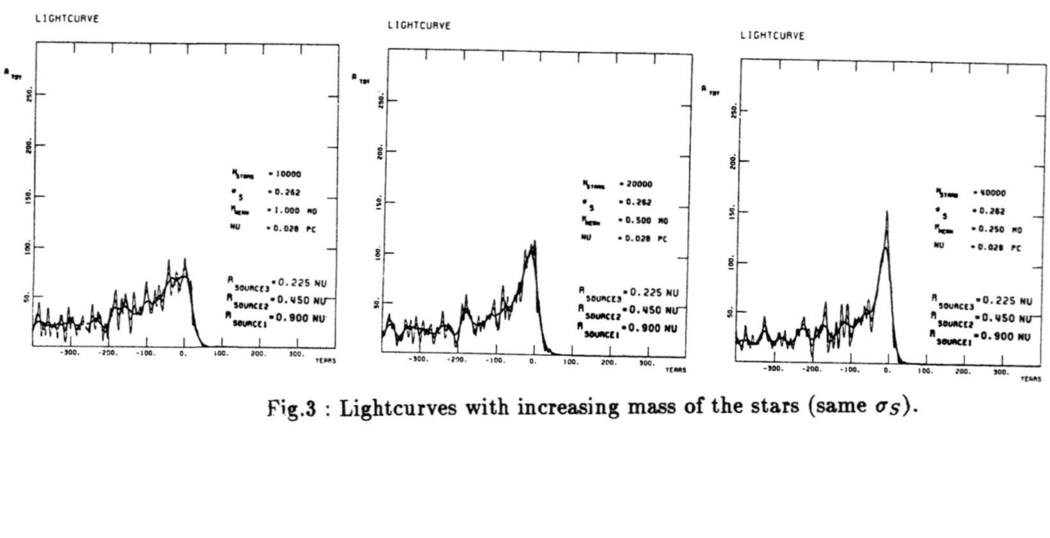

Fig.3 : Lightcurves with increasing mass of the stars (same σ_S).

STATISTICS OF LENSING ON EXTENDED SOURCES

Stéphane Marchandon

Observatoire de Meudon , place Janssen , 92195 Meudon

Abstract : *We study here the influence of the source size on the statistics of gravitational lensing, and predict that the distribution of the probability of magnifications is two-peaked: (i)the first lies at weak amplification, while its position doesn't depend on the source size; (ii)the second corresponds to the largest allowed value of the amplification; this is at variance with the A^{-3} law which is found for point sources until now admitted for $A >> 1$ .The increase of the source size progressively transfers a fraction of the probability from the first to the second peak.*

1. A reminder on point sources.

Studies done by several authors[1,2,3,4] concerning point sources show a highly non-gaussian amplification distribution having a maximum at $A_{peak} < \overline{A}$ (where $\overline{A}$ is the mean amplification) and a weak large amplifications contribution : $\mathcal{P}(A) \propto A^{-3}$ for $A >> 1$.

2. The influence of the source size.

We suppose that a light beam emitted from an extended source crosses a "d" plane where point lenses with mass M are randomly distributed (let μ be their mean number density). The mean $\overline{A}$ of the resulting shear-amplification must be equal[1,2,5,6] to the matter-amplification A_κ :

$$\overline{A} = A_\kappa = (1 - \kappa)^{-2} \tag{1}$$

with: $$\kappa = (4\pi G / c^2) \cdot M\mu \cdot \frac{D_d \cdot D_{dS}}{D_S} = \mu \cdot \pi f^2 \tag{2}$$

where D_d , D_{dS} and D_S are the usual angular-diameter distances and f is the critical radius of each lens. In the following all the distances will be expressed in f unit.

If both f and the typical projected (in the "d" plane) source dimension r are small with respect to the mean inter-lenses distance, i.e. if :

$$\kappa < 1 \quad \text{and} \quad r < \kappa^{-1/2} \quad , \tag{3}$$

then we can take only into account the influence of the nearest lens L_1 to the undeflected beam; under this hypothesis, the amplification writes for a circular source in Einstein linearised theory:

$$A(l_c,r) = \frac{1}{\pi r^2} \cdot \int_S \frac{l^2 + 2}{l \cdot (l^2 + 4)^{1/2}} \cdot l \cdot dl \cdot d\theta \quad , \tag{4}$$

where (l,θ) are polar coordinates centered on L_1 ; the integration is done all over the projected source surface $S = \pi r^2$. l_c is the distance between L_1 and the projected source center C .
The probability for the distance L_1 - C to lie between l_c and $l_c + dl_c$ writes :

$$P(l_c)\cdot dl_c = 2\kappa\cdot l_c \cdot \exp(-\kappa\cdot l_c^2) \cdot dl_c \qquad (5)$$

In fact, a study done in the point source case yields the conclusion that the amplification distribution doesn't hardly depend on the number of lenses taken into account [7].
Since the integration (Eq.4) can not be carried on analytically for all l_c values, we consider separately the cases of small, then of large amplifications.

2.1 Small amplifications ($A \approx 1$)

In that case $l >> 1$ in (Eq.4) , which becomes:

$$A(l_c,r) \approx \frac{1}{\pi r^2} \cdot \int_S (1 + \frac{2}{l^4}) \cdot l \cdot dl \cdot d\theta \;=\; 1 + \frac{2}{(l_c^2 - r^2)^2} \qquad (6)$$

Thus, the probability for the amplification to lie between A and A + dA is :

$$\mathcal{P}(A)\cdot dA = \frac{\kappa}{4} \cdot (2/(A - 1))^{3/2} . \exp[-\kappa \cdot (2/(A - 1))^{1/2}] \cdot \exp(-\kappa\cdot r^2)\cdot dA \qquad (7)$$

The striking fact is that the source size appears only inside the $\exp(-\kappa\cdot r^2)$ factor; therefore:

* The maximum is obtained for :

$$A_{peak} = 1 + \frac{2}{9} \cdot \kappa^2 \qquad (8)$$

Thus, the most probable amplification is small with respect to the average one $\overline{A} \approx 1 + 2\kappa$ and it doesn't depend on the source size (in fact, one can show that the formula (Eq.8) is rigorous at order κ^2; at higher order, A_{peak} increases very slowly with r) [7]. The intensity $\mathcal{P}(A_{peak})$ of this peak is proportionnal to $\exp(-\kappa\cdot r^2)$.

* The overall (integrated) weak amplifications contribution is also proportionnal to $\exp(-\kappa\cdot r^2)$, thus demonstrating that a probability transfer from the small to the large amplifications occurs as r increases. Now the question is: how does it occur ?

2.2 Large amplifications ($A \approx A_{max}$)

In that case $l_c << r$ in (4), which becomes :

$$A(l_c,r) \approx A_{max} \cdot [1 - \frac{4\cdot(l_c/r)^2}{(r^2 + 4)^2}] \qquad (9)$$

where: $$A_{max}(r) = (1 + \frac{4}{r^2})^{1/2} \qquad (10)$$

is the largest amplification obtained for $l_c = 0$.
Thus, the probability for the amplification to lie between A and A + dA is:

$$\mathcal{P}(A)\cdot dA = \mathcal{P}(A_{max}) \cdot \exp[\mathcal{P}(A_{max}) \cdot (A - A_{max})] \cdot dA \qquad (11)$$

with: $$\mathcal{P}(A_{max}) = \frac{\kappa}{4} \cdot [r\cdot (r^2 + 4)^{1/2}]^3 \qquad (12)$$

Then $\mathcal{P}(A)$ increases exponentially, reaching a maximum $\mathcal{P}(A_{max})$ described by (Eq.12) at $A=A_{max}$. The $\mathcal{P}(A_{max})$ increase with r confirms the probability transfer as r increases. The exponential growth law (Eq.11) contrasts with the A^{-3} decreasing law - due to the point sources hypothesis - until now admitted [3,4].

3. Conclusion.

We predict a distribution of amplification probability with two peaks: the first at $A_{peak} < \overline{A}$ described by (Eq.8), the second at $A_{max} > \overline{A}$ described by (Eq.10). The influence of the source size is to transfer a fraction of the probability from the vicinity of the first peak to the vicinity of the second as it increases (see Figure).

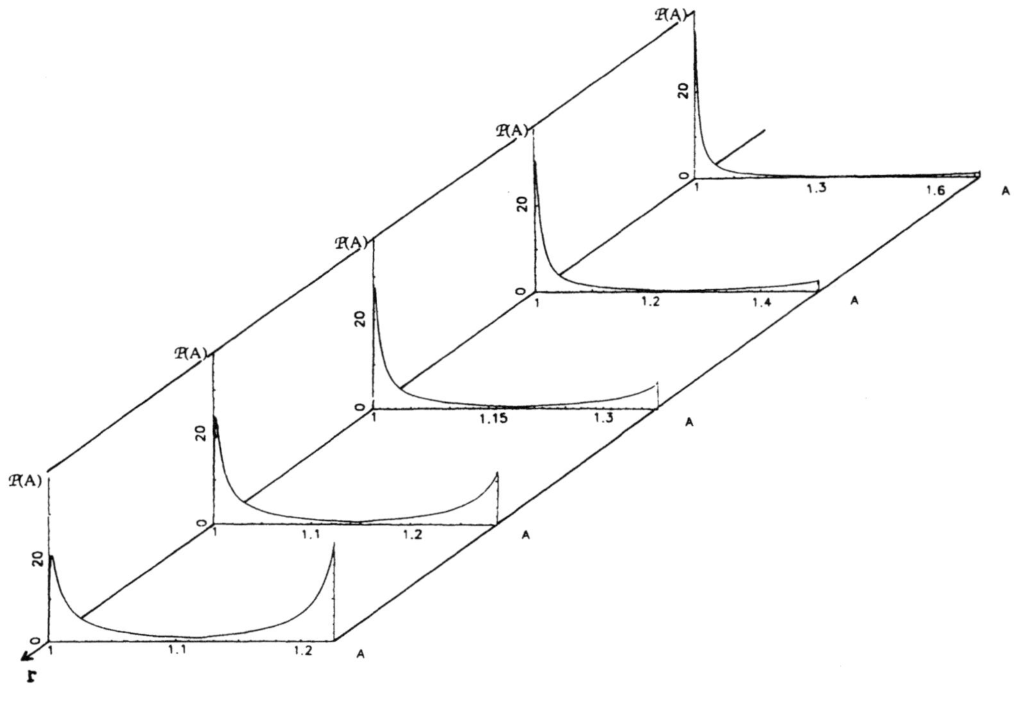

Figure: these curves show the function $\mathcal{P}(A)$ for $\kappa = 0.1$ and r = 1.437 ; 1.782 ; 2.127 ; 2.472 ; 2.817 (in f unit).

References:

[1] Nottale, L.: *Ann. Phys. Fr.*, **13** , 223 (1988)
[2] Peacock, J.A.: *Monthly Notices Roy. Astron. Soc.*, **223** , 113 (1986)
[3] Vietri, M., Ostriker, J.P.: *Astrophys. J.*, **267** , 488 (1983)
[4] Young, P.: *Astrophys. J.*, **244** , 756 (1981)
[5] Dyer, C.C., Roeder, R.C.: *G.R.G.*, **12** , 1157 (1981)
[6] Weinberg, S.: *Astrophys. J.*, **208** , L1 (1976)
[7] Marchandon, S., Nottale, L.: *in preparation.*

II

Observations

Observational aspects of gravitational lensing

Jean Surdej

Institut d'Astrophysique, Université de Liège

Chercheur Qualifié au FNRS (Belgique)

Abstract

In this review on the recent observational aspects of gravitational lensing, I first present the various systematic optical surveys for multiply lensed extragalactic objects that are being carried out at several observatories enjoying good seeing conditions. I then summarize the updated observational status of all proposed gravitational lens candidates, concentrating mainly on multiply lensed QSOs (the cases of giant luminous arcs, radio rings and distant 3C radio galaxies are described elsewhere in this volume). Finally, I give a general outline of all the additional pieces of observational evidence suggesting that gravitational lensing may (i) perturb significantly our view of the distant Universe and (ii) affect our physical understanding of various classes of extragalactic objects.

1. Introduction

The possible perturbations of our view of the distant Universe by gravitational lensing effects are strongly linked to the detailed distribution of matter at various scales. A theoreticians' approach usually consists in making use of all our present (but of course, limited) knowledge on the distribution of matter to predict the importance of such perturbations. Given all the modelling complications, limitations, as well as all yet unknown biases in our observations of distant objects, it is not surprising that the conclusions of such studies do oscillate between various degrees of pessimism or of optimism. There is a citation by Peebles, quoted in Blandford and Kochanek (1987), saying that: "Gravitational lenses provide a theorists' heaven and an observers' hell". I rather believe that the contrary is true and that the path through the examples will always remain shorter and more secure than that through the theories. This is also probably why, at least recently, observers have been more optimistic about the possible importance of gravitational lensing than theoreticians. Of course, there are exceptions and in this context, it is only fair to consider Sjur Refsdal as the leader of the optimistic supporters of gravitational lensing. The situation about "gravitational lensing" is a bit reminiscent of that associated with "atmospheric mirages". What does the average man think about the importance of atmospheric lensing effects on our view of distant earth-sources? (cf. the distant car lights seen along a straight road in Figure 1).

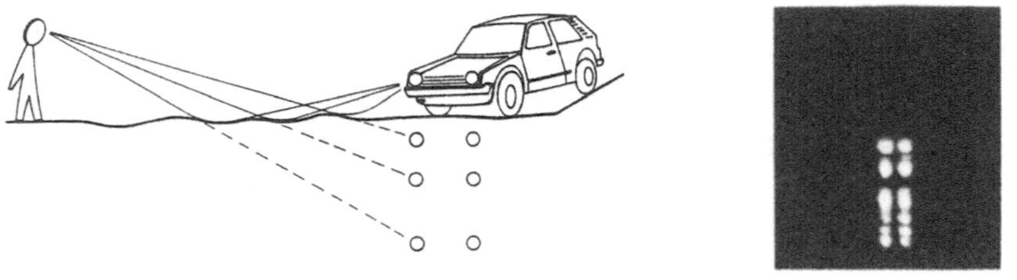

Figure 1: The figure at left gives a schematic representation of the light ray paths from a distant car when the ground turns out to be somewhat hotter than the ambiant air. Because air refraction always leads to a bending of light rays towards regions of colder air, several lower and somewhat deformed images of a distant source (the car lights in this example) may result. The figure at right represents the multiple images due to the lights from a distant car, as photographed by J. Lehár and the author, along the US 60 road, between Magdalena and Datil near the VLA (New Mexico) on the night of 19 January 1989. The distance between the car and the observers was estimated to be about 10 miles. Such terrestrial mirages, usually made of two single images, can actually be seen everyday, almost from anywhere. In addition to significantly affect our view (image deformation, multiplication, etc.) of distant resolved earth-sources, atmospheric lensing is also often responsible for the light magnification of distant unresolved objects located along straight and long roads or across flat countrysides. Other examples of terrestrial mirages are illustrated in Surdej et al. (1989).

The answer to the analogous question for "gravitational lensing" is of course not so trivial. Ten years after the serendipitous discovery of the double quasar Q0957+561 A and B by Walsh, Carswell and Weymann (1979), both theoretical and observational researches on gravitational lensing have led to a formidable burst of published work. Partly guided by this monument of information but also trying not to cover topics already addressed in recent reviews on the subject of gravitational lensing (e.g. Canizares (1987), Blandford and Kochanek (1987), Nottale (1988), Turner (1989a, b), Barnothy (1989) and references therein), I have decided to focus the present paper on a general discussion of the various known (or only hypothesized) pieces of observational evidence which suggest that gravitational lensing may affect significantly our view and physical understanding of the distant Universe, and of its major constituents.

In section 2, I present an overview of the various systematic optical surveys that are presently being carried out all over the world in order to search for multiply lensed distant objects. In section 3, I have listed the updated observational status of the best recognized cases of multiply lensed objects. I also discuss individually the objects for which recent and outstanding observations have led to a better understanding of the gravitational lenses. Suggestions are also given as to which complementary observations shoud be carried out in order to still improve some of the lens models. Section 4 summarizes all the additional observational evidence on statistical effects due to gravitational lensing and I also discuss there the various classes of extragalactic objects which may actually consist of lensed sources. The last section deals with general conclusions.

Before closing this introduction, let me recall the challenging goals of studying known gravitational lens systems as well as of discovering new ones. First of all, it is known that a statistical evaluation of the occurrence of gravitational lensing within well defined samples of extragalactic objects is of prime importance in order to improve our knowledge on quasars and distant radio galaxies (luminosity function, source counts, true part of their cosmic evolution). Detailed studies of identified gravitational lens systems are also important to test cosmological models (Refsdal 1964, 1966), to set constraints on the size and structure of the lensed source (Grieger et al. 1986, 1988) and to probe the luminous and dark matter distributions on various scales in the Universe (Refsdal 1964, 1966). In particular, gravitational lens studies can be used to set limits on populations of dark massive objects (Press and Gunn, 1973; Hewitt et al. 1987a; Webster et al., 1988a). Finally, analysis of narrow absorption lines observed in the spectrum of multiply lensed QSO images are of considerable interest for studies of the spatial structure of the inter-galactic medium (cf. Foltz et al. 1984 and the paper by Smette et al. in these proceedings).

2. Optical surveys for multiply lensed extragalactic objects

Whereas the first gravitational lens systems have been discovered by chance, systematic searches for lenses have proved to be very successful in identifying new systems. I shall describe hereafter several of the on-going optical surveys for multiply lensed extragalactic objects (I refer to the article by B. Burke in this volume for a description of the radio surveys).

2.1 Highly Luminous Quasars as gravitationally lensed objects

Considering the canonical log(N)-B relation for the count number of quasars per unit area brighter than a given magnitude B (cf. Boyle et al. 1988), it is easy to calculate the number density enhancement $q(M,B_0)$ of quasars in a flux limited sample (B_0) subject to a magnification M. Following Narayan (1989), one finds that $q(M,B_0) = (N(<B_0+2.5\log M)/N(<B_0))/M$. Referring to Figure 1 in Narayan (1989) where the author has illustrated the dependence of q versus both M and B_0, one immediately sees that the magnification bias works preferentially well for a bright flux limited sample of QSOs. There are at least three optical surveys for multiply lensed QSO images which benefit from this magnification bias:

1. The Liège/ESO/Hamburg survey, initiated in November 1986 and terminated in November 1988. A description may be found in Surdej et al. (1988a-c) and an update is presented in these proceedings by Swings et al. and Magain et al. I briefly recall that it has consisted in a high resolution imaging of apparently (m_V < 18.5) and intrinsically (M < -29) bright quasars observable from ESO (Chile). Out of 111 quasars observed under an average seeing FWHM = 1.2″, 25 turned out to be interesting candidates (i.e. 23% of HLQs showing elongated, multiple, or fuzzy images), 5 of which constitute very good lens candidates. The cases of UM673 and H1413+117 are described in Surdej et al. (1987) and Magain et al. (1988), respectively.

2. The Djorgovski and Meylan (1989a, b) optical survey for bright distant quasars that has been carried out during the past few years. It also consists in an optical (CCD) imaging search for gravitational lenses among a sample of (≈ 300) high redshift QSOs having an apparently large absolute luminosity. So far, they found one new gravitational lens candidate UM425 (Meylan and Djorgovski, 1989), two probable binary quasars PKS1145-071 (Djorgovski et al., 1987) and PHL1222 (Meylan and Djorgovski, these proceedings), and several other promising candidates. As in the Liège/ESO/Hamburg survey, they also identified several cases of QSOs with foreground galaxies within a few arcsec. I refer to the paper by Meylan and Djorgovski in this volume for more detailed information on their survey.

3. The Crampton et al. (1989) optical survey. Using a new-image stabilizing camera at the CFHT, these authors have also made a direct imaging search for closely spaced gravitationally lensed QSO components. Out of 32 quasars with z > 1.6, m_V < 19 (i.e. M < -27.8), seven were found to be good gravitational lens candidates, two of these having sub-arcsec. angular separations. This fraction (22%) of interesting HLQs is essentially the same as the one (23%) reported by Surdej et al. (1988c).

2.2 Quasars as gravitationally lensed objects

Whereas it is expected that for a fainter limiting magnitude survey the quasar number-density enhancement q(M,B_0) will get smaller, such surveys have the advantage of dealing with a much larger number of objects. One such sample of fainter quasars (m_B < 19) has been used as a basis for the automated survey for gravitational lenses described by Webster et al. (1988a). With the help of the Automated Plate Measuring (APM) facility at Cambridge, these authors have scanned both broadband direct and objective-prism Schmidt plates resulting in a survey of 2500 quasars, covering 130 sq. deg. of sky. Whereas this technique is only effective for separations greater than 2 - 3″, it enables one to quantify different parameters (separation, magnitude difference, lens brightness, etc.) in the lens survey. Since the frequency of lensing for a given separation of lensed QSO images may be predicted, it allows one to set interesting constraints on the mass distribution in the Universe. The new wide separation gravitational lens candidate Q1429-008 recently reported by Hewett et al. (1989) has been found via this survey. The detection of statistical gravitational lensing by foreground mass distributions described in Webster et al. (1988b) is also based upon this survey. It is discussed at length by R. Webster in this volume.

2.3 Distant 3C radio sources as gravitationally lensed objects

The magnification bias has also motivated Hammer, Nottale and Le Fèvre (1986) to assume that the distant (z > 1) and powerful (P(178MHz) > 10^{28} W/Hz) radio sources (hereafter DPRSs) constitute some of the best extragalactic candidates to search for the presence of gravitationally lensed images at arcsec./sub-arcsec. angular scale resolutions and/or for an excess of foreground objects (galaxies, clusters) in the vicinity of the relevant targets. Approximately 75% of the high redshift 3C sources observed by Hammer and Le Fèvre (1989) in optimal seeing conditions with the CFHT at Mauna Kea have been resolved into multiple (2 to 6) components. Strong arguments supporting the mirage hypothesis have been obtained for 3C324, by high resolution imagery through narrow band filters (Le Fèvre et al. 1987; Hammer and Le Fèvre 1989). High spatial resolution imaging and spectroscopy obtained for 3C208.1 definitely prove that the optical appearance of this source is due to the close projection (3.9″) of the optical counterpart of a radio loud quasar at z = 1.01 and a foreground AGN at z = 0.159 which gravitationally magnifies, by more than 0.5 mag., the background object (Le Fèvre and Hammer, 1989). Of special interest is that a detailed comparison between CCD frames obtained from a sample of 27 3C distant radio sources with z > 1 and selected blank fields indicates a significant excess of foreground bright galaxies (up to m_R ≈ 21) and Abell/Zwicky clusters near the 3C sources (Hammer and Le Fèvre, 1989). It therefore seems that gravitational magnification by foreground galaxies and rich clusters is at least partly responsible for the observed radio and optical luminosities of the bright 3C sources. In particular, the following DPRSs: 3C194 and 3C225A (Le Fèvre and Hammer, 1988), 3C238, 3C241, 3C266 and 3C305.1 (Le Fèvre et al. 1988b), 3C13 and 3C256 (Le Fèvre et al. 1988a), 3C239, 3C252, 3C267, 3C322, 3C230, 3C297 and 3C469.1 (Hammer and Le Fèvre, 1989) have been proposed to be either gravitationally magnified and/or multiply imaged. High spatial resolution narrow band imaging and/or deep 2D spectroscopy of most of these candidates are badly needed in order to further test the lensing hypothesis.

In closing I wish to mention here that the surveys described in 2.1.1, 2.1.2 and 2.3 constitute parts of an ESO key-programme that is being presently conducted at ESO-La Silla (see the description of the programme as well as the names of all participants in Surdej et al. 1989).

3. Gravitational lens candidates

Table 1 (see the appendix at the end of these proceedings) presents an updated list of the accepted and proposed candidates of multiply lensed extragalactic objects. Separate comments on individual gravitational lens systems follow:

1. 0957+561: The most extensive photometric monitoring of a gravitationally lensed quasar has of course been carried out for this famous double quasar. Tentative time delays of 1.55 +/- 0.1 years by Florentin-Nielsen (1984), 1.2 years by Schild and his collaborators and, very recently, 1.14 +/- 0.06 years by Vanderriest et al. (1989) have been reported. The radio time delay of < 500 days derived by Lehár et al. (1989) essentially corroborates the optical value. Both the groups of Vanderriest and of Schild find however that the observed lightcurves of 0957+561 A and B cannot be fully interpreted in terms of intrinsic brightness variations of the quasar alone. Micro-lensing seems to contribute to the variations of the B component. Note however that Falco et al. (cf. these proceedings) cast some doubt on this whole interpretation. A spectrophotometric monitoring of the two image components of 0957+561 would certainly help in disentangling the effects due to intrinsic and/or extrinsic variability. Whereas the estimate of the mass of the lensing galaxy is fairly secure (cf. Borgeest, 1986; see Table 1), a more precise modelling of the mass distribution in the lensing galaxy and its attendant cluster is necessary to derive a safe estimate for the Hubble parameter. A deep mapping in luminosity and velocity (velocity dispersion + redshift) of most of the foreground galaxies is necessary in order to better constrain the free parameters of the mass distribution. Very high angular resolution radio observations of images A and B at several epochs would also be extremely valuable.

2. 1115+080: Following the photometric variability study reported by Vanderriest et al. (1986), it is likely that micro-lensing effects are responsible for the brightness variations observed between the A1 and A2 image components. Spectrophotometry of the whole system at various epochs should provide us with a definite answer as to the reality of these effects. Note that in addition to the lensing galaxy detected by Shaklan and Hege (1987) between the A and B images, Henry and Heasley (1986) found that there was also a galaxy centred approximately midway between the two A components. These authors find that the properties of this galaxy (G4, $M_V \approx -23.3$) are consistent with it being the brightest member of a small group at $z = 0.305$ (cf. galaxies G1 and G2 in Young et al. 1981). If real, this would lend support to the micro-lensing induced variability suspected for the A twin components. By means of the pupil segmentation technique used with the CFHT (see Lelièvre et al. 1988), high angular resolution observations supporting this picture for PG1115+080 have been obtained by Arnaud et al. (1989). However, using speckle interferometric observations, Foy et al. (1985) have detected an elongation of image A2 and they suggested that it was caused by a fifth lensed QSO image, located at 0.04" from A2. It is therefore clear that image A is made of more than simply two lensed QSO images and it is likely that further high angular resolution imaging will be necessary in order to definitely settle this point. Although the redshift of the galaxy detected between the A and B images has not yet been directly measured, it is quite possible that for $z = 0.305$, the observed reddening of the B component relative to A, as seen in the spectra published by Young et al. (1981), is actually due to the contamination by the foreground galaxy. In any case, any serious modelling of this lens should take into account the several (5?) galaxies detected in the nearby field.

3. 2345+007: Nieto et al. (1988) have obtained very high angular resolution images of this system using a photon counting detector in its resolved imaging mode. By using recentring and selection algorithms plus image restoration techniques, they have achieved good S/N images with resolution of $FWHM \approx 0.25''$. These observations lend a good support to the gravitational lensing hypothesis of 2345+007 A and B because they reveal that the fainter (B) component actually consists of a double image (B1, B2 with roughly equal brightnesses), separated by 0.36″ and roughly aligned along the same direction as the A and B images, and that there is also good evidence for the outer of these two sub-images to be resolved along a direction perpendicular to the line between B1 and B2. Furthermore, comparison of the ratio A/B at the time of their observations with previously reported values suggests some variability in this system. A spectrophotometric monitoring of the A and B components would also be of great scientific interest to search for micro-lensing effects and/or for a measurable time delay between the two brighter image components.

4. 2016+112: This gravitational lens system appears to be very complex because it consists of at least three detected lensed images A-C′, two foreground galaxies C and D, including one with a measured redshift $z = 1.01$ and two diffuse narrow line emission regions, that appear to be physically distinct, and located near images A and B (see Figure 1 in Schneider et al. 1986). Observational evidence for the possible occurrence of micro-lensing effects has also been reported for this system and a spectrophotometric follow-up of the lensed QSO images would be of great value. See the contribution in this volume by Heflin et al. related to the interesting constraints that their VLBI observations impose on the proposed gravitational lens models (cf. those by Narasimha et al. 1987).

5. 1635+267: The detailed spectroscopic study of 1635+267 A and B by Turner et al. (1988) has led to the good conviction that this double quasar constitutes a good case of gravitational lensing. Indeed, not only were they able to show that the wavelengths, strengths, widths and profiles of different lines were the same, after a proper scaling and to within measuring errors, in the two image spectra but they also found that the excess of red light in the bright component ressembles the continuum emission of a $z \approx 0.57$ galaxy. Independently, the modelling of this system by Narasimha and Chitre (1988) also led to the prediction that a lensing galaxy should be located at 0.75″ from component A. One should of course try to image this system in red light and under very good seeing conditions in order to confirm the presence of the lensing galaxy ... nearly on top of the A (multiple?) image(s).

6. 2237+0305: First of all, superb high angular resolution imagery (Yee, 1988 and Schneider et al. 1988) and spectroscopy (De Robertis and Yee, 1988 and Adam et al. 1989) have convincingly demonstrated that the Einstein cross consists of four lensed images plus a central galaxy nucleus. Kayser and Refsdal (1989) have pointed out the uniqueness of this gravitational lens system to display micro-lensing effects. Indeed, i) because the expected time delays are so short, intrinsic variability should show up almost simultaneously in the four images so that any difference may be attributed to micro-lensing, ii) due to the large distance ratio between the lens and the source, micro-lensing should lead to more frequent and rapid high amplification events (HAEs) and iii) the expected number of HAEs should be large (about 0.3 events per year and per image). It was therefore not a surprise when Irwin et al. (1989) announced the brightening by 0.5 mag. of component A on CCD frames obtained during the 1988 summer. CCD frames taken at La Silla by Remy et al. (1989) in april 1989 in the framework of the "Gravitational Lensing" ESO key-programme indicate that the relative brightening of image A was still about 0.3 mag. in blue light. Integral field spectrophotometric monitoring of this system, similar to the *one epoch* observations reported by Adam et al. (1989) with the TIGER spectrograph (Courtès et al. 1987), would also be extremely valuable. Such observations would present the additional potential of measuring simultaneously the rotational velocity field of the deflecting spiral galaxy. Note that simulations of micro-lensing effects for 2237+0305 by Wambsganss et al. (1989) predict unfortunately that the time scale and the amplitude of HAEs should not only depend on the source size and the relative transverse velocity between the source, the lens and the observer but that, due to the very strong effect of the shear by the galaxy, it should also depend on the direction of this velocity.

7. 0142-100: Since the lens of this system appears to be made of a single isolated galaxy at a redshift $z = 0.49$ (the redshift of the galaxy D in Surdej et al. 1987 has been measured to be $z = 0.17$), it constitutes a very good candidate to attempt an independent measurement of H_0 . A photometric monitoring of 0142-100 A and B has been initiated two years ago at ESO. It should nevertheless be mentioned that we have failed to detect the A and B images with the VLA at 6cm in the A configuration (0.3 mJy r.m.s. noise).

8. 3C324: Spectroscopic confirmation of this first galaxy-galaxy gravitational lens system is strongly awaited. This optical source should also be monitored photometrically for the possible detection of a SN event in the different images.

9. 1413+117: This gravitational lens system has been resolved at 3.6 cm with the VLA in the A configuration. A detailed modelling of these observations has been made by Kayser et al. (1989) and supports very nicely the optical observations. High angular resolution integral spectroscopy of this system with the integral field spectrograph SILPHID (Vanderriest and Lemonnier, 1987) has enabled us to resolve the spectra of the four individual images (Angonin et al., these proceedings). Spectroscopic and photometric indications of micro-lensing effects in the D component have possibly been found.

10. 1120+019: Except for images A and B, no spectra have yet been obtained for the many other objects in the field of this gravitational lens candidate (see Meylan and Djorgovski, 1989).

11. 0414+0534: It would be very important to make a new attempt to determine the redshift of the optically detected object (source?). All four components of this system have been detected with VLBI to show compact flux at 18cm (Hewitt, 1989). There is no doubt that this gravitational lens system turns out to be a very interesting one.

12. 1429-008: Because of the slight, although probably real, spectroscopic differences seen between the two image components, better S/N spectra are mandatory in order to confirm or reject the lensing hypothesis for this system.

13. M82 quasars: This is just one example of QSO images with large separations and all general remarks which follow could apply equally well to many other cases of quasars having very similar redshifts and spectra but large angular separations. Paczynski and Gorski (1981) did actually first suggest that the unusual grouping of quasars near M82 (Burbidge et al. 1980) may consist of another case of gravitational lensing. In the lensing scenario, one or several of the observed QSO images could actually be multiple (due to macro-lensing by a galaxy located in one of the two hypothetical clusters) and one could try to detect i) the presence of a cluster by direct imaging (direct detection or search for distorted background galaxies), ii) the possible multiplicity of the QSO images, iii) the presence of additional magnified QSOs, iv) the similarity or, on the contrary, noticeable differences between the redshifts of the three identified quasars.

4. Further observational evidence for gravitational lensing

4.1 Statistical gravitational lensing

4.1.1 Galaxies near flat radio spectrum quasars: using an automatic search and classification technique for counting galaxies on CCD frames, Fugmann (1988, 1989) has reported, at typically a 97.5% significance level, an increase in the number density of relatively bright ($r < 21.5$) galaxies towards distant ($z > 1.7$) quasars. The increased number of galaxies seems to pertain mainly to fields of (12) flat-spectrum quasars with a much smaller increase near steep-spectrum and radio-quiet quasars. Fugmann invokes gravitational lensing effects, specially micro-lensing to account for these observations and he concludes that gravitational lensing contributes strongly to observed counts of distant sources, particularly to the statistics of flat-spectrum quasars. Whereas Tyson (1986) had also reported a QSO-galaxy correlation for moderately distant quasars ($1 < z < 1.5$), Yee and Green (1984) did not. It is most likely that these different results arise because of the different fractions of radio quiet, steep and flat spectrum quasars constituting those different samples. Fugmann further postulates that gravitational lensing may be responsible for the appearance of different sub-classes of quasars (e.g. the optically quiet compact radio sources (OQCRSs) including objects such as AO0235+164, 0406+121, 0500+01, 1413+135, etc.). Since X-rays detected from selected quasars are supposedly emitted from a very compact core, one may naturally wonder whether X-ray selected quasars could also reveal observational evidence for statistical gravitational lensing.

4.1.2 Galaxies near X-ray selected quasars: during the course of an extensive program of spectroscopic identification of faint X-ray sources discovered serendipitously with the Einstein satellite, 10 X-ray selected AGNs have been discovered by Stocke et al. (1987) to lie within three optical diameters of bright ($m_v < 18$) foreground galaxies . These authors report that, at a confidence level > 97.5%, these AGNs have significantly higher redshifts than X-ray selected ones in general. Stocke and his collaborators have interpreted their findings in terms of micro-lensing in which stars in the foreground galaxy significantly brighten the X-ray emission from these higher redshift AGNs, allowing them to be detected. However, Rix and Hogan (1989) have recently re-investigated this problem by taking deeper CCD frames of the complete sample of the 56 X-ray flux limited AGN fields (MSS initial sample) in order to enlarge the subsample of AGNs with foreground galaxies ($R < 20.5$) from 3 to a total of 8. On the basis of these new data, Rix and Hogan conclude that there is no longer any evidence for lensing effect by the galaxies. The presence of an excess of high redshift X-ray selected quasars near bright foreground galaxies remains still a subject of debate.

4.1.3 Galaxies near APM quasars: at a significance level greater than 99.99%, Webster et al. (1988b) have presented evidence for the detection of an excess of distant quasars ($z > 0.5$, $B_J < 18.7$) associated with foreground galaxies ($r < 6''$, $B_J < 21$) and they invoke gravitational lensing effects in order to account for such a large number of associations. However, they report that the mass of the foreground galaxies must be substantially greater than is conventionally attributed to a luminous galaxy and its halo. I believe that this apparent problem could be easily overcome if one could prove that the principal deflector lies much closer to the line-of-sight (e.g. $r < 1.5''$) than it has been claimed. Several such tight associations have been identified in the HLQ samples described above (see Magain et al. in these proceedings and also section 4.1.12). Given the excess of 4.4 times more APM quasars near galaxies, which corresponds to an average magnification of 2.7, Narayan (1989) has shown that, if this effect is really caused by gravitational lensing, then the quasar-galaxy correlation found by Webster et al. must be smaller than the reported one by a factor of about 2 and/or the magnitude limit of their survey should be brighter by a few tenths of magnitude than the one reported ($B_J = 18.7$). Narayan also demonstrates that the observed effect does not require halos of foreground galaxies to be composed of micro-lenses (stars, black holes, etc.); the quasar-galaxy correlation works equally well with smooth halos. All these conclusions directly follow from the quasar luminosity function and are independent of the lens structure. This result shows how important it is to know accurately what the unlensed luminosity function of quasars is (see the contribution in this volume by R. Webster for an update of their findings on the APM QSO-galaxy associations).

4.1.4 The giant luminous arcs: see the updated report by B. Fort in these proceedings.

4.1.5 Distorted background galaxies: using a similar technique to that of Tyson and his collaborators, Elston et al. (1989) have carried out searches for statistical gravitational lensing effects using background galaxies projected on selected rich, massive clusters. They found that the background galaxies are preferentially aligned at 90° to a radial vector from the cluster center. Some good agreement is reached between the observed distributions of position angles and those of lensing models. They consider this finding as evidence that the background galaxies are being lensed by the foreground clusters (see also the contributions by Tyson and Grossman in these proceedings).

4.1.6 Anomalous quintets of galaxies: Hammer and Nottale (1986a) have shown that the well known quintets of galaxies such as VV172, VV115, etc. which contain a discrepant redshift member may be fully understood in terms of the effects of gravitational magnification by the halo of the foreground quartet on the more distant galaxy.

4.1.7 Gravitational lensing magnification of the bright cluster galaxies (BCGs): Hammer and Nottale (1986b) have presented good evidence that the BCGs of the Kristian et al. (1978) sample lie in regions of the sky containing about two times more foreground Zwicky clusters, the latter clusters being approximately 5 times richer than the mean value for the whole sky. The authors suggest that the gravitational lens magnification of the BCGs by the foreground clusters has induced a strong selection effect in defining the Kristian et al. sample, artificially increasing the deceleration parameter q_0 from 0.2 to about 1.7, as measured from the Hubble diagram of these objects.

4.1.8 The Arp QSO-galaxy associations: Nottale (1988) also presents good evidence that the Arp QSO-galaxy associations may be the result of the combined lensing effects of several superposed galaxies, groups and clusters near their lines-of-sight.

4.1.9 Quasars with 2.45 < z < 3.8: speculations have been made by Ph. Véron that the second observed rise in the space comoving density of quasars from z = 2.45 up to at least z = 3.8 (Véron, 1986) could be due to statistical gravitational lensing effects by foreground objects located along their lines-of-sight.

4.1.10 Gravitationally magnified narrow absorption line quasars: to my knowledge, Nottale (1987) has reported the first observational evidence for gravitational lensing magnification within a sample of absorption line quasars. His conclusions are based upon a comparison between observational data on absorption line quasars with 1.6 < z < 2.2 from the catalogue of Barbieri et al. (1982) with predictions from the theory of multiple gravitational lensing (Nottale and Chauvineau, 1986). Nottale finds that the luminosity of a large fraction of known quasars with a redshift z > 1.6 is enhanced by a factor up to about 10 because of the action of intervening matter (clusters of galaxies, superclusters) along their lines-of-sight. More recently, Thomas and Webster (1989, see also their contribution in these proceedings) have presented tests of both evolution and bias due to gravitational lensing in the number density of QSO metal absorption line systems. Their study is mainly based on sets of CIV and MgII absorption line systems observed and compiled by Sargent et al. (1988a-c). They do not find any evidence for evolution except perhaps in the high equivalent width systems, where gravitational lensing may affect the statistics. They rather propose a model in which the distribution of high equivalent width systems reflects clustering of an unevolving low equivalent width population. Note however that, unlike Nottale (1987), Thomas and Webster assume that gravitational lensing resulting from two or more independent systems is unlikely.

4.1.11 Gravitationally magnified quasars with damped Lyα absorption line(s): in studies of the formation of disk galaxies at high redshift, various authors have carried out a search for damped Lyα absorption in the spectra of high redshift QSOs. Smith et al. (1987) report that the observed number of those absorption lines is at least a factor five in excess of the number of galaxy disks expected to be intercepted by lines-of-sight to this sample of QSOs (cf. their Fig. 3). The derived column densities of those damped Lyα lines are found to be well in excess of $10^{21} cm^{-2}$ (cf. their Fig. 4). Although these authors suggested that gaseous disks may form relatively early in galaxies and that they may be considerably larger at early epochs, I naturally suggest that gravitational lensing magnification could as well account for these observations in terms of induced observational biases, i.e. the quasars from this sample have been observed because they are preferentially magnified due to galaxy disks intercepting their lines-of-sight.

4.1.12 Quasar-galaxy superpositions: among the 23% of interesting HLQs found in the above surveys for gravitational lenses, quite a few of these objects turned out to show that a bright galaxy is superimposed on the central image of the QSO (see also the contribution by Magain et al. in these proceedings). We believe that this form of gravitational lensing magnification of background quasars (HLQs, etc.) could actually be more common than the formation of multiple images. One should note that a lensing galaxy has been identified on top, or very near to one of the lensed images in the cases of 0957+561, PG1115+080, 2016+112, and UM673. Recently, Arnaud et al. (1988) have published another nice case of a galaxy superimposed on an HLQ image. These authors report the detection of a galaxy (z = 0.63, M_R = -21.7) at 1.3″ from the line-of-sight to the QSO 1209+107 (z = 2.19). Another galaxy (z = 0.39), lying at 7″ from the QSO had previously been identified by Cristiani (1987). Arnaud et al. conclude that the QSO is certainly gravitationally magnified by the first and second galaxy. One should also note that the distant flat-spectrum (Fugmann, 1988) and APM (Webster et al., 1988b) quasars are also good candidates to this form of lensing. In any case, one should derive the redshifts of the galaxies found in the vicinity of those distant quasars in order to have a definite proof that they constitute foreground objects. Indeed, at low redshifts (up to about z = 0.7), most of us would tend to agree that a galaxy superimposed on a QSO image does actually consist of its host galaxy or a galaxy in interaction whereas for higher redshifts, we would be tempted to call it a foreground (lensing) object. Could it be for instance that some of the reported low redshift host galaxies are in fact foreground objects, and vice-versa at high redshifts?

4.2 Gravitational lensing and AGN segregation

4.2.1 Gravitationally lensed high luminosity AGNs: in a spectrophotometric monitoring of high luminosity AGNs, Pérez et al. (1989) report that significant emission line variations take place in high luminosity, high redshift objects (cf. 3C 446, PKS 2134+004, etc.). These variations are found to occur on timescales much shorter than those expected from ionization models and therefore the authors invoke beaming of the optical and ultraviolet continuum towards the observer, leading to an anisotropic broad line emission region and a selection effect in favour of cases in which the axis of this anisotropy is close to the line-of-sight (cf. their Fig. 7). Since the phenomenological effects of gravitational lensing are very similar to those expected from beaming, it is equally plausible that the spectroscopic variations observed in high luminosity AGNs are caused by (micro-?) lensing effects. Let us mention that metal absorption line systems, at optimal redshifts for a lens, have been detected in the spectrum of most these objects.

4.2.2 *Rapid variability of extragalactic sources due to lensing effects:* Quirrenbach et al. (1989) have reported radio variability of several flat-spectrum compact sources with amplitudes in the range 10-20% and timescales of hours (cf. their Fig. 1). Although they discuss intrinsic effects and scattering in the interstellar medium as possible explanations for the observed rapid variability, they also mention that it could be due to micro-lensing effects in a foreground object. They searched for the presence of a lensing galaxy in the fields of 0716+71 and 0917+62 but did not find any object on the POSS plates. They conclude on this basis that the other interpretations are more likely to be the right ones. Besides taking deeper CCD frames of the above fields ... one should certainly postpone any sound and firm conclusion.

4.2.3 OVVs as micro-lensed objects: Chang and Refsdal (1979, 1984) have first proposed that the main observable effect of lensing by stars or compact objects in galaxies and their halos will be a noticeable photometric variability of the distant source, due to the passage of the micro-lens in front of its line-of-sight. Ostriker and Vietri (1985), Nottale (1986) and Schneider and Weiss (1987) have independently proposed that micro-lensed objects were to be searched among the Optically Violently Variable (OVV) extragalactic objects, particularly those behaving like BL Lac objects at maximum brightness and like quasars when fainter. In the case of 0846+51w1, Nottale (1986) has shown that all its observed properties (sudden brightening by 4 mag. in less than one month, changes of its spectrum between that of a BL Lac object and that of a QSO, etc.) could be explained by the effects of gravitational lensing due to a compact -Jupiter like- object (10^{-2} M_o), located within a galaxy at z = 0.072 near the line-of-sight and that the core size of the central region emitting the QSO continuum is about $2.10^{-4}\,pc$. Stickel et al. (1989) recently reported that this BL Lac object displays a noticeable elongated appearance due to an intervening galaxy at z = 0.235 located nearly exactly on the line-of-sight. This observation therefore strengthens the hypothesis by Nottale (1986) that the observed properties of 0846+51w1 are markedly influenced by gravitational micro-lensing.

Stickel et al. (1988a) have also suggested that the correlated dramatic flux variability in the optical and radio wavebands of the BL Lac object AO0235+164 (z = 0.94) are most likely the result of gravitational lensing of the compact non thermal BL Lac core in the intervening galaxies at z = 0.524, seen at 2″ to the south and at 0.5″ to the east from the background source. Note that Yanni et al. (1989) locate this second newly detected foreground object at 1.3″ east of AO0235+164. Surdej and his collaborators found recently good evidence that these two claimed galaxies are actually forming a single object. Kayser (1988) has analysed the plausibility of a micro-lensing model to explain the observed characteristics of AO0235+164. Although he finds that several observational data are in apparent contradiction with the micro-lensing hypothesis, he does not totally exclude that micro-lensing may contribute in part to the observed variability of AO0235+164. More investigations on this particular object are badly needed.

Stickel et al. (1988b) have presented convincing arguments that the southern blazar PKS0537-441 could be another example of a gravitationally lensed object. This blazar has been found by these authors to have an extended, spatially resolved appearance, despite its high redshift of z = 0.894. Using image decomposition techniques, Stickel and his collaborators have shown that a foreground disk galaxy (z = 0.186?), seen nearly face on, lies very near the line-of-sight to the blazar (cf. their Figure 2). Stickel et al. derive $M \approx -22.7$ and an exponential disk scale length of 4.3 kpc for the foreground galaxy. These authors propose that micro-lensing effects are responsible for the rapid and large optical variations (up to 5 mag.) previously observed for this object and that macro-lensing magnification by the foreground galaxy could account for the large luminosity radiated by this blazar. During an active state of the object in february 1985, Tanzi et al. (1986) found that PKS 0537-441 brightened by a similar factor ($\approx$ 2) in the IR, optical, UV and X-ray frequencies, suggesting that a same spatial region may be responsible for the emission in the whole spectral range observed. It is of course not difficult to account for these observational trends with a micro-lensing model that magnifies a very compact and distant source.

3C279, as one of the most variable ($\delta B > 6.7$) and possibly most luminous QSO yet observed (M = -31.4), constitutes another excellent candidate for micro- and/or macro- lensing effects (see the historical lightcurve of 3C279 in Figs. 1 and 2 of Eachus and Liller, 1975). Additional good micro-lensed candidates may be found among the following OVVs: PKS0215+015, 0420-014, 0454-234, 0511-220, 0735+178, PKS0823-23, 1308+326, 1638+398, 3C345 (1641+399) and 3C446 (2223-052). Some of these candidates are known during different brightness phases to have a spectrum turning from that of a QSO to that of a BL Lac object. Others undergo sudden light brightening reminiscent of magnification by micro-lensing effects, and some are located very near a diffuse object and/or show narrow metallic absorption lines at an optimal redshift for a lens.

4.2.4 Variability of some BAL profiles induced by micro-lensing effects: variability of some broad absorption line (BAL) profiles have been reported for several objects: UM232 (Barlow et al., 1989), Q1246-057 (Smith and Penston, 1988), Q1303+308 (Foltz et al., 1987) and the well known cloverleaf H1413+117 (Turnshek et al., 1988). In the latter case, it has been recently possible to obtain with the CFHT + SILFID spectrograph very high angular resolution spectra of each of the four QSO lensed images (cf. Angonin et al. in this volume). The spectrum of component D turns out to be markedly different from the other ones and I suggest that micro-lensing might be responsible for this. Other evidence for this effect comes from the photometric variability observed for that same image (see Kayser et al. 1989). The possible implication of this is that if an observer takes an integrated spectrum of H1413+117 under average seeing conditions (FWHM > 1″), he may detect variability in the trough or emission line component(s) of the SiIV, CIV, etc. BAL profiles not only depending on when (time dependence of micro-lensing) he took the spectrum but also on how he set the instrument (width, orientation and exact location of the slit with respect to each of the four image components). It could therefore very well be that some of the variability of BAL profiles reported for other BAL QSOs is induced by micro-lensing effects ... and that, maybe, the BAL phenomenon itself could be closely related to gravitational lensing effects.

5. Conclusions

From an observational point of view, it has been shown in the previous sections that the challenge of further studying "gravitational lenses" is great, essential and manifold. Indeed, observations taken with the best performing instruments, and under the best seeing conditions, will certainly contribute: (i) to a better understanding of the already known examples of gravitational lensed objects (cf. section 3), (ii) to the discovery of new interesting cases (cf. section 2), (iii) to test the various proposed suggestions that gravitational lensing may significantly perturb our view of the distant Universe (cf. section 4.1) as well as to check how much gravitational lensing has induced an apparent segregation among the various specific classes of extragalactic objects (cf. section 4.2). In summary, there is no doubt that future observations of very remote objects will help us in assessing the real importance of the various known, recently hypothesized or yet unknown gravitational lens effects which seem to condition our view of the distant Universe (see Figure 2).

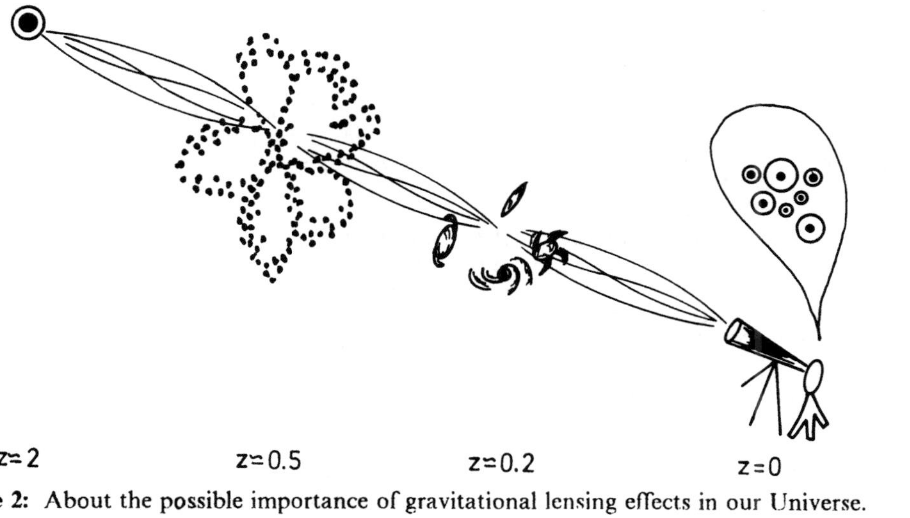

Figure 2: About the possible importance of gravitational lensing effects in our Universe.

Acknowledgements: It is a pleasure to thank my friends and colleagues Pierre Magain, Alain Smette and Jean-Pierre Swings for their careful reading of the manuscript. I also wish to thank them as well as George Djorgovski, François Hammer, Géo Meylan, Laurent Nottale, Olivier Le Fèvre, Sjur Refsdal and Peter Schneider for interesting discussions. Finally, my thanks also go to Armand Kransvelt for drawing the figures.

References:

- Adam, G., Bacon, R., Courtès, G. et al.: 1989, Astron. Astrophys 208, L15.
- Arnaud, J., Hammer, F., Jones, J., Le Fèvre, O.: 1988, Astron. Astrophys. 206, L5.
- Arnaud et al.: 1989, private communication.
- Barbieri, C., Capaccioli, M., Cristiani, S. et al.: 1982, Mem. S. A. It. 53, 511.
- Barlow, T.A., Junkkarinen, V.T., Burbidge, E.M.: 1989, preprint.
- Barnothy, J.M.: 1989, in "Gravitational Lenses", eds. J.M. Moran et al., (New York: Springer-Verlag), p. 23.
- Blandford, R.D., Kochanek, C.S.: 1987, in "Jerusalem Winter School for Theoretical Physics": Vol. 4, "Dark Matter in the Universe", (World Scientific), Singapore, eds. J. Bahcall et al., p. 133.
- Borgeest, U.: 1986, Astrophys. J. 309, 467.
- Boyle, B.J., Shanks, T., Peterson, B.A.: 1988, Mon. Not. Roy. Astr. Soc. 235, 935.
- Burbidge, E.M., Junkkarinen, V.T., Koski, A.T. et al.: 1980, Astrophys. J. 242, L55.
- Canizares, C.R.: 1987, in "Observational Cosmology": Proceedings of IAU Symp. N° 124, eds. A. Hewitt et al. (Reidel: Dordrecht), p. 729.
- Chang, K., Refsdal, S.: 1979, Nature 282, 561.
- Chang, K., Refsdal, S.: 1984, Astron. Astrophys. 132, 168.
- Courtès, G., Georgelin, Y., Bacon, R. et al.: 1987, in "Instrumentation in Astronomy": Proceedings of the 1987 Santa Cruz Workshop, p. 266.
- Cowie, L.L., Hu, E.M.: 1987, Astrophys. J. 318, L33.
- Crampton, D., McClure, R.D., Fletcher, J.M. et al.: 1989, DAO preprint.
- Cristiani, S.: 1987, Astron. Astrophys. 175, L1.
- De Robertis, M.M., Yee, H.K.C.: 1988, Astrophys. J. 332, L49.
- Djorgovski, S., Spinrad, H.: 1984, Astrophys. J. 282, L1.
- Djorgovski, S., Perley, R., Meylan, G. et al.: 1987, Astrophys. J. 321, L17.
- Djorgovski, S., Meylan, G.: 1989a, in "Active Galactic Nuclei": Proceedings of the IAU Symp. N° 134, eds. D. Osterbrock and J. Miller, (Dordrecht: Kluwer), p. 269.
- Djorgovski, S., Meylan, G.: 1989b, in "Gravitational Lenses", eds. J.M. Moran et al., (New York: Springer-Verlag), p. 173.
- Eachus, L.J., Liller, W.: 1975, Astrophys. J. 200, L61.
- Elston, R., Grossman, S., Zaritsky, D.: 1989, preprint.
- Florentin-Nielsen, R.: 1984, Astron. Astrophys. 138, L19.
- Foltz, C.B., Weymann, R.J., Röser, H.-J. et al.: 1984, Astrophys. J. 281, L1.
- Foltz, C.B., Weymann, R.J., Simon, L.M. et al.: 1987, Astrophys. J. 317, 450.
- Foy, R., Bonneau, D., Blazit, A.: 1985, Astron. Astrophys. 149, L13.
- Fugmann, W.: 1988, Astron. Astrophys. 204, 73.
- Fugmann, W.: 1989, preprint.
- Grieger, B., Kayser, R., Refsdal, S.: 1986, Nature 324, 126.
- Grieger, B., Kayser, R., Refsdal, S.: 1988, Astron. Astrophys. 194, 54.
- Hammer, F., Nottale, L.: 1986a, Astron. Astrophys. 155, 420.
- Hammer, F., Nottale, L.: 1986b, Astron. Astrophys. 167, 1.
- Hammer, F., Nottale, L., Le Fèvre, O.: 1986, Astron. Astrophys. 169, L1.
- Hammer, F., Le Fèvre, O.: 1989, CFH preprint.
- Henry, J.P., Heasley, J.N.: 1986, Nature 321, 139.
- Hewett, P.C., Webster, R.L., Harding, M.E. et al.: 1989, preprint.
- Hewitt, J.N., Turner, E.L., Burke, B.F. et al.: 1987a, in "Observational Cosmology": Proceedings of IAU Symp. N° 124, eds. A. Hewitt et al. (Dordrecht: D. Reidel), p. 751.
- Hewitt, J.N., Turner, E.L., Lawrence, C.R. et al.: 1987b, Astrophys. J. 321, 706.
- Hewitt, J.N., Turner, E.L., Schneider, D.P. et al.: 1988, Nature 333, 537.

- Hewitt, J.N., Burke, B.F., Turner, E.L. et al.: 1989, in "Gravitational Lenses", Eds. J.M. Moran et al. (New York: Springer-Verlag), p. 147.
- Hewitt, J.N.: 1989, private communication.
- Huchra, J.P., Gorenstein, M., Kent, S. et al.: 1985, Astron. J. 90, 691.
- Irwin, M.J., Webster, R.L., Hewett, P.C. et al.: 1989, CITA preprint.
- Kayser, R.: 1988, Astron. Astrophys. 206, L8.
- Kayser, R., Surdej, J., Condon, J.J. et al.: 1989, preprint.
- Kayser, R., Refsdal, S.: 1989, Nature 338, 745.
- Kristian, J., Sandage, A., Westphal, J.A.: 1978, Astrophys. J. 221, 383.
- Langston, G.I., Schneider, D.P., Conner, S.: 1989, preprint N° 346 from the MPI für Radioastronomie.
- Lawrence, C.R., Schneider, D.P., Schmidt, M. et al.: 1984, Science 223, 46.
- Le Fèvre, O., Hammer, F., Nottale, L. et al.: 1987, Nature 326, 268.
- Le Fèvre, O., Hammer, F., Nottale, L. et al.: 1988a, Astrophys. J. 324, L1.
- Le Fèvre, O., Hammer, F., Jones, J.: 1988b, Astrophys. J. 331, L73.
- Le Fèvre, O., Hammer, F.: 1988, Astrophys. J. 333, L37.
- Le Fèvre, O., Hammer, F.: 1989, CFH preprint.
- Lehár, J., Hewitt, J.N., Roberts, D.H.: 1989, in "Gravitational Lenses", eds. J.M. Moran et al. (New York: Springer-Verlag), p. 84.
- Lelièvre, G., Nieto, J.L., Salmon, D. et al.: 1988, Astron. Astrophys. 200, 301.
- Magain, P., Surdej, J., Swings, J.-P. et al.: 1988, Nature 334, 325.
- Meylan, G., Djorgovski, S.: 1989, Astrophys. J. 338, L1.
- Narasimha, D., Subramanian, K., Chitre, S.M.: 1987, Astrophys. J. 315, 434.
- Narasimha, D., Chitre, S.M.: 1988, preprint.
- Narayan, R.: 1989, Astrophys. J. 339, L53.
- Nieto, J.-L., Roques, S., Llebaria, A. et al.: 1988, Astrophys. J. 325, 644.
- Nottale, L., Chauvineau, B.: 1986, Astron. Astrophys. 162, 1.
- Nottale, L.: 1986, Astron. Astrophys. 157, 383.
- Nottale, L.: 1987, Ann. Phys. Fr. 12, 241.
- Nottale, L.: 1988, Ann. Phys. Fr. 13, 223.
- Ostriker, J.P., Vietri, M.: 1985, Nature 318, 446.
- Paczynski, B., Gorski, K.: 1981, Astrophys. J. 248, L101.
- Paczynski, B.: 1986, Astrophys. J. 308, L43.
- Pérez, E., Penston, M.V., Moles, M.: 1989, preprint.
- Press, W.H., Gunn, J.E.: 1973, Astrophys. J. 185, 397.
- Quirrenbach, A., Witzel, A., Krichbaum, T. et al.: 1989, Nature 337, 442.
- Refsdal, S.: 1964, Mon. Not. Roy. Astr. Soc. 128, 295 and 307.
- Refsdal, S.: 1966, Mon. Not. Roy. Astr. Soc. 132, 101.
- Remy et al.: 1989, private communication.
- Rix, H.-W., Hogan, C.J.: 1989, preprint N° 883 of the Steward Observatory.
- Sargent, W.L.W., Boksenberg, A., Steidel, C.C.: 1988a, Astrophys. J. Suppl. 68, 539.
- Sargent, W.L.W., Steidel, C.C., Boksenberg, A.: 1988b, Astrophys. J. 334, 22.
- Sargent, W.L.W., Steidel, C.C., Boksenberg, A.: 1988c, RGO preprint.
- Schneider, D.P., Gunn, J.E., Turner, E.L. et al.: 1986, Astron. J. 91, 991.
- Schneider, D.P., Turner, E.L., Gunn, J.E. et al.: 1988, Astron. J. 95, 1619; plus erratum in Astron. J. 96, 1755.
- Schneider, P., Weiss, A.: 1987, Astron. Astrophys. 171, 49.
- Shaklan, S.B., Hege, E.K.: 1987, preprint N° 609 of the Steward Observatory.
- Smith, H.E., Cohen, R.D., Bradley, S.E.: 1987, preprint.
- Smith, L.J., Penston, M.V.: 1988, Mon. Not. Roy. Astr. Soc. 235, 551.
- Stickel, M., Fried, J.W., Kühr, H.: 1988a, Astron. Astrophys. 198, L13.
- Stickel, M., Fried, J.W., Kühr, H.: 1988b, Astron. Astrophys. 206, L30.
- Stickel, M., Fried, J.W., Kühr, H.: 1989, preprint.
- Stocke, J.T., Schneider, P., Morris, S.L. et al.: 1987, Astrophys. J. 315, L11.
- Surdej, J., Magain, P., Swings, J.-P. et al.: 1987, Nature 329, 695.
- Surdej, J., Magain, P., Swings, J.-P. et al.: 1988a, Astron. Astrophys. 198, 49.
- Surdej, J., Magain, P., Swings, J.-P. et al.: 1988b, in "Large Scale Structures: Observations and Instrumentation": Proceedings of the First DAEC Workshop (Paris-Meudon), eds. C. Balkowski and S. Gordon, p. 95.
- Surdej, J., Swings, J.-P., Magain, P. et al.: 1988c, Astr. Soc. Pac. Conf. Ser. 2, 183.

- Surdej, J., Arnaud, J., Borgeest, U. et al.: 1989, The Messenger 55, 8.
- Tanzi, E.G., Barr, P., Bouchet, P. et al.: 1986, Astrophys. J. 311, L13.
- Thomas, P.A., Webster, R.L.: 1989, CITA preprint.
- Turner, E.L., Hillenbrand, L.A., Schneider, D.P. et al.: 1988, Astron.J. 96,1682.
- Turner, E.L.: 1989a, in "Gravitational Lenses", eds. J.M. Moran et al. (New York: Springer-Verlag), p. 69.
- Turner, E.L.: 1989b, in "The 14th Texas Symp. on Relativistic Astrophysics", preprint.
- Turnshek, D.A., Foltz, C.B., Grillmair, C.J. et al.: 1988, Astrophys. J. 325, 651.
- Tyson, J.A.: 1986, in "Quasars", Proceedings of the IAU Symp. N° 119, eds. G. Swarup and V.K. Kapahi (Dordrecht: Reidel), p. 551.
- Vanderriest, C., Wlérick, G., Lelièvre, G. et al.: 1986, Astron. Astrophys. 158, L5.
- Vanderriest, C., Lemonnier, J.P.: 1987, in "Instrumentation in Astronomy": Proceedings of the 1987 Santa Cruz Workshop, p. 304.
- Vanderriest, C., Schneider, J., Herpe, G. et al.: 1989, Astron. Astrophys. 215, 1.
- Véron, P.: 1986, Astron. Astrophys. 170, 37.
- Walsh, D., Carswell, R.F., Weymann, R.J.: 1979, Nature 279, 381.
- Wambsganss, J., Paczynski, B., Katz, N.: 1989, Preprint POP-315.
- Webster, R.L., Hewett, P.C., Irwin, M.J.: 1988a, Astron. J. 95, 19.
- Webster, R.L., Hewett, P.C., Harding, M.E. et al.: 1988b, Nature 336, 358.
- Weedman, D.W., Weymann, R.J., Green, R.topet al.: 1982, Astrophys. J. 255, L5.
- Weymann, R.J., Latham, D., Angel, J.R.P. et al.: 1980, Nature 285, 641.
- Yanny, B., York, D.G., Gallagher, J.S.: 1989, Astrophys. J. 338, 735.
- Yee, H.K.C.: 1988, Astron. J. 95, 1331.
- Yee, H.K.C., Green, R.F.: 1984, Astrophys. J. 280, 79.
- Young, P., Deverill, R.S., Gunn, J.E. et al.: 1981, Astrophys. J. 244, 723.

QUASAR-GALAXY ASSOCIATIONS

Rachel L. Webster
Canadian Institute for Theoretical Astrophysics
60 St George St., Toronto, Ontario, Canada, M5S 1A1

Paul C. Hewett
Institute of Astronomy
Madingley Road, Cambridge, UK, CB3 0HA

Abstract

Evidence for an overdensity of galaxies around high-redshift quasars is reviewed and possible explanations for such an overdensity are considered. Published work has supported the existence of an overdensity, however there is no consensus about either the amplitude or the cause of this effect. We present arguments which favour gravitational lensing as the correct model for the overdensity, and discuss the implications of such a model for the determination of the quasar luminosity function and the total distribution of mass around luminous galaxies.

1. Introduction

Deep CCD imaging and spectroscopy of faint galaxies around low-redshift ($z \lesssim 0.5$) quasars have established that quasars are located in groups and small clusters of galaxies (Stockton 1980, Yee and Green 1987). More controversial has been the claim that there is a statistical excess of close associations between galaxies at faint apparent magnitudes and quasars with much higher redshifts ($z \gtrsim 1$) - *e.g.* Webster *et al.* 1988a). Establishing an excess of associations between quasars and galaxies is difficult; the procedure is statistical in nature and since the amplitude of any effect is likely to be small, large samples of objects are required. This review summarises recent results concerning the incidence of quasar–galaxy associations and considers the evidence supporting the gravitational lensing and galaxy evolution hypotheses invoked to explain the claimed excess of such associations.

Studies of quasar–galaxy associations have been hampered by the lack of suitable catalogues of both quasars and galaxies. Large samples of objects selected in a quantitative and homogeneous fashion are essential. Until recently the only catalogues that came close to satisfying these criteria were the radio-selected quasar catalogues, such as the 3C and 4C samples, and the visual compilations of bright galaxies, for example the Zwicky and ESO B catalogues. Little weight can be given to the anomalous associations between individual bright galaxies and nearby quasars such as those studied by Arp (1970). After decades of work there has been no estimate of the significance of these associations which incorporates sufficient control fields, realistic estimates of quasar surface densities or properly treats the consequences of using quasar samples with poorly determined magnitude limits and large magnitude errors for individual objects. Careful analyses involving large visually compiled galaxy catalogues have been attempted (*e.g.* Seldner and Peebles 1979, Nieto and Seldner 1982) but results are uncertain because of the unknown selection biases involved in the compilation of the catalogues.

Two principal hypotheses have been proposed to explain any excess of associations between high-redshift quasars and faint galaxies: (i) the mass associated with galaxies gravitationally lenses the quasar. The apparent brightness of quasars is enhanced when the line of sight passes close to a foreground galaxy, thereby increasing the probability that such quasars are included in flux limited samples. For a given quasar luminosity function (QLF) the number of associations and the distribution of quasar–galaxy separations constrains the total amount of mass associated with the galaxies. A consequence of this model is a steepening of the bright tail of the observed QLF. (ii) the galaxies are at the same redshift as the quasar. This model requires substantial luminosity evolution amounting to a two or three magnitude brightening of galaxies by redshift $z \sim 1$, compared to a passive evolution model. Obtaining spectroscopic redshifts of the galaxies in the associations provides a clear discriminatory test of the two hypotheses.

In section 3 we describe many of the published imaging studies of high-redshift quasar samples, but the most recent study, that of Magain *et al.*, can be found in these proceedings, where they discuss the number of close galaxy–quasar associations found in their high-resolution CCD data of a sample of very luminous, high-redshift quasars.

2. Observational Data

The detection of quasar–galaxy associations depends strongly on the selection biases involved in the compilation of quasar catalogues. Most search techniques consider only stellar objects as quasar candidates and have consequently been biased against the detection of quasars lying within a few galaxy radii of galaxies visible on photographic plate or CCD material used in the quasar identification. Some radio selected quasar samples were not entirely free of such problems, although more recent surveys with high-precision source positions and complete optical identifications are unbiased with respect to resolved objects. In some cases the selection effects depend on the cause of the excess. If the region producing the radio or emission-line radiation is larger than the continuum source, quasar samples selected according to their radio power or emission-line strength may be biased against inclusion of objects affected by microlensing, compared to samples selected by continuum colour. The existence of these, and other, unquantifiable selection biases makes the statistical analysis of samples drawn from the heterogeneous compilations of Hewitt and Burbidge (1987) and Véron-Cetty and Véron (1987) problematic at best.

Most recent analyses have concentrated on samples derived from two different observational strategies. Fast automated densitometers (*e.g.* APM and COSMOS) can now produce positions, magnitudes and morphological information for hundreds of thousands of sources to faint apparent magnitudes ($m_B \sim 22$) from high-quality wide-field plate material. Using these facilities it is feasible to generate homogeneous galaxy (Maddox *et al.* 1989) and quasar (Foltz *et al.* 1989) catalogues of unprecedented size according to precisely defined selection criteria. The availability of such samples does not eliminate all selection effects in identifying quasar–galaxy associations but the selection function can be determined to arbitrary precision. Model predictions may be convolved with the selection function and compared directly to the data. Such a procedure is discussed by Webster *et al.* 1988b for the case of detection of multiply imaged quasars from digitised photographic plates.

The second approach involves obtaining deep CCD frames, centred on known quasars, in order to examine the distribution of faint galaxies close to the quasars. It is possible to obtain high-quality CCD images reaching galaxies of magnitude $m_R \sim 23.5$ (equivalent to $z_{mean} \sim 0.4$), for large samples of quasars using relatively short exposure times. It is assumed that there is no coupling between the number of faint galaxies and the procedure used to identify the quasar in the first place. The strong clustering of galaxies together with their extended luminosity function means that the density of galaxies in a given field can be correlated over many magnitudes. Thus care must be taken in the choice of quasar sample and any selection-related correlation (or anti-correlation) with galaxies within $\delta m \sim 3$ magnitudes of those to be examined.

Detailed procedures must be carried out to determine the amplitude of the galaxy overdensity from the photographic or CCD data. Good magnitude calibration for both the target quasars and control fields is essential as any derived overdensity depends critically on the level of the adopted galaxy background counts. Background counts can be estimated from: (i) published galaxy counts, (ii) deliberately chosen control fields or (iii) areas, such as the edge of CCD frames, a specified distance from the target quasars. In the latter case it is important to ensure that the coherence effects due to galaxy clustering are fully taken into account. Some form of image classification must be undertaken to eliminate stellar images that will dilute the amplitude of the overdensity shown by the galaxy population. The stellar sample then forms an excellent control for testing the analysis procedures.

Observational studies often image a sample of quasars and count the number of surrounding galaxies, while theoretical calculations are performed the other way around. Matching the measured statistics to the theoretical calculations requires that the detection methodology be symmetric. Schneider (1989) has suggested that selecting quasars and looking for associated galaxies is different from selecting galaxies and searching for associated quasars. In practice different biases may be introduced depending on the search procedure used to identify each population, but if the selection procedure is free from bias, then the two methodologies will give the same result. Any bias involved in the observational determination of the number of associations must be quantified so that theoretical predictions can be convolved with the selection function allowing direct comparison of models with the data.

3. Recent Results

Stocke *et al.* (1987) have examined the distribution of bright ($m_V \lesssim 18$), low-redshift ($z \leq 0.15$) galaxies near a complete sample of 56 X-ray selected, predominantly low-redshift ($z_{median} \sim 0.3$), active galactic nuclei (AGN). They find that the 3 closest quasar–galaxy pairs contain quasars with redshifts amongst the highest in the sample. The small numbers involved preclude a definitive statement on the significance of the result but Stocke *et al.* interpret the result in terms of microlensing by compact objects associated with the galaxies. Rix and Hogan (1989, preprint) have obtained R band CCD images of the 56 AGNs, and performed a similar analysis to Stocke *et al.* using a deeper sample of galaxies ($m_R \lesssim 20.5$). They find no significant correlation between the quasar X-ray luminosities or redshifts and the presence of a nearby galaxy, concluding that the Stocke *et al.* result was a statistical fluctuation. The significance of the latter result is also unclear because at $m_R \sim 20.5$ their galaxy sample can be expected to contain galaxies in the foreground, background and at the same redshift as the AGNs.

In the Yee and Green (1987) CCD imaging study (to $m_r \sim 23$) of low-redshift ($z \lesssim 0.5$) quasar environments there is an excess of galaxies clustered around the quasars, showing that the quasars are located in groups or poor clusters of galaxies. At higher redshifts ($0.5 \lesssim z \lesssim 0.65$), there is a general increase in the amplitude of the quasar–galaxy covariance function as a function of quasar redshift and absolute magnitude. There are surprisingly rapid changes in the galaxy richness of the quasar environment; for radio-loud quasars, an increase of a factor ~ 3 in the amplitude of the quasar–galaxy covariance function between redshifts $z \sim 0.4$ and $z \sim 0.6$ (Yee and Green 1987). They interpret the change as substantial evolution of galaxy luminosities (0.6 ± 0.4 magnitudes brighter at redshift $z \sim 0.6$ (Yee 1989)). Gravitational lensing effects might also account for the evolution at higher redshifts. Qualitative evidence which favours such an interpretation includes: (i) some very high-redshift quasars show excess galaxy counts (Yee and Green 1984), and (ii) the observed trend of increased galaxy counts as a function of quasar absolute magnitude and of quasar redshift.

Tyson's (1986) CCD imaging study, to $m_R \sim 21$, of a sample of low-redshift ($0.1 \lesssim z \lesssim 0.5$) and high-redshift ($1.0 \lesssim z \lesssim 1.5$) quasars also proposed strong luminosity evolution of galaxies associated with the quasars. Direct comparison with Yee and Green is difficult since Tyson's observations were taken in white light and subsequently matched to the R band. However, Tyson found an excess of galaxies around low-redshift quasars at small angular separations, which, after some adjustments in background counts, is similar to that of Yee and Green (1984) and can be interpreted as spatial clustering of galaxies in the vicinity of the quasar. The excess in the high-redshift quasar sample is less significant, but if interpreted as showing galaxies clustered at the quasar redshift, implies evolution in galaxy luminosities of more than 2 magnitudes at redshifts $z = 1 - 1.5$. If the high-redshift quasars in Tyson's sample are separated into two groups by radio spectral index, the flat-spectrum quasars ($\alpha > -0.5$) show all the excess of close galaxies.

Hintzen, Romanishin and Valdes (1989 preprint) have undertaken a study similar to that of Tyson (1986), and have imaged 16 radio selected quasars with redshifts $0.9 < z < 1.5$

in the R and I passbands. The minimum separation at which they could distinguish a faint galaxy is ~ 4 arcsec. A statistically significant excess is found within 15 arcsec of the quasars. The R band data shows 32 quasars where 18 ± 5 would have been expected at a limiting magnitude of $m_R = 23$. At $m_R = 22$, 15 galaxies are found where 9 ± 3 were expected. Hintzen *et al.* interpret the excess galaxies as part of the galaxy cluster at the quasar redshift. Using a very conservative method, they calculate that on average each galaxy is 1.1 ± 0.8 magnitudes brighter than a first-ranked cluster galaxy, representing an increase of $1 - 3$ magnitudes in luminosity by redshifts $z = 0.9 - 1.5$. In Figure 1, the apparent R band magnitudes of all galaxies within 15 arcsec of the quasars are plotted as a function of quasar redshift. No trend with redshift is seen; thus either the luminosity evolution exactly cancels the cosmological corrections (distance plus k-corrections) or the galaxies are in the foreground.

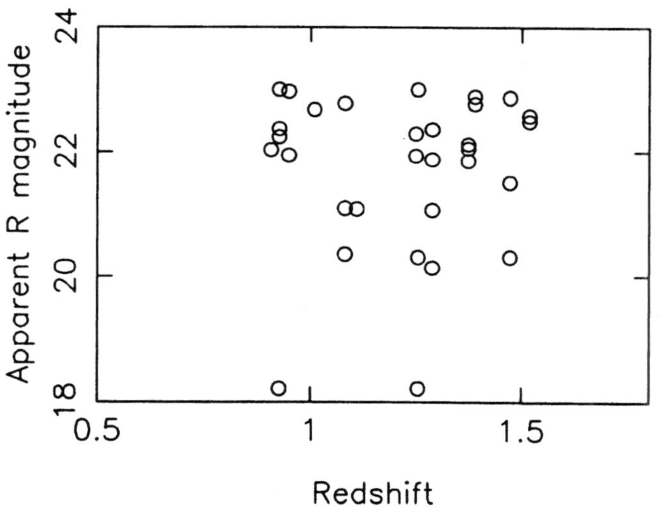

Figure 1: Apparent R band magnitude for each of the galaxies within 15 arcsec of the Hintzen *et al.* quasars, plotted against the quasar redshift.

Fugmann (1988) has imaged a total of 29 strong radio sources in the r band. For a subsample of 12 high-redshift ($z \geq 1.7$), flat-spectrum quasars he finds an excess of associated galaxies ($m_r < 21.5$) on scales of $6.7 - 15$ arcsec. Five galaxies were observed where 1 ± 1 were expected, based on background counts obtained from 5 control fields.

Webster and collaborators (Webster *et al.* 1988a, 1988b) have investigated the incidence of close ($\sim$ 6 arcsec) quasar–galaxy associations in a sample of $\sim$ 850 bright ($m_{B_J} \lesssim 18.7$), high-redshift ($z > 0.5$), optically selected quasars. The quasar selection was specifically developed to be independent of the source morphology. A preliminary analysis of a subsample of the survey shows that for 334 quasars there are 7 certain and a further 2 probable quasar–galaxy associations with separations $\theta \leq 6$ arcsec. In a second subsample of 296 quasars (Webster *et al.* 1988a) there are 9 certain quasar–galaxy associations with separations $\theta \leq 6$ arcsec and several more possible associations. A total of 5 redshifts for galaxies, with magnitudes $19.5 < m_{B_J} < 21.5$, in quasar–galaxy associations with separations $4.5 < \theta < 8$ arcsec have been obtained. The redshifts span the range $0.1 < z < 0.35$ in excellent agreement with the results of the field-galaxy survey of Broadhurst *et al.* (1988). If we consider only those associations containing images identified as certain galaxies, there are 16 associations compared to 5.5 ± 2.3 expected. The galaxy surface density used in this calculation is obtained directly from the plates on which the quasars are identified. The surface density agrees with King and Ellis (1985) who obtained $\sim 10^{3.1}$ galaxies per square degree with $m_{B_J} \leq 21.5$. The relatively bright magnitudes ($m_{B_J} \sim 20 - 20.5$), normal surface-brightness properties and the 5 galaxy redshifts all support the hypothesis that the galaxies in this sample are foreground objects.

It is not easy to compare the results from different samples. There are significant differences in: (i) the radio and optical properties of the quasars, (ii) the quasar sample redshifts, (iii) the colour-band and magnitude limit used to define the galaxy samples, (iv) the procedures used to provide control samples and/or define the background galaxy counts. There is a consensus of opinion that for low-redshifts ($z \lesssim 0.5$) the observed excess of galaxies results from the detection of galaxies at the same redshifts as the quasars. For quasars at high-redshifts an excess of galaxies is still seen, but there is no agreement on the relative importance of gravitational lensing effects and strong luminosity evolution of the galaxy population. In the case of high-redshift radio-loud quasars there appears to be a greater overdensity of galaxies about flat-spectrum compared to steep-spectrum sources.

Discriminating between the gravitational lensing and strong luminosity evolution hypotheses is difficult since for most samples redshifts are not available. In addition, there is uncertainty about the amplitude of the galaxy excess, which is dependent on the determination of the background number density of galaxies. Fugmann (1988) and Yee *et al.* (1986) disagree over the size of the background correction at $m_r \sim 21.5$ by a factor ~ 3 for example! The only galaxy redshifts to be obtained are those of the Webster *et al.* group (unpublished) which do provide substantial evidence against the galaxy luminosity evolution model. Publication of the full catalogue of Webster *et al.*'s associations together with complete redshift information demonstrating that the galaxies are foreground, would be powerful support for the lensing hypothesis.

4. Discussion

Gravitational lensing provides a natural explanation for excess numbers of quasar–galaxy associations. Mass associated with the galaxy magnifies quasar images seen close to the line-of-sight and these quasars are preferentially included in flux-limited samples. Stimulated by new observations, a number of authors (Narayan 1989, Kovner 1989 and Schneider 1989) have calculated the amplitude of the excess for various lensing models and intrinsic QLFs. Narayan's (1989) elegant calculation illustrates the main features of this work. Calculations are presented in terms of the enhancement factor, q, defined as the ratio of the observed number of quasar–galaxy associations to the number predicted to occur by chance. For a given lensing model the value of q is sensitive to the form of the QLF. Narayan formulates this dependence in terms of the observed quasar number counts as a function of apparent magnitude. The critical parameters are (i) the slope of the bright end of the intrinsic number counts, without the effects of lensing, and (ii) the magnitude difference $\delta B = B_{break} - B_0$, between B_0, the limiting apparent magnitude of the quasar sample, and B_{break} the magnitude where the intrinsic quasar number counts change from a steep to a flatter slope. It is then a simple matter to deduce that

$$q(M, B_0) = \frac{1}{M}\frac{N(< B_0 + 2.5 \log M)}{N(< B_0)}$$

where M is the average magnification in the sample and N are the cumulative number counts. The observed differential number counts show a sharp turnover at $B \sim 19.3$ (Boyle *et al.* 1988). At brighter magnitudes, $d \log N/dB \sim 0.9$, while at fainter magnitudes the slope is ~ 0.3. For intrinsic number-count slopes < 0.4, $q \sim 1$ but as the slope increases above 0.4, q becomes significantly larger than unity. The slope of the intrinsic number counts is not known but may be appreciably steeper than the observed slope. Figure 2 shows a plot of $\log q$ as a function of B_0 for different values of the intrinsic bright-end slope of the number counts. For the sample parameters of Webster *et al.* and an intrinsic number count slope equal to the observed slope, the maximum attainable value of q is ~ 2. For intrinsic number count slopes steeper than that observed, the value of q can be larger.

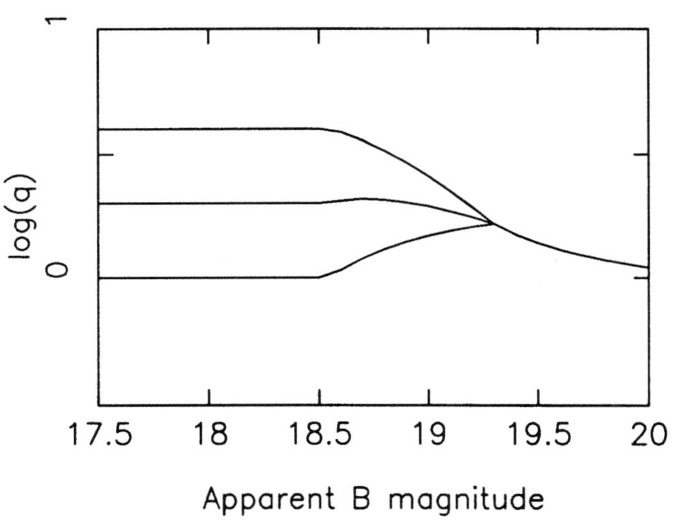

Figure 2: For an average magnification $M = 2$, $\log q$ is plotted as a function of the apparent magnitude m_B. The slope of the faint end of the QLF is fixed at 0.3, while the bright end slopes are 0.4, 0.9 and 1.4 from top to bottom.

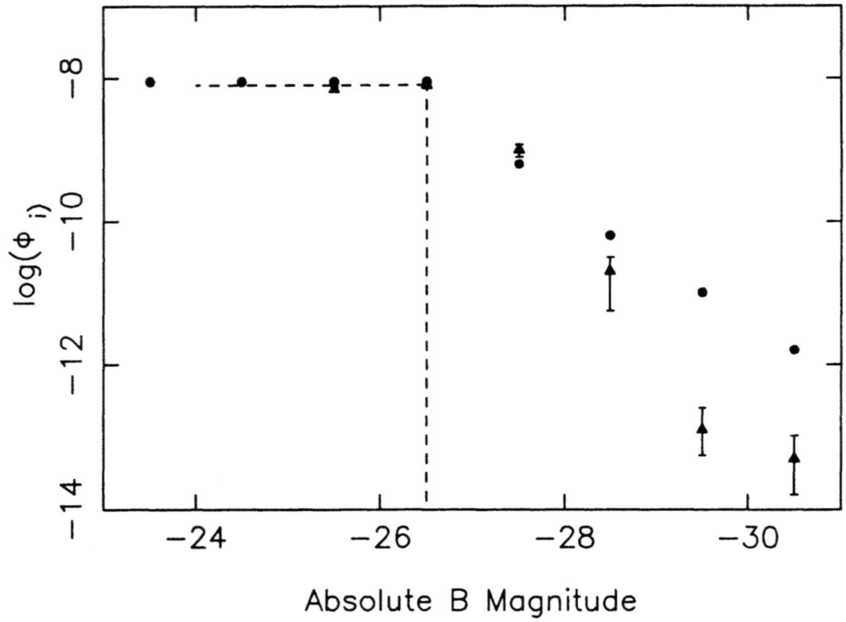

Figure 3: Triangles represent the observed QLF in the redshift range $1.7 \leq z \leq 2.2$ taken from the Boyle *et al.* (1988) sample. The dotted line is the assumed intrinsic QLF, which represents the most conservative hypothesis about the number of high-luminosity quasars. The circles are the result of convolving this intrinsic QLF with a realistic amplification function for an $\Omega = 1$ universe filled with compact objects capable of lensing.

The size of the difference between the observed QLF and the intrinsic QLF caused by the effects of gravitational lensing depends strongly on the form of the intrinsic QLF. The quasar number counts represent an integration of the QLF, over redshift and absolute magnitude, which involves a substantial loss of information. The QLF and its evolution is conventionally represented by a two power-law form (*e.g.* Boyle *et al.* 1988) which retains its shape but moves to brighter absolute magnitudes with increasing redshift. To understand the effects of gravitational lensing on the QLF a redshift dependent convolution is required. Webster (in preparation) has used an analytic form for the amplification, which conserves flux, has a properly defined minimum amplification, and an arbitrary maximum amplification to calculate the expected effects. For an intrinsic QLF represented by a power-law, $\Phi_i(> L, z) \propto L^{-\alpha}$, where $\Phi_i(> L, z)$ is the intrinsic cumulative space density, and L is the quasar luminosity, the value of the overdensity q is ~ 1 for power-law slopes $\alpha \lesssim 1$. As the index α increases, the value of the maximum amplification plays a more important role. For example, if the intrinsic

QLF is represented by a step-function and convolved with the amplification function appropriate to a total mass $\Omega = 1$ distributed in compact objects, it is possible to produce more high-luminosity quasars than observed in the high-redshift sample of Boyle *et al.* (1988).

While a model with $\Omega = 1$ in compact objects may be extreme, this calculation demonstrates that the bright tail of the observed QLF may be entirely due to the gravitational lensing effects of dark matter. Once the shape of the observed QLF is known as a function of redshift, this knowledge can be used to constrain cosmological parameters such as the amount of clumped dark matter, even if the value of $q \sim 1$, and there is no overdensity of galaxies near high-redshift quasars. Under the gravitational lens model for the overdensity, the value of q will depend on the quasar redshift, the galaxy redshift, δB, the surface mass distribution around the galaxy and the shape of the QLF. The first three parameters can be measured observationally. Thus the measured overdensity will be a function of the shape of the intrinsic QLF and the total mass around galaxies. Observations of the QLF combined with a better theoretical understanding of the gravitational lensing phenomenon, offer the possibility of determining the total mass associated with galaxies, an outstanding goal of the search for dark matter.

References

Arp H C 1970 *Astr. J.* **75** 1

Boyle B J, Shanks T and Peterson B A 1988 *Mon. Not. R. astr. Soc.* **235** 935

Broadhurst T J, Ellis R S and Shanks T 1988 *Mon. Not. R. astr. Soc.* **235** 827

Foltz C B, Chaffee F H, Hewett P C, Weymann R J, Anderson S F and MacAlpine G M 1989 *Astr. J.* 98 in press

Fugmann W 1988 *Astr. Astrophys.* **204** 73

Hewitt A and Burbidge G 1987 *Astrophys. J. Suppl.* **63** 1

King C R and Ellis R S 1985 *Astrophys. J.* **288** 456

Kovner I 1989 *Astrophys. J. Lett.* **341** L1

Maddox S J, Sutherland W J, Efstathiou G and Loveday J 1989 *Mon. Not. R. astr. Soc.* in press

Narayan R 1989 *Astrophys. J. Lett.* **339** L53

Nieto J-L and Seldner M 1982 *Astr. Astrophys.* **112** 321

Schneider P 1989 *A A* **221** 221

Seldner M and Peebles P J E 1979 *Astrophys. J.* **227** 30

Stocke J T, Schneider P, Morris S, Gioia I, Maccacaro T and Schild R 1987 *Astrophys. J. Lett.* **315** L11

Stockton A 1980 in *Objects of High redshift, IAU Symp. 92* eds G O Abell and P J E Peebles (Reidel: Dordrecht)

Tyson J A 1986 *Astr. J.* **92** 691

Véron-Cetty M -P and Véron P 1987 ESO Scientific Report No 5

Webster R L, Hewett P C, Harding M E and Wegner G A 1988a *Nature* **336** 358

Webster R L, Hewett P C and Irwin M J 1988b *Astr. J.* **95** 19

Yee H K C and Green R F 1984 *Astrophys. J.* **280** 79

Yee H K C and Green R F 1987 *Astrophys. J.* **319** 28

Yee H K C, Green R F and Stockman H S 1986 *Astrophys. J. Suppl.* **62** 681

Yee H K C 1989 in *Evolution of the Universe of Galaxies Astr. Soc. Pacific Conf. Series* in press

A search for gravitational lensing among highly luminous quasars : observational strategy and survey

J.P. Swings, P. Magain[1], M. Remy[2], J. Surdej[1]
Institut d' Astrophysique
Université de Liège, Belgium

A. Smette
European Southern Observatory
Garching, F.R.G.

D. Hutsemekers, E. van Drom
European Southern Observatory
La Silla, Chile

Abstract

The observational technique used for our search for gravitational mirages is presented : recent results, as well as future observations, are outlined.

Introduction

In order to explain the puzzle of the very high luminosity observed for the most luminous and most distant objects in the Universe, our team has suggested that many, if not all (!), of the brightest quasars could, in fact, be gravitational mirages. The apparent brightness of these objects would thus be nothing more than an illusion. An observational research project was first proposed to be performed with the Hubble Space Telescope, but ...; because of the Challenger disaster it was decided to attempt to do it from the ground, in the best possible seeing conditions, with large optical and/or radio telescopes. In the following, we briefly outline the strategy of our observations : it should be stated that our program actually started in 1986, and that preliminary reports about it were delivered in January 1988 (Tucson, see Surdej et al., 1988a) and in June 1988 (Paris, see Surdej et al., 1988b).

----

[1] Chercheur Qualifié au Fonds National Belge de la Recherche Scientifique

[2] Aspirant au Fonds National Belge de la Recherche Scientifique

Recently our program has been "upgraded" to an ESO key-program which is described in Surdej et al., 1989 (where all the names of the participants to the consortium are given).

Since gravitational lensing may play a double role, i.e. that of multiplication and/or that of magnification of quasar images, it is clear that a statistical evaluation of the occurrence of such effects, within a well defined sample of quasars, is of prime importance to better understand the QSO luminosity function and possibly the QSO phenomenon itself, to further test cosmology and to improve our knowledge on the hidden content of the Universe.

The scientific background of our search for gravitational mirages may thus be formulated as follows (see Surdej et al., 1988a, b) : the answer to the question "what fraction of extragalactic objects are gravitationally lensed ?" is very closely related to that to "how do the visible and dark matter distributions look at different scales in the Universe ?". It results that any prediction made for the expected number of gravitationally lensed quasars is bound to be very model dependent. It was therefore essential to carry out a systematic observational search and study of gravitational lensing effects in order to understand better the luminosity function of quasars, their observed number counts, their apparent cosmic evolution and the basic physical mechanism(s) powering these energetic objects. Such observations should also allow one to determine the Hubble parameter as well as galaxy masses on account of the expected time delay between the brightness variations of multiply lensed QSO images using, in particular, methods developed in Hamburg and in Meudon. More generally we may, in such a way, obtain information on the distribution of luminous and dark matter at various scales in the Universe. Furthermore, data on the size and structure of quasars ought to be derived from the observation of micro-lensing effects. Several communications at this meeting will deal with the latter points.

The selection of gravitational lens candidates

First of all, we considered that the apparently ($m_V < 18.5$) and intrinsically ($M_V < -29$) highly luminous quasars (hereafter HLQs) consitituted the best extragalactic candidates to search for the presence of gravitationally lensed images at arcsec./sub-arcsec. angular scale resolutions and/or for a brightness magnification due to an excess of foreground objects (galaxies, clusters) in the vicinity of the relevant targets*. The technical arguments leading to this assumption are the following :

----

* In the present contribution we do not consider giant radio galaxies nor arcs, since they are dealt with by several other authors.

(i) the HLQs form particularly high flux limited samples of extragalactic objects for which the probability of detecting multiply lensed images is higher than in volume limited ones;

(ii) the HLQs are the most likely objects for which one may assume that their intrinsic brightness is partially due to amplifcation by gravitational lensing;

(iii) the large cosmological distances suggested by the higher redshift values observed for the HLQs imply a high probability for gravitational lenses to be located along their lines of sight. This is also suggested by the rich absorption systems at redshifts $z_a < z_q$ recorded in the optical spectrum of most HLQs;

(iv) it is likely that the paucity of known cases of multiply lensed HLQ images with angular separations in the range below 2-3 arcsec is caused by strong observational selection effects.

Observations

a) In the optical

About 150 objects selected from the Véron and Véron (1989) catalog have been observed under good to excellent seeing conditions since 1986, many of them with the ESO/MPI 2.2m telescope (La Silla), the others with the ESO 3.6m telescope and/or with the CFHT (Hawaii). Wide-band filters (B, V, R, or I), and, whenever possible, narrow-band filters chosen to isolate one of the bright redshifted emission line of the QSO (Ly α or occasionally CIII]) as well as a nearby portion of the QSO continuous spectrum were used to observe the QSOs as well as the objects (galaxies ?) in the field. Subsequent spectra were obtained for the most interesting objects, essentially with the EFOSC spectrograph attached to the ESO 3.6m telescope.

b) In the radio

A radio detection survey of a few tens of optically selected objects has been carried out with the VLA, and a few sources have been detected and/or resolved (see Surdej, these proceedings).

c) Preliminary statistical results

As of now, the histograms and statistics given in Surdej et al. (1988a, b) remain unchanged, and they can be summarized as follows :

(i) the redshifts of the objects fall in the range between 1 and 3.6 (i.e. we are dealing here with fairly high redshift objects);

(ii) the visual apparent magnitudes of the QSOs are between 16 and 18.5, while

(iii) their absolute visual magnitude is located between - 29 and - 30.6.

Of all the objects observed so far, 23 % turn out to be "interesting" and definitely deserve further attention. By "interesting", we mean that the images show either multiple components, elongated structures, the presence of jet-like feature(s), the presence of a (or more) faint nearby galaxy(ies) or of some fuzz, etc.

d) Illustrations

In the recent years we have shown that our search for gravitational lensing from a survey of optically selected HLQs turned out to be quite successful and rewarding. To illustrate this, it suffices to show images and spectra of the double quasar UM 673, and its "overlying" lens-galaxy (see Surdej et al., 1987), of the "clover leaf" H 1413 + 117 (see Magain et al., 1988) or of UM 425 (Meylan and Djorgovski, 1988). Table 1 in Surdej et al. (1988a) actually clearly shows that most of the accepted cases of quasars being multiply lensed are of highly luminous quasars ($M_V < -29$).

Present and future observations and analyses

Since "magnification" of quasar images via gravitational lenses does not necessarily imply "multiplication" of images, it could very well be that many more HLQs are gravitationally lensed. We are therefore presently :

(i) performing a systematic subtraction of star-like images on our CCD frames in order to search for "overlying" galaxies (see reports by P. Magain et al. and J. Surdej at this meeting);

(ii) attempting to determine the frequency of galaxies in fields "near" HLQs (on the CCD frames) compared to that in random fields : the apparent excess in the spatial vicinity of HLQs is also discussed by P. Magain et al. (these proceedings);

(iii) acquiring, through a "test program" short exposure CCD frames of HLQs under superb seeing conditions (< .7 arcsec) with the newly installed New Technology Telescope at La Silla.

Furthermore, in order to determine the size of Ly α clouds in the surroundings of quasars, we are presently analyzing high resolution spectra of e.g. UM 673 (see report by A. Smette et al. at this meeting).

As far as the (near) future is concerned, we are :

(i) actively participating in a 3 year key-program for imagery and spectroscopy of many HLQs, as well as for monitoring gravitationally lensed objects we discovered, such as UM 673 and the cloverleaf H 1413 + 117;

(ii) monitoring other lenses, such as the "Einstein cross" Q 2237 + 031 for which variations in the brightness of one of the components has been observed (the aim here is to distinguish between micro-lensing effects and macro-lensing effects that could lead to cosmological implications, e.g. the determination of the Hubble constant);

(iii) planning further observations at high spatial resolution with the CFHT and/or the VLA, as well as speckle observations with the ESO/MPI 2.2m telescope;

(iv) looking forward to observations of a selection of HLQs with the Hubble Space Telescope.

* * * * * *

Acknowledgement

This paper was presented on behalf of the consortium involved in the ESO key program on Gravitational Lenses and on that of those who collaborated with us in the recent years.

References

Magain, P., Surdej, J., Swings, J.P. et al.: 1988, Nature, **334**, 325.

Meylan, G. and Djorgovski, S.: 1989, Astrophys. J., **338**, L1.

Smette, A., Surdej, J., Shaver, P. et al. : 1989, these proceedings.

Surdej, J., 1989: these proceedings.

Surdej, J., Swings, J.P., Magain, P. et al., 1987: Nature, **329**, 695 (and additional observations reported in 1988: Astron. Astrophys., **198**, 49)

Surdej, J., Swings, J.P., Magain, P. et al.: 1988a, Astron. Soc. Pac. Conf. Series, **2**, 183.

Surdej, J., Magain, P., Swings, J.P.: 1988b, in "Large Scale Structures : Observations and Instrumentation", Proceedings of the 1st DAEC Workshop, eds. Balkowski, C., and Gordon S., p. 95.

Surdej, J., Arnaud, J., Borgeest, U. et al.: 1989, The Messenger, 8.

Véron, M.-P. and Véron, P.: 1989, ESO Scientific Report n° 7.

A search for gravitational lensing among highly luminous quasars: new results *

P. Magain † M. Remy ‡ J. Surdej †
J.-P. Swings
Institut d'Astrophysique, Université de Liège
5, avenue de Cointe, B-4200 Ougrée-Liège

A. Smette
European Southern Observatory
Karl-Schwarzschild-Str. 2, D-8046 Garching bei München

1 Introduction

A number of recent investigations (e.g. Webster et al., 1988; Fugmann, 1988) have presented evidence in favour of an excess of galaxies in the fields of distant QSOs. This excess is generally interpreted as a manifestation of the gravitational magnification of the background QSO by matter associated with the foreground galaxies.

Having collected a significant number of images of highly luminous QSOs, we decided to analyse them in order to search for the presence of such an effect in our sample (for a description of our survey, see Surdej et al., 1988a, b and Swings et al., 1989). To improve the contrast in favour of galaxy detections, we kept only the images taken through a red filter (Bessel R – ESO serial number 421 or 447). 83 images were selected: 40 obtained with EFOSC at the ESO 3.6 m telescope and 43 with a direct CCD camera at the ESO/MPI 2.2 m telescope. The limiting R magnitude of these images generally ranges, for stellar objects, between 22.5 and 23.0.

2 Galaxy counts

The exposure time having been chosen in order to reach an optimum S/N for the QSO image, it varies from frame to frame. In order to ensure meaningful statistical checks, a comparison field had thus to be chosen on the same frame as that of each QSO.

*Based on observations collected at the European Southern Observatory

†Chercheur Qualifié au Fonds National Belge de la Recherche Scientifique

‡Aspirant au Fonds National Belge de la Recherche Scientifique

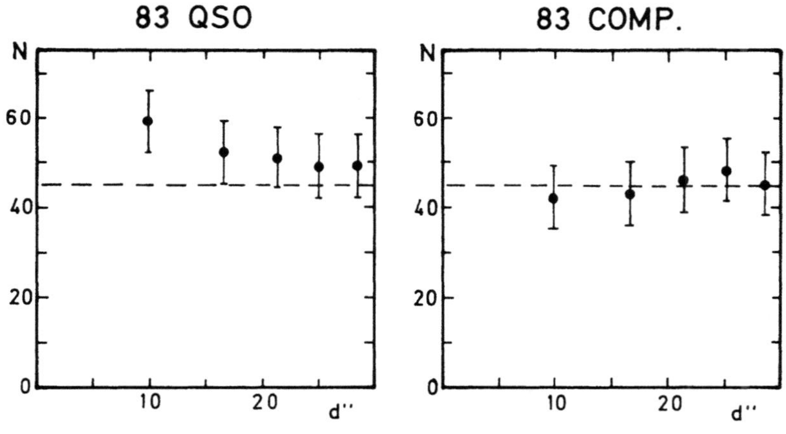

Figure 1: Counts of galaxies in 5 concentric rings around the QSOs (left) and in the comparison fields (right). One sigma error bars are indicated.

These fields were defined a priori in the following way. The QSO field is bounded by a circle of 30 arcsec radius centered on the QSO. The comparison field has the same size and shape and is centered on the same x column as the QSO, with a y position as far as possible from that of the QSO. The separation between the QSO and the center of the comparison field ranges from 1 to ~1.5 arcmin. In both the QSO and the comparison fields, the inner 3 arcsec are excluded in order to avoid bias against galaxy detection due to the light from the QSO.

All objects – stellar as well as extended – were first detected by eye. Two–dimensional gaussians were then fitted to these objects, which led to a determination of their extension and brightness. This procedure allowed the detection of most objects down to a S/N per pixel less than 2 in the center of the object. In order to ensure completeness of the sample and to avoid confusion between stars and galaxies, the objects were rejected if their mean S/N per pixel in the inner circle of 0.5 arcsec radius was less than 3. Each field was then divided into 5 concentric rings of equal surface and the total number of galaxies in each ring was determined.

The results are presented in Fig. 1. The expected number of galaxies in each ring for all the frames, as derived from the comparison fields, amounts to 45. The only likely significant deviation from this expectation occurs for the inner ring (3.0 to 13.7 arcsec) in the QSO field, where an overdensity of a factor 1.3 is obtained. The probability that this is just a statistical fluctuation from a uniform density of galaxies amounts to 1.5 %.

Our sample was then divided in two sub–samples, the first one containing 24 radio–loud QSOs, and the second one the remaining 59 optically selected (radio–quiet) QSOs.

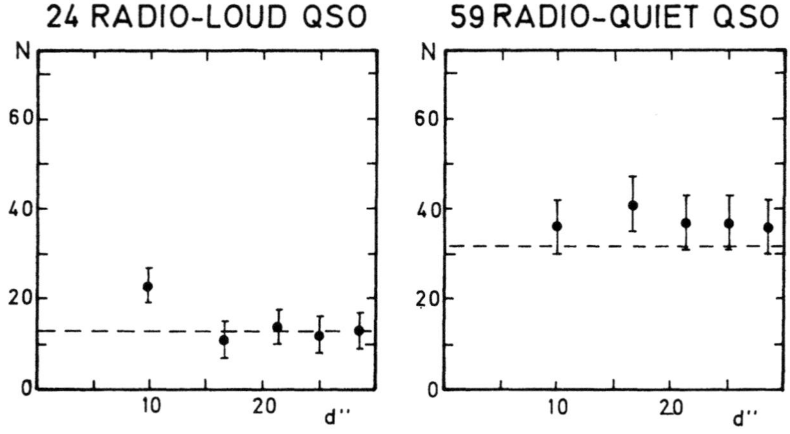

Figure 2: Counts of galaxies in 5 concentric rings around 24 radio QSOs (left) and 59 optical QSOs (right).

It can be seen from Fig. 2 and Table 1 that the galaxy overdensity in the inner ring is significant only for the case of the radio QSOs. The overdensity factor around the radio QSOs is 1.8 and the probability that this is due to a random fluctuation around uniformity is less than 1 %.

Table 1: Galaxy counts in the inner ring (3.0 – 13.7 arcsec) around the QSOs.

QSO sample	all (83)	radio (24)	optical (59)
number expected	45	13	32
number detected	59	23	36
overdensity	1.3	1.8	1.1
confidence level (%)	98.5	99.2	<75

Since most of the radio QSOs in our sample have flat radio spectra, the present result at first sight seems to confirm Fugmann's (1988) conclusion that the density of relatively bright galaxies around flat–spectrum radio QSOs is significantly higher than average. However, the overdensity found by Fugmann is much more pronounced and extends to larger distances. If Fugmann's result is to be taken at face value (which is not necessary the case as, in our opinion, his treatment of the comparison fields is far from optimum), our comparison field might also present a galaxy density above average. In this case, the discrepancy between Fugmann's results and ours would be significantly reduced. As a partial check for such an effect, we compared the galaxy density in the two halves of the comparison field, towards the QSO and opposite to it, and found no difference. This rules out the presence of a meaningful gradient in the density of comparison field galaxies. Any overdensity should then extend to distances exceeding a few arc minutes, which is, however, not incompatible with the size of large galaxy clusters.

A comparison can also be made with the results of Webster et al. (1988) who found an overdensity of galaxies located within 6 arcsec from optically selected QSOs, by a factor 4.2. In our 83 QSO fields, 7 galaxies would be expected in the 3 to 6 arcsec ring on the basis of a uniform distribution, while 29 galaxies should be found if the density was 4.2 times higher than in the comparison field. 10 galaxies are detected, in agreement with a more or less uniform distribution, but in disagreement with an enhancement by a factor 4.2 at the 99.998 % confidence level. Putting the S/N cutoff at brighter magnitudes in order to better match the magnitude limit of the Webster et al. sample does not change the result although its significance is lowered as a result of the smaller numbers involved.

It should be noted that the Webster et al. density enhancement has now been revised downwards on the basis of better statistics (see Webster, this volume). This downward revision is not sufficient, however, to bring their results in agreement with ours.

Such a comparison between widely different samples may seem unsecure. However, our sample being biased towards large QSO brightness magnifications, one would naively expect that it is also biased towards more detections of foreground galaxies. This is in apparent contradiction with the observations, and the reason for this has to be understood. A possible explanation would again be that our comparison field has more galaxies than average. This would mean that galaxy clusters are important sources of light magnification for the highly luminous QSOs. This possibility certainly deserves further investigation.

3 Subtraction of the point spread function

Another possibility is that a significant fraction of galaxies that are responsible for the QSO light magnification are in fact so close to the line–of–sight that they are within the inner 3 arcsec circle. To investigate this, we made a systematic subtraction of the point spread function (PSF) from the QSO images, in search for overlying objects. The PSF was determined whenever possible from stellar images of comparable brightness on the same frame. An analytic function, of the form of a generalized Moffat (1969) profile, was fitted to these stellar images. The PSF parameters were used in a similar fit of the QSO image from which the resulting profile was then subtracted.

An overlying galaxy was detected in three cases (with multiple QSO images in two of these). The faintest of these galaxies would have, in the absence of the QSO, a S/N per pixel in the central part somewhat higher than 25. The comparison fields were then used to predict the number of galaxies with S/N $\geq$ 25 which should be found within a circle of 3 arcsec radius for the 83 fields considered. The expected number turns out to be 0.16, which is 18 times lower than the observed number. Considering the fact that some PSF subtractions could not be carried out properly because of a lack of adequate stellar images or because of divergence of the actual PSF from the adopted analytical function, some overlying galaxies of similar brightness may have been missed. Moreover, if the comparison field would present some galaxy overdensity, the expected number of overlying galaxies would be even lower. All this means that the enhancement factor of 18 for relatively bright galaxies within 3 arcsec from the QSOs should probably be considered as a lower limit.

Our results thus suggest that galaxies which are very close to the line–of–sight probably contribute significantly to the magnification of highly luminous QSOs.

References

Fugmann, W.: 1988, Astron. Astrophys. **204**, 73

Moffat, A. F. J. : 1969, Astron. Astrophys. **3**, 455

Surdej, J., Swings, J.-P., Magain, P. et al.: 1988a, Astron. Soc. Pac. Conf. Ser. **2**, 183

Surdej, J., Magain, P., Swings, J.-P. et al.: 1988b, in "Large Scale Structures: Observations and Instrumentation": Proceedings of the First DAEC Workshop (Paris–Meudon), eds. C. Balkowski ans S. Gordon, p. 95

Swings, J.-P., Magain, P., Remy, M. et al.: 1989, this volume

Webster, R.L., Hewett, P.C., Harding, M.E. and Wegner, G.A.: 1988, Nature **336**, 358

Gravitational magnification effects on distant 3CR galaxies

O. LE FEVRE
Canada-France-Hawaii Telescope Corporation, P.O. Box 1597, Kamuela, HI96743, USA.

F. HAMMER
D.A.E.C. Observatoire de Meudon, 92195 Meudon Principal Cedex, France

Abstract. A photometric survey of 27 3CR galaxies with z>1 is presented. Unprecedented image quality of the data (mean FWHM=0"75 for the sample) have allowed us to find that 75% of the galaxies are made of 2 to 6 components in a typical 50(H_0/50) kpc size.

We present direct as well as statistical evidences showing that the sample is subjected to significant gravitational magnification effects. In 4 cases where a foreground bright galaxy has been observed near the 3C object, the geometric configuration is well known and the basic parameters can be estimated; this allows gravitational magnification calculations to be made and show that these galaxies are magnified by 0.5 to 1.6 mag. Four galaxies over 27 are found at less than 4 arcsec from R<21 foreground galaxies, 9 times more than expected from a random distribution, and shows that the sample is affected in a statistical way by the foreground matter. We finally emphasize the importance of taking into account the gravitational magnification effects before estimating the amount of evolution in these systems.

I. Introduction

The 3CR sample with 30 galaxies at z>1 is by far the largest and best studied sample of galaxies at these redshifts, and is thus the corner stone of most of the studies on galaxy evolution and cosmological tests (see e.g. Spinrad, 1986; Dunlop et al., 1989). However, all these galaxies are extremely powerful radio sources ($P_{178MHz}>10^{28}$ W.Hz^{-1}.ster^{-1}), and it seems dangerous to derive conclusions on the normal population of galaxies from the properties of such peculiar objects. Selection effects are expected to affect the properties of this sample: the 30 3CR z>1 galaxies have been selected over one half of the sky from their high radio power, because they are either very powerful sources ("monsters") or are gravitationally magnified by foreground matter (Hammer et al., 1986). Since the 3CR catalog is a flux limited sample (9 Jy), it has a higher probability to contain gravitationally magnified sources. This is even more relevant to the z>1 galaxies which all have their radio fluxes just above the flux limit. Some of them could be gravitationally magnified radio sources which originally belonged to a population of radio sources of lower radio power, typically a few Jy, but with a much higher number density than the 3CR sources. With such considerations in mind, we undertook a high spatial resolution imaging survey of the high redshift 3CR galaxies, in order to carefully evaluate the existence of gravitational lensing.

II. High quality imaging data: galaxies morphologies

All observations have been made with a double density RCA CCD at the prime focus of the Canada-France-Hawaii telescope on Mauna Kea, were consistently good seeing conditions prevail. The mean image quality on our sample is 0.75 arcsecond (σ=0"15), which is a must at these redshifts were 1"=12(H_0/50) kpc at z=1.5. We observe that 75% of the galaxies have 2 to 6 components detected 4σ above the sky, with total sizes for the galaxies in the range 15(H_0/50) to 120(H_0/50) kpc, and component sizes on the order of 10(H_0/50) kpc. The red magnitudes for single components have a mean of 23.2, i.e. a mean absolute magnitude M_R=–22.5+5log(50/H_0), which, computed without k correction or evolution correction is comparable to an elliptical galaxy's luminosity. These basic parameters show how peculiar the morphologies are, the galaxies have more the appearance of compact groups of bright galaxies than single ellipticals. 3/4 of our sample shows multiple components against 1/4 of the z<0.6 3CR galaxies. This is a very fundamental property. A description of individual galaxies is given in Le Fèvre et al., 1987, 1988a,b; Le Fèvre and Hammer, 1988, 1989; Hammer and Le Fèvre, 1989, A complete discussion on the morphologies of distant radio-galaxies is given in a review by Hammer (1989).

III. Gravitational magnification/lensing by foreground matter

1. Excess of foreground bright galaxies on the line of sight. At least 4 sources are close in projection to a bright foreground galaxy (R<21, see Table I). From red counts of galaxies (see e.g. Tyson, 1988), the probability to find 4 3CR sources in such a configuration is 10^{-3}. The location of high redshift 3CR galaxies with respect to foreground galaxies is therefore not due to a random distribution at the 99.999% confidence level. It is of interest to compare our result with the study of Webster et al. (1988) on 285 QSOs, who found that the surface density of QSOs around $B_J<21$ galaxies is 4.4 times higher than would be expected from a random distribution. A similar reasoning shows that the density of 3CR high z galaxies at less than 5 arcseconds from a R<21 foreground galaxy is *9 times higher* than is expected from a random distribution.

Moreover, a case by case analysis shows that all these bright foreground galaxies actually gravitationally magnify the background 3CR galaxy. The redshifts of the background galaxy and of the 3CR object are known and estimates of the velocity dispersions of the background galaxies have been obtained from their luminosities. This allowed gravitational magnification calculations to be performed and showed that the foreground 3CR galaxies are gravitationally magnified by 0.5 to 1.6 magnitudes (Table I). See Hammer and Le Fèvre (1989) for a complete analysis.

2. Excess of foreground distant and rich clusters of galaxies on the line of sight. Over 23 3CR galaxies with z>1 and a galactic latitude > 18°, there is 5 with a line of sight close to a foreground distant cluster of galaxies. The probability to obtain such a configuration at random is only 6.10^{-2}, therefore the high z 3CR galaxies are not distributed at random with respect to foreground clusters of galaxies.

IV. Conclusion

There is strong direct and statistical evidences that the z>1 3CR galaxy sample is affected by gravitational magnification. With 4 galaxies observed at foreground redshifts close in projection to 4 3CR objects, there is 9 times more foreground galaxies with R<21 close to galaxies in this sample than would be expected from a random distribution. Moreover, there is 5 3CR objects close to distant foreground clusters of galaxies. Therefore, the z>1 3CR galaxies are not distributed at random with respect to the foreground matter and one could expect some gravitational lensing effects to be present. Indeed, it is possible to compute gravitational magnifications for the 4 cases with close projections of foreground galaxies and they are found to be very significant, in the range 0.5 to 1.6 magnitudes. Within the 75% of the galaxies showing multiple components, it is to be checked that all of them are at the same redshift. It is possible, in at least one case, 3C324, that some of them are foreground and act as gravitational deflectors to produce lensing effects.

This gravitational magnification/lensing effects are a further complication in the process of interpreting the properties of these high redshift galaxies. They have to be precisely estimated and removed before any attempt of an interpretation in terms of physical processes internal to the galaxies.

Table I

3CR	z 3CR	z and mag R foreground galaxy	Dist. (arcsec) between 3CR and foreground galaxy	Gravitational magnification (in mag)
13	1.351	0.477; 20.3	3.9	1.2
194	1.185	0.312; 19.3	4.0	1.6
208.1	1.02	0.158; 20.9	3.9	0.5
225A	1.565	0.134; 18.8	4.9	0.6

References

Dunlop, J.S., Guiderdoni, B., Rocca-Volmerange, B., Peacock, J.A., and Longair, M.S., 1989, M.N.R.A.S., in press.

Hammer, F., 1989, proceedings of the XI European Astronomy Meeting of the IAU, Cambridge University Press, in press.

Hammer, F., Le Fèvre, O., 1989, Ap.J., in press.

Hammer, F., Nottale, L., Le Fèvre, O., 1986, Astr. Ap., **169**, L1.

Le Fèvre, O., Hammer, F., Nottale, L., Mathez, G., 1987, Nature, **326**, 268.

Le Fèvre, O., Hammer, F., Nottale, L., Mazure, A., Christian, C., 1988a, Ap.J., Lett., **324**, L1.

Le Fèvre, O., Hammer, F.,, Jones, J., 1988b, Ap.J. Lett., **331**, L63.

Le Fèvre, O., Hammer, F., 1989, Ap.J., Lett., in press.

Spinrad, H., 1986, Pub.A.S.P., **98**, 269.

Tyson, T., 1988, A.J., **96**, 1.

Webster, R., Hewett, P.C., Harding, M.E., Wegner, G.A., 1988, Nature, **336**, 358.

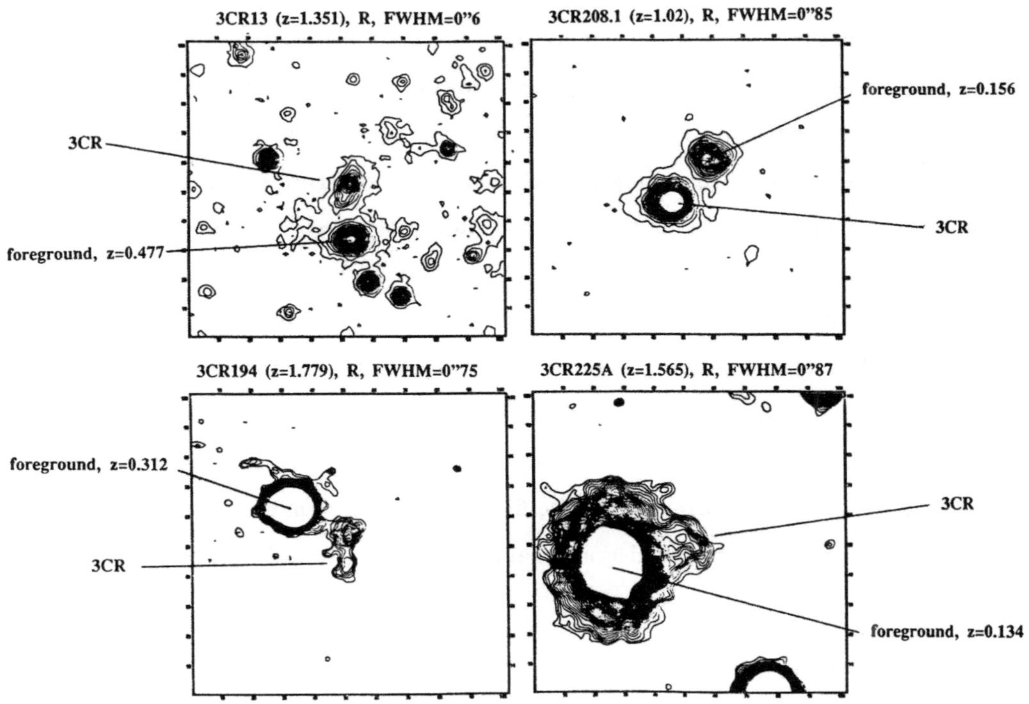

Figure 1: Images of the 4 cases of contamination by foreground galaxies of the line of sight to distant 3CR galaxies. All images obtained with the CFHT and a RCA CCD.

0957+561: THE TIME DELAY REVISITED

E. E. Falco and I. I. Shapiro
Harvard-Smithsonian Center for Astrophysics
Cambridge, MA, USA

J. H. Krolik
Johns Hopkins University
Baltimore, MD, USA

I. Introduction

The optical fluctuations of the A and B images of the 0957+561 gravitational lens system were observed for nearly ten years by Schild in the R band (Schild and Cholfin 1986; these proceedings) and by Vanderriest and collaborators in the B band (Vanderriest *et al.* 1989; these proceedings). The goal was to determine $\Delta\tau_{BA}$, the difference in light propagation times from the source to the observer via A and B, by cross-correlating the light curves. Schild's data set (hereafter RS) and Vanderriest *et al.*'s (hereafter CV) have long, periodic gaps due to occultation by the sun for 4-5 months every year, and shorter, random gaps of up to three months due to the usual problems of ground-based astronomy. Here we reanalyze RS and CV, discuss requirements for a reliable determination of $\Delta\tau_{BA}$, and apply the Edelson and Krolik (1988) method (hereafter EK) of discrete correlations. We find that the unevennesss of the sampling and possible systematic errors render unreliable any inference of $\Delta\tau_{BA}$ from these data.

II. RS and CV data sets

Figures 1 and 2 show, respectively, RS and CV, with fluxes in each normalized to A$\equiv$ 1 at time zero. The figures also show the least-squares fits of straight lines to the fluxes of A and B.

We assume, as in Vanderriest *et al.* (1989), that the light flux from the main lensing galaxy, G1, did not vary significantly over the interval covered by these data. In RS, B contains light from G1; in CV, the estimated effect of G1 was subtracted. Contamination of the light from B by G1 could affect the cross-correlation, even if the light from

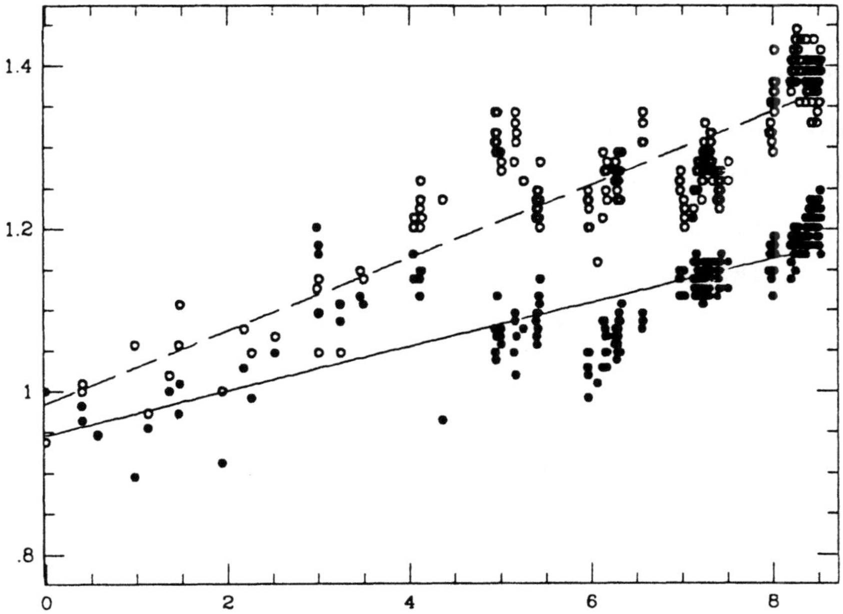

Figure 1. RS data set. The abscissa is time (years), the ordinate is R-band flux for A (filled) and B (open), normalized to A≡ 1 at time zero (16 November 1979).

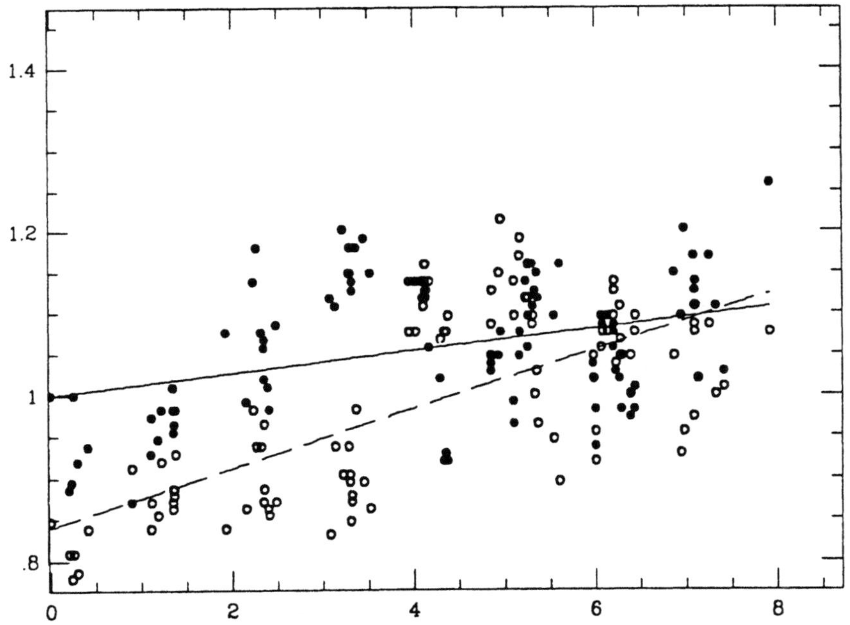

Figure 2. CV data set. The abscissa is time (years), the ordinate is B-band flux for A (filled) and B (open), normalized to A≡ 1 at time zero (19 December 1979).

G1 were strictly constant, and must be considered carefully before drawing (positive) conclusions about $\Delta\tau_{BA}$.

The ratios of the rms fluctuations to the mean A and B fluxes are: 5.9% (7.8%) in A and 7.3% (8.2%) in B for RS (CV), compared to the stated measurement errors of 1-2%. The ratios of the slope of the straight-line fit to the fluxes from A to that from B are 1.7 ± 0.1 (RS) and 2.6 ± 0.3 (CV). Since RS and CV correspond to R and B bands, respectively, these two sets of slopes need not coincide, but the ratios of slopes should, within the noise. The inferences made by the observers of values for $\Delta\tau_{BA}$ rely on interpolation of the light curves and on removal of long-term trends. Since the fluctuations in the light curves are near the noise level, and the trends may be systematically biased, the significance of the value deduced for $\Delta\tau_{BA}$ must be examined carefully.

Given our assumptions, we believe there are three requirements for estimating $\Delta\tau_{BA}$ reliably:

1. Cross-correlation must yield a statistically significant peak (*e.g.*, the probability of it being due to noise alone should be $\lesssim 1\%$), if microlensing effects are negligible.
2. A and B should be statistically indistinguishable in RS and in CV, up to a multiplicative factor due to the relative magnification for A and B, if microlensing effects are negligible.
3. If microlensing effects are not negligible, they must be accounted for, because they could mask temporal variations of the quasar if the times scales were comparable.

III. Discrete correlations

We apply EK because it avoids interpolation of the time series, and thus possible biases (Scargle 1982). EK yields correlations in discrete time bins, with error bars for each bin. Figures 3 and 4 show EK cross-correlations of A and B, for RS and CV, respectively. RS yields a peak, for delay ~ -0.25 years. CV yields only a very broad, low-amplitude "bump", certainly not a statistically significant peak. The peak yielded

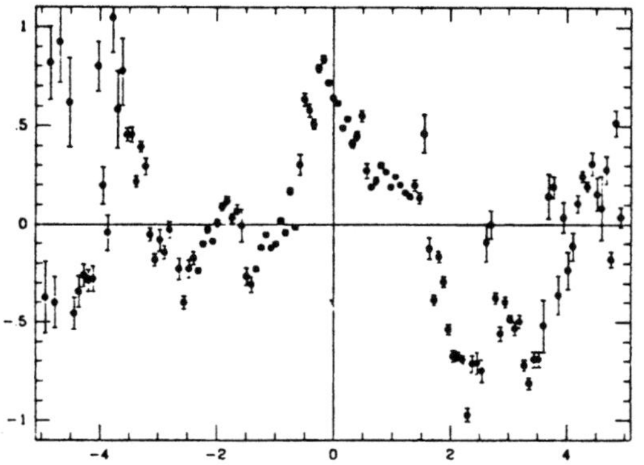

Figure 3. RS cross-correlation; lags in years.

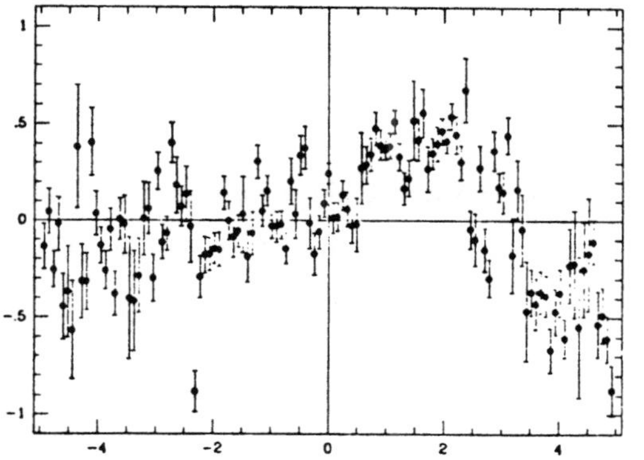

Figure 4. CV cross-correlation; lags in years.

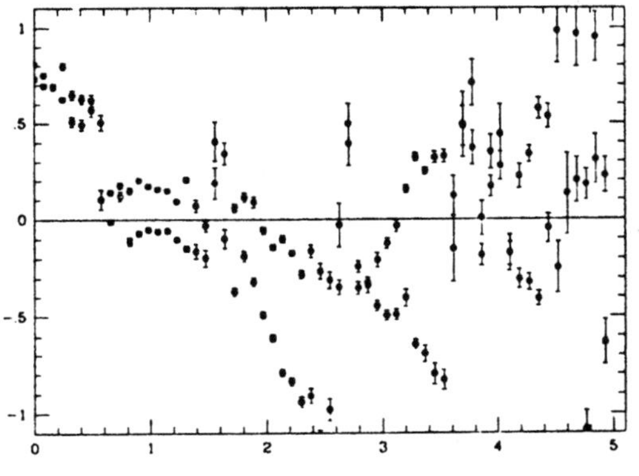

Figure 5. RS auto-correlations for A (filled) and B (open); lags in years.

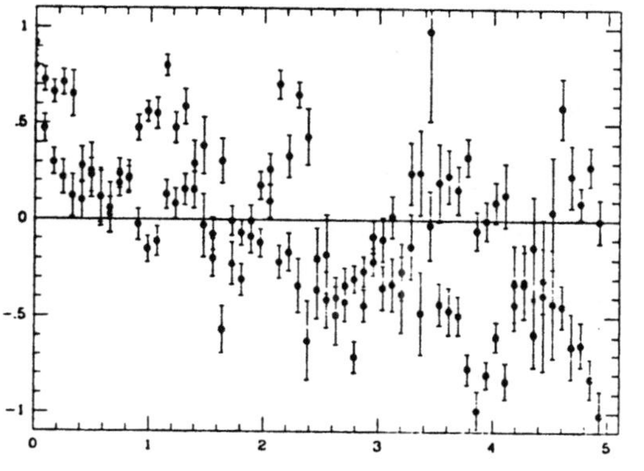

Figure 6. CV auto-correlations for A (filled) and B (open); lags in years.

by RS may or may not be significant, but it does violate expectations of a positive delay (*e.g.*, Dyer and Roeder 1981); CV violates the first criterion. In any event the two cross-correlations seem to be mutually inconsistent.

Figures 5 and 6 show auto-correlations of A and B, for RS and CV, respectively. The plots demonstrate that A and B are statistically different (by ten error bars per bin, on average). However, A and B should be just time-delayed views of the same object with different magnifications, so their auto-correlations should be indistinguishable within the noise, save for any microlensing, on either image or on both. The second criterion is therefore also violated.

Could microlensing save the day? Since B is nearer G1's center of brightness than is A, we could assume B fluctuated on a time scale of ten years due to a single, simple microlensing event; such a fluctuation of light intensity would be approximately parabolic in time. To check, we divided the light curve of B by a parabola, auto-correlated, and compared the result with the auto-correlation of A. We searched the space of the three parabola parameters to find the locus of the smallest difference between auto-correlations of A and of B for RS and CV. The search yielded no microlensing model that even came close to saving the day: the mean difference between the auto-correlations could not be reduced below ten correlation error bars.

IV. Conclusions

Application of EK implies that the optical flux data for A and B do not yield a reliable value for $\Delta\tau_{BA}$. Possible reasons, barring 0957+561 not being a lens system, are:

1. Any fluctuations of the quasar's flux, compared to the measurement errors, are too small to yield a significant result. This possible problem is compounded by the relatively poor sampling of the light curves.

2. Microlensing is affecting one or both light curves on time scales of weeks to months, a possibility that we plan to examine by detailed simulations (*e.g.*, Wambsganss, Paczyński and Katz 1989). If so, it is doubtful that the effects could be estimated and subtracted with sufficient accuracy to allow a reliable value of $\Delta\tau_{BA}$ to then be extracted.

3. G1 has varied significantly over the ten years of observations.
4. The measurements suffer from unknown systematic effects.

Finally, we stress that the "standard" auto- and cross-correlations, based on, for example, linear interpolation between data points usually yield dramatically different results. These differences must be understood in detail before the results from any of the techniques can be considered definitive.

References

Dyer, C. C., and Roeder, R. C. 1981, *Ap. J. Letters*, **174**, L115.

Edelson, R. A., and Krolik, J. H. 1988, *Ap. J.*, **333**, 646.

Scargle, J. D. 1982, *Ap. J.* **263**, 835.

Schild, R. and Cholfin, B. 1986, *Ap. J.*, **300**, 209.

Schild, R. 1989, These Proceedings.

Vanderriest, C., Schneider, J., Herpe, G., Chevreton, M., Moles, M., and Wlérick, G. 1989, *Astron. Astrop.*, **215**, 1.

Vanderriest, C. *et al.*, These Proceedings.

Wambsganss, J., Paczyński, B., and Katz, N. 1989, Springer-Verlag: Proceedings of "Gravitational Lenses", Cambridge, Mass., June 1988, 209.

THE TIME DELAY OF Q0957+561 A,B
FROM 10 YEARS OF OPTICAL MONITORING

R. Schild

Center for Astrophysics

Cambridge, MA 02138, U.S.A.

Abstract: Brightness monitoring of Q0957+561 A,B since its discovery in 1979 permits a time delay of 1.1 years to be established with 99 percent confidence. A best value for the time delay is now 404 days. Comparison of the brightness records shifted for the time delay shows an important discrepancy which we attribute to microlensing by a star or stars in the lens galaxy.

I. Introduction

Brightness monitoring of the gravitational mirage object Q0957+561 A,B (TwQSO) has continued intensively because a demonstration that the pattern of the QSO's brightness fluctuations seen in the two images with a time delay provides the most conclusive proof that the two images are two views of the same object. In addition, determination of the time delay between observation of the QSO's intrinsic brightness fluctuations provides important information about the gravitational lens creating the mirage. Because relativity theory predicts the amount of image displacement as a function of observed image properties, the measurement of the image separations and time delay permits a redshift-independent determination of the lens distance. This can be combined with the measured redshift of the lens galaxy and cluster to determine a value of the Hubble constant which is independent of the distance to the Hyades. Finally, comparison of the brightness records of the two image components time-shifted for the time delay allows determination of the microlensing which results from the passage of a star through the beam of an individual image component.

Since our first publication of the TwQSO time delay of 1.03+/- 0.1 years (Schild and Cholfin 1986, Schild 1986), other determinations have been attempted, as summarized in Table 1.

II. Observations

As soon as the time delay was recognized, the brightness records of the two mirage images were carefully compared and it was realized that microlensing

TABLE I

Time Delay Measurements

Reference	Year	Time Delay
Florentin-Nielsen	1984	1.55
Schild & Cholfin	1986	1.03 +/- 0.1 y
Gondhalekar	1986	1.8 +/- 0.2 y
Gorenstein et al.	1987	4 y
Lehar, Hewitt, Roberts	1988	1.31 +/- 0.25 y
Vanderriest et al.	1989	1.14 +/- .05 y
Schild	1989	1.11 +/- .01 y

effects being seen were at the level of a few percent per year. Because such small effects would be swamped by the larger intrinsic brightness fluctuations of the QSO, a highly refined time delay measurement was needed; the goal was to determine the time delay to the nearest day. Early observations demonstrated that nightly brightness variations occur at the level of a few percent in brightness, with the largest nightly variation being probably less than 10 percent; in addition the QSO seems to have intervals of quiescence with nightly brightness fluctuations of less than 1 percent (Vanderriest et al, 1982; Schild and Weekes,. 1984; Schild and Cholfin 1986). Thus it would be necessary to monitor nightly brightness variations at some initial epoch for at least two weeks to have a reasonable chance of detecting an event. Follow-up observations would of course need to extend over a much longer time interval in the year following because of the large uncertainty in the value of the time delay.

During the 1985-86 season of the current campaign, monitoring was possible on 46 of the 57 nights scheduled for the 61 cm telescope of the F. L. Whipple (Mount Hopkins) Observatory. No large nightly brightness fluctuations were found but some trends lasting over several days were noted, and a second season of monitoring resulted in observations on 79 of 116 scheduled nights. Analysis of this data gave the suggestion of a 406 day time delay, which was used to schedule 138 nights in the 1987-88 observing season. Observing was successful on 109 of these nights. No particularly prominent brightness change like the one seen in winter 1984 was found and refinement of the time delay could be done only with a cross-correlation analysis.

Our observations and reductions in the current campaign were as described in Schild and Cholfin (1986) except that all observations were made with a standard red (F) filter described by Schild and Kent(1981). This compromise was made because although the contribution of lens galaxy G1 to the QSO images is greater in the red, the background sky illuminated by moonlight is much fainter in the red and many of the observations had to be made on nights of full moon.

III. The Time Delay

Because early calculations indicated that the time delay would be nearly 5 years, sampling of the brightness was infrequent for the first several years. As it became clear that the time delay was only 1 year, the density of sampling increased, but unfortunately long intervals exist when the object is not observable because of the sun's northerly declination at 9 to 10 hours right acension.

A cross-correlation of the two 9-year data records is shown as Figure 1. The calculation was done using the standard software which has been in use at CfA for years for the determination of redshifts (Tonry and Davis 1979). This calculation gives a time delay of 404 +/- 10 days with a level of significance r=8.1, corresponding to greater than 99 percent confidence. This value is consistent with the Vanderriest et al result of 415 +/- 20 days. Furthermore the cross correlation peak is the same one as originally found by Schild and Cholfin (1986), which was 376 +/- 37 days.

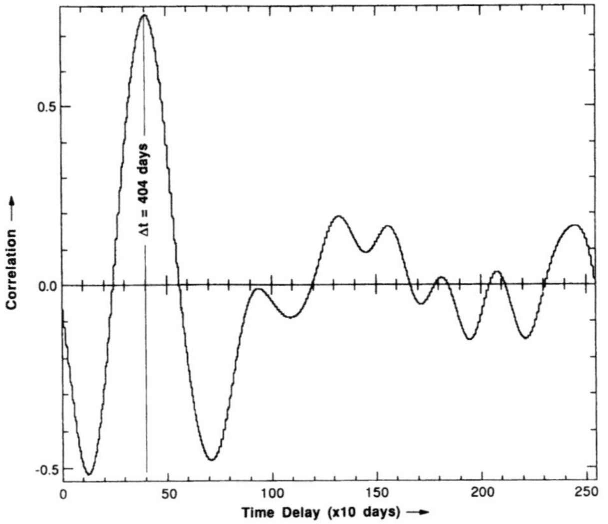

Figure 1: The cross-correlation for the 10-year records of brightness monitoring of the image components.

As noted above, an intensive monitoring campaign was started in 1985 to refine the time delay determination to the nearest day. From the first two seasons of monitoring, a provisional result of 406 days was used to optimize the dates of observation in the third season. A cross-correlation of the data from the second and third seasons is shown as Figure 3. We find a peak at 1 year and 39 days, or 404 days. It is the only peak compatible with the three previous determinations. It is the more believable in that the peak is single and relatively symmetrical. Other peaks seen in figure 3 at 1 year plus 60+ days are unacceptable because they are not compatible with the long-term monitoring record. The relatively high plateau which occurs for 1 year + 60 days is apparently the result of the microlensing, which will be discussed below. Because of the microlensing, which was noted already in 1986, we have not attempted to cross-correlate together all of the data from the three seasons of intensive monitoring.

We adopt a value of 404 +/- 1 day as the Q0957+561 A,B time delay. This can be remembered as one year plus one month plus one week plus two days. The value is convenient to astronomers in that an event seen in the first arriving A (northern) image will be seen 1.1 years later in the second arriving B image; a time delay of 1.5 years would be more problematic. The time delay of 404 days corresponds to 13 sydonic months plus 20 days. This means that an event seen at new moon first in image A will be seen near last quarter lunar phase the next year in image B.

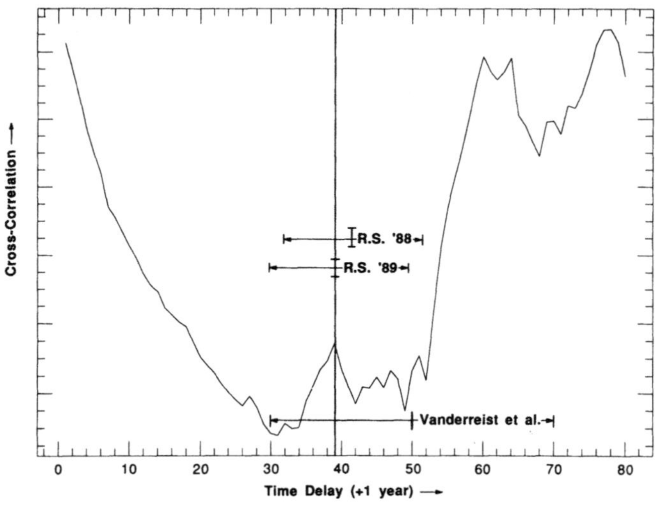

Figure 2: The cross-correlation for the 1-year intensive monitoring records. Also shown are the correlation peaks determined from long- term data records by the author and from Vanderriest et al (1989).

IV. Microlensing

With our 404-day time delay we can phase the observations together and look for the microlensing by individual stars in the lens galaxy G1. This would be seen as a change in the brightness ratio of the two mirage components, which we show as a change in the magnitude difference. As noted above, brightness monitoring was relatively infrequent during the first four years of the program and we plot in Figure 3 the mean difference between the two mirage image components averaged over an entire season's observations. For years after 1984 when monitoring was more intense, we average the results for the first half (Nov-Feb) and the second half (Mar-Jun) separately.

We see in Figure 3 that a 10 percent change in the brightness was observed during 1985. Relative to the brightness observed in spring 1984 in the A component, a 10 percent brightening was seen in the spring 1985 data for the later arriving B component. This change in brightness level was apparently preceded by an overshoot of the kind predicted to occur at the onset of microlensing by a single star in the lens galaxy G1 according to the models of Kayser, Refsdal, and Stabell (1986). These conclusions regarding microlensing are not sensitive to small uncertainties in the measured time delay.

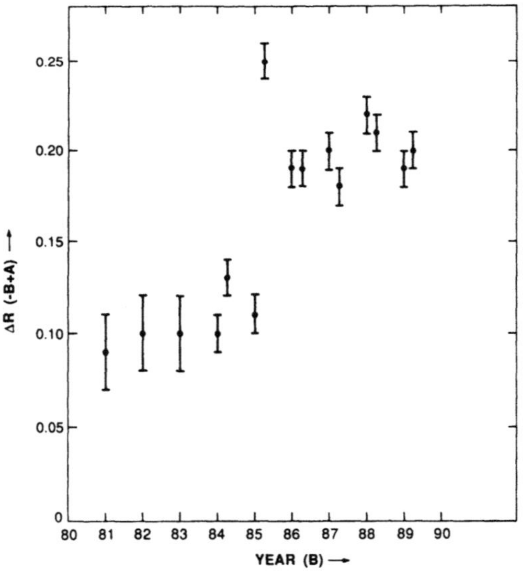

Figure 3: The brightness ratio of the two image components shown as a magnitude difference. Note that since magnitudes are defined as minus the log of the ratio, we plot (-B+A). The plot shows that relative to the brightness of the northern (A) component, the southern (B) component brightened by 10 percent in 1985.

Also seen in Figure 3 are small, possibly significant microlensing effects in the brightnesses observed since 1985. As soon as the time delay is considered to be certainly known to the nearest day, it will be possible to re-examine the brightness records with higher time resolution and look for microlensing effects with much shorter time scales. This is equivalent to looking for structure in the hypothetical accretion disc of the lensed QSO.

REFERENCES

Florentin-Nielsen, R. Astronomy and Astrophysics 138 (1984):119.

Gondhalekar, P.M.; Wilson, R.; Dupree, A.K.; and Burke, B.F. Joint NASA/ESA/SERC Conference, University College, London. ESA SPP-263 (1986):715.

Gorenstein, M.V.; Bonometti, R.J.; Cohen, N.L.; Falco, E.E.; Shapiro, I.I.; Bartel, N.; Rogers, A.E.E.; Marcaide, J.M.; and Clark, T.A. IAU Symposium 129 (1988):201.

Kayser, R.; Refsdal, S.; and Stabell, R. Astronomy and Astrophysics 166 (1986):36.

Lehar, J.; Hewitt, J.N.; and Roberts, D.H. Gravitational Lenses. J. Moran, J. Hewitt, and K.Y. Lo, eds. (1989):84.

Schild, R.E., and Kent, S. In Solid State Imagers for Astronomy. J.C. Geary and D.W. Latham, eds. Proceedings Soc. Photo-Opt. Instru. Eng. 290 (1981):186.

Schild, R.E., and Weekes, T. Astrophys. Journal 277 (1984):481.

Schild, R.E., and Cholfin, B. Astrophys. Journal 300 (1986):209.

Schild, R.E. In Quasars. G. Swarup and V.K. Kapahi, eds. Proceedings of IAU Symposium 119 (1986):549.

Tonry, J., and Davis, M. Astronom. Journal 84 (1979):1511.

Vanderriest, C.; Schneider, J.; Herpe, G.; Chevreton, M.; Moles, M.; and Wlerick, G. Astronomy and Astrophysics 215 (1989):1.

Application of Image Processing Techniques in a Search for Gravitational Lenses and Quasar Pairs

S. Djorgovski and N. Weir
Palomar Observatory, California of Technology
Pasadena, CA 91125, USA

G. Meylan
Space Telescope Science Institute
3700 San Martin Drive, Baltimore, MD 21218, USA

Abstract. We discuss application of two image processing techniques to high-resolution imaging of quasars, in a search for multiple images or companion objects: PSF subtraction, and MEM seeing deconvolution, on the examples of QSO 2300–352, and QQ 2345+007.

For the past few years, we have been conducting an optical imaging search for gravitational lenses and quasar pairs, using CCD imaging (Djorgovski and Meylan 1989ab; Meylan *et al.*, this conference). Good seeing is the most important factor in this kind of work. However, one can go beyond the immediately obvious, and search for an interesting morphology well within the seeing limit, by using a variety of image processing methods. Here we illustrate applications of two reasonably understood and extensively tested techniques, viz., the PSF subtraction and the Maximum Entropy (MEM) deconvolution, on the example of QSO 2300–352, and the probable lens QQ 2345+007. Other techniques and objects will be explored in a future publication.

There are two aspects of the problem: both small separation of components and/or a large intensity ratio make the detection of a multiple-image morphology difficult. For most methods, the S/N and the quality of the seeing can be traded off, but ultimately the pixel sampling is the limit. The choice of the optimal technique will depend on the data in hand. For example, we find that the Maximum Entropy Method, albeit very powerful in separating close components with comparable intensity ratios, is not very good for separating close objects with very large intensity ratios, as it tends to "smooth away" the fainter components. The PSF subtraction technique is complementary to it: it can bring out a faint object near a bright one, but is not very useful in separating close pairs with comparable intensities. Most Fourier-based methods (e.g., Wiener filtering) tend to introduce "ringing" in the image, and may create artifacts.

The quasar 2300–352 was discovered by Osmer (1982), who remarked upon the "discrete absorption features" in its spectrum. It was selected for our CCD imaging on the basis of its brightness ($R \simeq 17.8$) and redshift ($z = 2.84$). Our initial images were obtained on the night of UT 1988 July 20, with the CTIO 60-inch telescope and a TI 800×800 CCD, in the 2-by-2 binning mode, giving the effective pixel scale of 0.552 arcsec. The seeing was mediocre, with $FWHM \simeq 2$ arcsec. In all three images (VRI) there was a hint of an extension or a

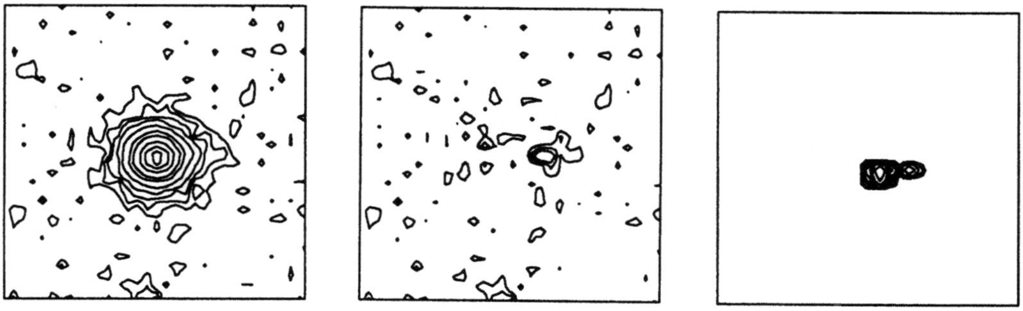

Figure 1. A section of the R band image of QSO 2300–352, obtained at CTIO. The field shown is 17.7 arcsec square, with N to the top, E to the left. The seeing was $FWHM \simeq$ 2 arcsec. Left: the original image. Only a slight appendage to the W of the QSO is evident. Center: the PSF-subtracted image, with the companion showing. Right: the MEM deconvolved image; the two objects are clearly separated. Images obtained at ESO in a better seeing confirm the existence of the companion.

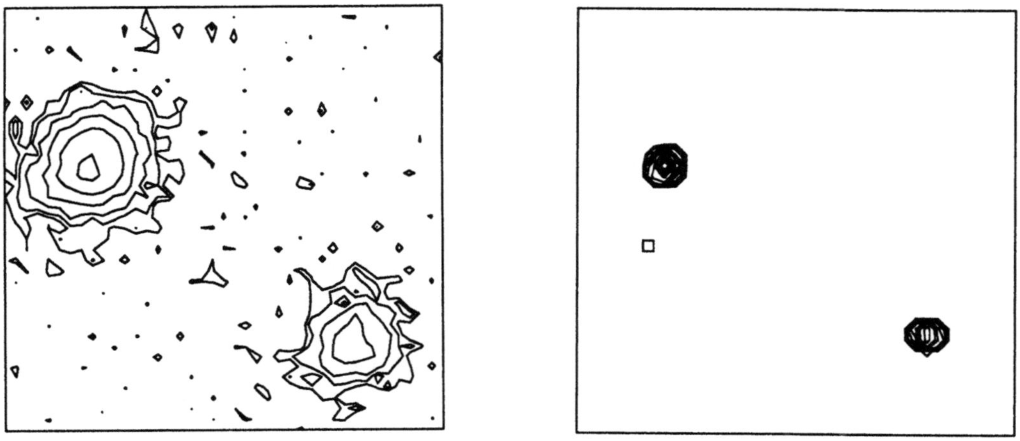

Figure 2. The original and the MEM-deconvolved images of QQ 2345+007, obtained at Palomar in the Gunn r band, in $FWHM \simeq 1$ arcsec seeing. The field shown is 10.1 arcsec square, with N to the top, E to the left. Neither component is resolved in our data, but our restored images are limited by the sampling, which is comparable to the component separation claimed by Nieto *et al.* (1988). Thus we cannot make any definitive statement at this time, but we hope to do so in a near future, as more data are being obtained. The results for the i band image are similar.

companion within about 2.5 arcsec to the west of the QSO. (Subsequently we learned that this companion was discovered independently by Surdej *et al.*)

Figure 1 shows the R band image of the quasar, original, with the scaled PSF subtracted from the quasar, and deconvolved using a MEM algorithm. PSF-subtracted images were generated using program *Daophot* (Stetson 1987). The companion is now clearly detected in both R and I. In the images deconvolved with a MEM algorithm the companion is probably over-resolved, but it is clearly detected. It is $\sim 3.7^m$ fainter than the QSO in R, giving the magnitude $R \simeq 21.5 \pm 0.5$, and is redder in color. It appears to be slightly non-stellar in morphology. Many other faint galaxies are present in the field, and it is possible that the companion is a member of a loose foreground cluster. It is also possible that it is physically associated with the QSO itself. Subsequent imaging of the object was done on the ESO 2.2-m telescope, using RCA 1024×640 CCD, with the pixel scale 0.175 arcsec, in fair seeing conditions ($FWHM \simeq 1.4$ arcsec). The companion is clearly seen in both V and R images. Thus, the better-seeing data to the first order confirm our image restorations.

The spectrum of the QSO is rather unusual: the object appears to be a transition case between an "ordinary" and a BAL QSO. It is tempting to speculate that the companion is responsible for some of the absorption, or that we are seeing an interaction which may be responsible for the QSO activity or its (incipient?) BALQ nature. Further spectroscopy and narrow-band imaging may help us resolve this question. Detecting such an interaction at a redshift where the comoving density of quasars begins to reach its maximum would be rather fascinating.

The lens candidate 2345+007 was discovered by Weedman *et al.* (1982), but to this date, no lensing cluster has been detected, in spite of a substantial effort (cf. Tyson *et al.* 1986). Nieto *et al.* (1988) obtained good-seeing images of the pair, using a seeing-compensating camera, and processed these images with a MEM-type algoritm. They claimed that the brighter component was resolved into at least two subcomponents. If true, this evidence would corroborate the lens hypothesis. We obtained a number of good-seeing images of this system at Palomar Observatory, in an attempt to check the Nieto *et al.* result using MEM deconvolution. The work is still in progress, and the results inconclusive at the time of this writing. As an illustration of the image sharpening power of the MEM technique, Figure 2 shows one of the Palomar images, before and after the seeing deconvolution.

References:

Djorgovski, S., and Meylan, G. 1989*a*, in Moran, J., *et al.* (eds.), *Gravitational Lenses, Lecture Notes in Physics* **330**, p. 173. Berlin: Springer.

Djorgovski, S., and Meylan, G. 1989*b*, in Osterbrock, D., and Miller, J. (eds.), *IAU Symp. #134*, p. 269. Dordrecht: Kluwer.

Nieto, J.-L., *et al.* 1988, *Astrophys. J.* **325**, 644.

Osmer, P. 1982, *Astrophys. J.* **253**, 28.

Stetson, P. 1987, *P. A. S. P.* **99**, 191.

Tyson, J. A., Seitzer, P., Weymann, R., and Foltz, C. 1986, *Astron. J.* **91**, 1274.

Weedman, D., Weymann, R., Green, R. J., and Heckman, T. M. 1982, *Astrophys. J. Letters* **255**, L5.

A SEARCH FOR GRAVITATIONAL LENSES AND QUASAR PAIRS AND THE DISCOVERY OF A NEW MULTIPLE QSO SYSTEM

G. Meylan
Space Telescope Science Institute
3700 San Martin Drive, Baltimore, MD 21218, USA

S. Djorgovski and N. Weir
Palomar Observatory, California Institute of Technology
Pasadena, CA 91125, USA

P. Shaver
European Southern Observatory
D-8046 Garching bei München, FRG

Abstract. We are conducting an optical imaging survey for lensed quasars, with a spectroscopic follow-up for the promising cases. A sample of known QSOs has been selected on the basis of apparently large absolute luminosities and high redshifts. So far, this survey has yielded one close pair of possibly interacting quasars, PKS 1145–071, one very probable gravitational lens, UM 425, several other promising lens candidates, and several cases of foreground or associated galaxies within a few arcsec from the quasars. Recently, we have discovered another interesting system, PHL 1222, probably a close physical pair of QSOs with similar redshifts, $z \simeq 1.91$, and separated by 3.3 arcsec. At least three other foreground faint objects ($R \sim 24$ mag) are observed inside a circle of 7 arcsec around the quasar. In PHL 1222 we may be witnessing a triggering event responsible for the nonthermal activity in both objects, i.e., the birth of a quasar pair at a redshift where the comoving density of quasars was close to its maximum.

The survey. The distribution of image separations for gravitationally lensed quasars is expected to peak at arcsecond or subarcsecond scales (for extensive modelling see: Turner, Ostriker, and Gott 1984, Ostriker and Vietri 1986, and Blandford and Kochanek 1987). Since most quasars are identified on Schmidt plate material (e.g., sky survey prints or grism plates), and thus with effective resolution of $\sim 3-5$ arcsec, one may expect to find some lens candidates simply by CCD imaging in good seeing conditions. However, one can improve the odds dramatically by making a careful selection of candidates. The luminosity function of quasars falls off very steeply at every redshift, and thus an apparently high-luminosity QSO has a higher *a priori* probability of being lensed. On account of larger intercept lengths, the quasars at large redshifts also have a higher *a priori* probability of being lensed. For the last few years, we have been conducting just such an imaging survey for gravitational lenses and other interesting systems. The survey is described briefly by Djorgovski and Meylan (1989*a*, *b*). The two selection criteria for quasars in our survey are:

- Large apparent luminosity ($M_V \leq -28$), and
- Large redshift ($z \geq 1$).

A QSO sample preselected in this way should have a considerably higher frequency of lensing events than a "random" QSO catalog. The exact probability enhancement is difficult to compute, since the biasing function depends on many unknowns (e.g., luminosity functions of QSOs at different redshifts, mass distribution and clustering on large scales, and cosmology). Our candidates are selected from the catalog by Hewitt and Burbidge (1987). We have surveyed about 300 QSOs so far. A similar survey is being conducted independently by Surdej *et al.* (1988). Our two surveys have now merged into the ESO Key Program for Gravitational Lensing (Surdej *et al.* 1989).

The efficiency of our lens candidate selection criteria is demonstrated by the two cases published by the Liège group, viz., UM 673 (Surdej *et al.* 1987), H1413+117 (Magain *et al.* 1988), and by UM 425 (Meylan and Djorgovski 1989). Several other promising candidates are now awaiting spectroscopy. Another interesting result of this survey is the discovery of two close pairs of quasars, viz., PKS 1145–071 (Djorgovski *et al.* 1987), and PHL 1222, described here and announced in Meylan *et al.* (1989). In addition, we found several cases of QSOs with foreground galaxies within a few arcsec, which should be useful for absorption-line studies, or monitoring for gravitational microlensing.

PHL 1222: the initial observations. The quasar PHL 1222 = UM 144 = QSO 0151+048 (Burbidge 1968) is one of the objects selected as potential lens candidates on the basis of the two criteria mentioned above. The first piece of evidence of the multiple character of PHL 1222 is found in the multicolor CCD frames obtained with the 40-inch telescope at Las Campanas Observatory (Chile) on UT 1988 October 21. These images show at least one close companion (which we denote as B), with the same colors (within the errors) as the bright QSO itself (denoted as A).

The confirmation of the interesting character of PHL 1222 is provided by observations taken at the European Southern Observatory (ESO), La Silla (Chile). Spectra of both components A and B were obtained with the ESO Faint Object and Spectrograph Camera (EFOSC) at the ESO 3.6-m telescope, on the nights of UT 1988 December 8, 9, and 10. Some deep imaging frames were obtained with a high resolution CCD camera at the ESO/PMI 2.2-m telescope, on the nights of UT 1988 December 11, 12, and 13. Subsequent images and spectra were obtained at Palomar Observatory using the 200-inch Hale telescope and the 4-Shooter imager/spectrograph, on the nights of UT 1989 September 6 and 7. Additional CCD images were obtained at the Palomar 60-inch telescope.

The contour map of the system is shown in Figure 1. Four close faint companions, labelled B, C, D, and E (in decreasing brightness) encircle the bright image of the quasar A. The separation between the quasar image A ($V = 17.60$ mag) and its brightest companion B ($V = 21.25$ mag) is 3.3 arcsec. Object C, at 6.8 arcsec from A, is unresolved and much redder than A or B; it might be a foreground star. Objects D and E are both resolved, at about 4 arcsec from A; they are very blue in color. Preliminary photometric information is summarized in Table 1, for both components A and B. The differences in *BVRI* magnitudes between A and B components are almost constant, $\Delta m \sim 3.55$. From the color indices, the companion B seems slightly bluer than the quasar A.

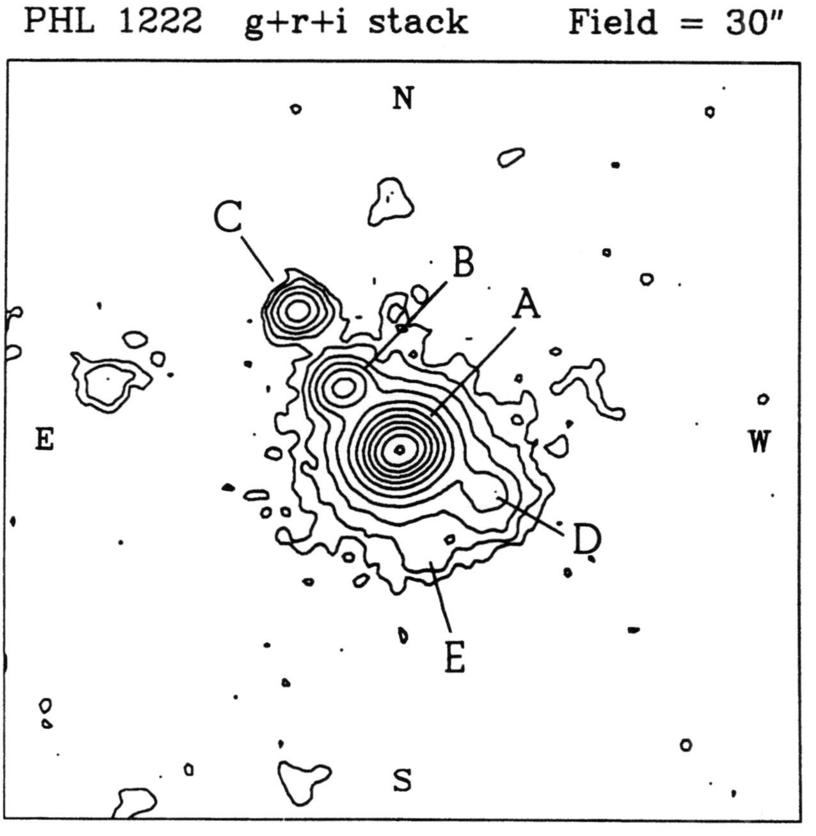

Figure 1: Image of the PHL 1222 = UM 144 = QSO 0151+048 system, from a stack of CCD *gri* frames obtained with the Palomar 200-inch Hale telescope. The field shown is 30×30 arcsec square, with north at the top and east at the left. The components of the PHL 1222 system are indicated with letters. The isophotal levels are spaced logarithmically in factors of 2.

PHL 1222: a gravitational lens or an interacting quasar pair? The spectra of the two components A and B (with a total exposure time of 3 hours in both B300 and R300 grisms) immediately confirmed that both objects are quasars, with the same emission lines (viz., C IV 1549, C III] 1909, and Mg II 2799) at about the same redshift, $z \simeq 1.91$ (Figure 2). Differences in velocity between the two spectra have been obtained from

Table 1. From the ESO December 1988 data:

PHL 1222 spectroscopic magnitudes

object	B	V	R	I	estimated errors
QSO A	18.05	17.60	17.30	16.95	~ 0.05
QSO B	21.50	21.25	20.90	20.50	~ 0.10
B–A	3.45	3.65	3.60	3.55	

PHL 1222 spectroscopic color indices

object	B–V	V–R	R–I	estimated errors
QSO A	0.45	0.30	0.35	~ 0.03
QSO B	0.25	0.35	0.40	~ 0.05

emission line redshifts ($\Delta V = 1380 \pm 240$ km s^{-1}) and from cross-correlation ($\Delta V = 520 \pm 160$ km s^{-1}). These ΔV values are typical of velocity dispersion in clusters of galaxies. The relative intensities of the C IV 1549 and C III] 1909 lines are reversed in the spectra of A and B. While the continuum of the faint quasar B is nearly flat, the continuum of the bright quasar A increases significantly from the blue to the red, so far as to have a flux level twice as high in the red than in the blue (thus the bluer $B-V$ color index of component B). The equivalent widths of the emission lines are much larger for the fainter component B. All these dissimilarities in the shapes of the emission lines and the continuum, as well as the differences in redshift favor the interpretation of PHL 1222 as being a physical quasar pair, rather than a gravitational lens.

The component C of the system (Figure 1), with very red colors, has been weakly detected in the R300 grism spectra. The very low S/N ratio hampers any clear determination of the nature of this object. It could be a foreground galaxy at $z \sim 0.8$ if we interpret the increase in intensity towards the red as the break at 4000 Å. It may also be a foreground galactic star, probably an M dwarf.

In spite of a few attempts about ten years ago, PHL 1222 has not been detected in radio. Sramek and Weedman (1980) summarize the flux density limits obtained so far: 20 mJy at 1415 MHz, 18.3 mJy at 2380 MHz, and 2.3 mJy at 4885 MHz. We are unaware of any other radio observations of this system.

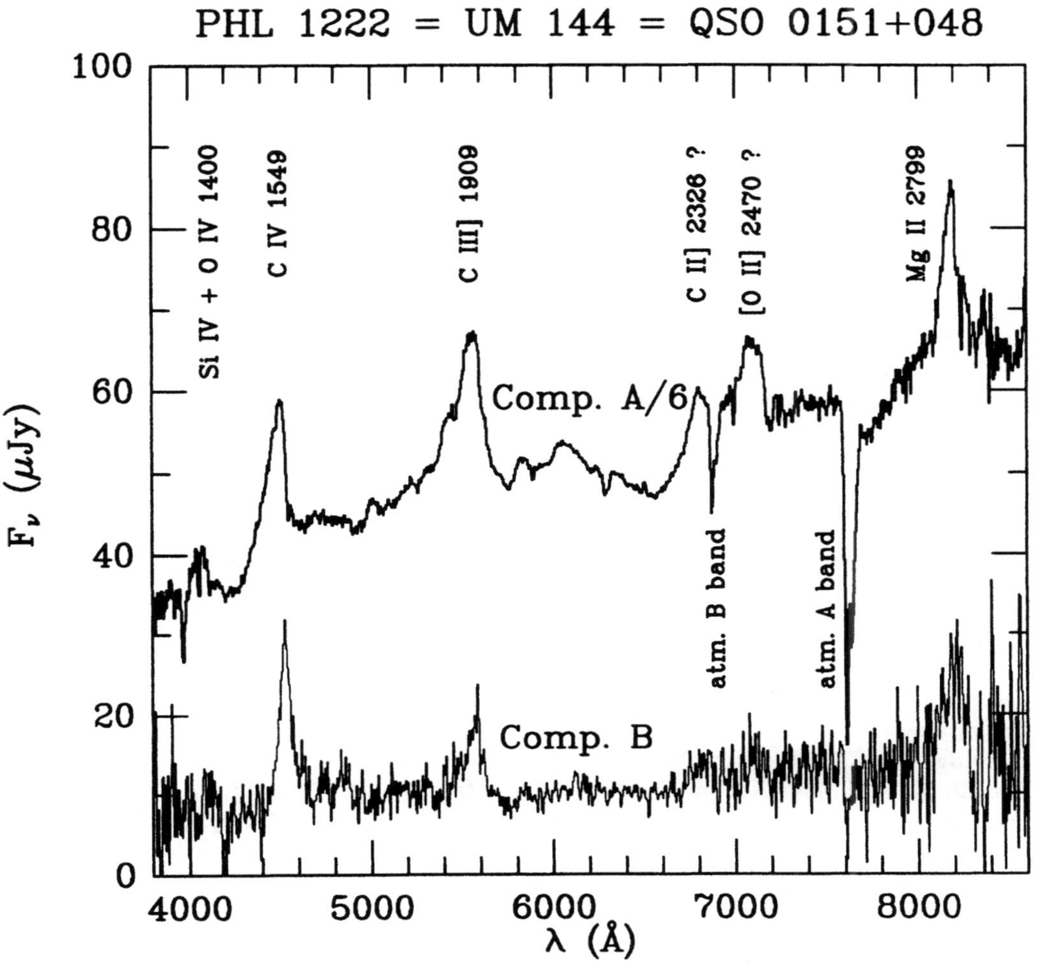

Figure 2: CCD spectra of components A and B, obtained at ESO, indicating that both objects are quasars. The spectrum of the brighter component A is scaled down by a factor of 6 for easier comparison. The two spectra show some dissimilarities in redshifts, in the shapes of the emission lines (viz., C IV 1549, C III] 1909, and Mg II 2799) and in the continuum, which favor the interpretation of PHL 1222 as being a physical pair of quasars rather than a gravitational lens.

PHL 1222: a tentative estimate of the total mass of the system. Under the assumption that the two QSOs are gravitationally bound, and that there is no other massive object in their vicinity, their projected separation and their velocity difference allow us to place a lower limit to the total mass of the system by using the Virial mass equation. At $z = 1.91$, with $H_0 = 75$ km/s/Mpc and $\Omega_0 = 0.0$, 3.3 arcsec correspond to 28 kpc, whereas with $\Omega_0 = 1.0$, 3.3 arcsec correspond to 18 kpc. Considering a minimum separation of about 20 kpc and a minimum velocity $\Delta V = 500$ km s^{-1}, the (minimum) virial mass amounts to about $M_A + M_B \simeq 1.7 \times 10^{11}\ M_\odot$, a reasonable value for normal galaxies. Allowing for the projection effects would suggest a true value of the total mass of the system several times larger than that.

PHL 1222: a possible interaction event. Recent numerical simulations show that gas distributed throughout a galaxy responds strongly to the tidal field of a close companion. In some cases, dynamical instability drives a large fraction of the gas into the inner regions of the galaxy (Hernquist 1989). A strong burst of star formation follows and subsequent evolution may lead to the formation of a black hole. Continued accretion of gas by the black hole may provide enough power to explain quasars and nuclear activity. From an observational point of view, the "interaction model" seems also to be the dominating paradigm for explaining the origin of nuclear activity in galaxies (cf. Fricke and Kollatschny 1989 for a recent review and further references).

PHL 1222 is already known to have an absorption system with $z_{abs} > z_{em}$ (Williams and Weyman 1977; Sargent, Boksenberg, and Steidel 1988). Associated *red*-shifted C IV 1549 absorption is present in the spectrum of the brighter component A, which corresponds roughly to the redshift of the fainter component B. This may be a signature of the ambient gas in a possibly interacting system. In PHL 1222 we may be witnessing a triggering event responsible for the nonthermal activity in both objects, i.e., the birth of a quasar pair at a redshift where the comoving density of quasars was close to its maximum.

References:

Burbidge, E.M. 1968, *Astrophys. J. (Letters)*, **154**, L109.
Blandford, R.D., and Kochanek, C.S. 1987, *Astrophys. J.*, **321**, 658.
Djorgovski, S., Perley, R., Meylan, G., and McCarthy, P. 1987, *Astrophys. J.*, **321**, L17.
Djorgovski, S., and Meylan, G. 1989*a*, in Moran, J. *et al.* (eds), *Gravitational Lenses, Lecture Notes in Physics*, bf 330, p. 173, Berlin: Springer.
Djorgovski, S., and Meylan, G. 1989*b*, in Osterbrock, D., and Miller, J. (eds), *Active Galactic Nuclei, IAU Symp.* # 134, p. 269, Dordrecht: Kluwer.
Fricke, K.J., and Kollatschny, W. 1989, in Osterbrock, D., and Miller, J. (eds), *Active Galactic Nuclei, IAU Symp.* # 134, p. 425, Dordrecht: Kluwer.
Hernquist, L. 1989, *Nature*, **340**, 687.
Hewitt, A., and Burbidge, G. 1987, *Astrophys. J. Suppl.*, **63**, 1.
Magain, P., *et al.* 1988, *Nature*, **334**, 325.
Meylan, G. and Djorgovski, S. 1989, *Astrophys. J. (Letters)*, **338**, L1.
Meylan, G., Djorgovski, S., and Shaver, P. 1989, in *B.A.A.S.*, in press..
Ostriker, J., and Vietri, M. 1986, *Astrophys. J.*, **300**, 68.
Sargent, W.L.W., Steidel, C.C., and Boksenberg, A. 1988, *Astrophys. J.*, **334**, 22.
Sramek, R.A., and Weedman, D.W. 1980, *Astrophys. J.*, **238**, 435.
Surdej, J., *et al.* 1987, *Nature*, **329**, 695.
Surdej, J., *et al.* 1988, in Osmer, P.S. *et al.* (eds), *Optical Surveys for Quasars, P.A.S.P. Conference Series* **2**, p. 183.
Surdej, J., *et al.* 1989, *ESO Messenger*, **55**, 8.
Turner, E., Ostriker, J., and Gott, J.R. 1984, *Astrophys. J.*, **284**, 1.
Williams, R.E., and Weymann, R.J. 1976, *Astrophys. J. (Letters)*, **207**, L143.

EVOLUTION AND GRAVITATIONAL LENSING IN QUASAR ABSORPTION LINE SYSTEMS

Peter A. Thomas and Rachel L. Webster
CITA, 60 St. George St., Toronto, Ontario, M5S 1A1, CANADA

Abstract

We present tests of both evolution and the bias due to gravitational lensing in the redshift distribution of quasar metal-absorption-line systems.

1 Introduction

Absorption line systems of heavy elements are thought to arise in gas associated with galaxies and their halos. Thus the detection of absorption line systems at high redshift may provide a tracer for the evolution of both number density and the chemical composition of galactic halos over redshift ranges not accessible by other means. However these results can be biased by gravitational lensing: if a quasar is intrinsically too faint to be included in the sample but is magnified into it by gravitational lensing , and if there is an absorption line associated with the halo which causes the magnification, then the distribution of these absorption lines will reflect the variation of the lensing cross-section with redshift.

Over the last ten years, a number of groups have compiled samples of metal line absorption systems. In this paper, we use the data from these groups, and in particular the extensive compilations of Sargent, Boksenberg and Steidel (1988, hereafter SBS), Sargent, Steidel and Boksenberg (1988a, hereafter SSB1) and Sargent, Steidel and Boksenberg (1988b, hereafter SSB2). Steidel, Sargent and Boksenberg (1988, hereafter SSB) have used these data-sets to support claims of an increasing comoving number density with redshift of magnesium absorption line systems (for $z \lesssim 1$) and decreasing comoving number density of carbon line systems (for $z \gtrsim 1$). This behaviour is similar to that which would be expected in the case of biasing by gravitational lensing (see Fig. 1).

Identification of lensed quasars can be difficult; micro-lensed quasars will not show multiple images and although the frequency of multiply-imaged quasars is relatively well-known, the percentage of bright quasars which might be micro-lensed is not (Webster *et al.* 1988). The fact that a quasar is gravitationally lensed does not necessarily invalidate its contribution to a random sample of absorption line systems. A bias will only arise if the object responsible for lensing the quasar into the sample also contributes a related absorption line system. Even so, the other absorption lines in the quasar spectrum may be treated as selected from a random line-of-sight. In practice this means that if the number of absorbers is large then lensing will not seriously affect their redshift distribution. We elaborate on this in §2.2 below.

There are a number of selection effects to be avoided when choosing a sample of absorption lines to test these hypotheses. For each quasar we must determine the redshift interval over which absorption lines with a given rest equivalent width, W_0, could have been detected. Because of quasar-cluster correlations that part of the redshift interval which corresponds to a recession speed

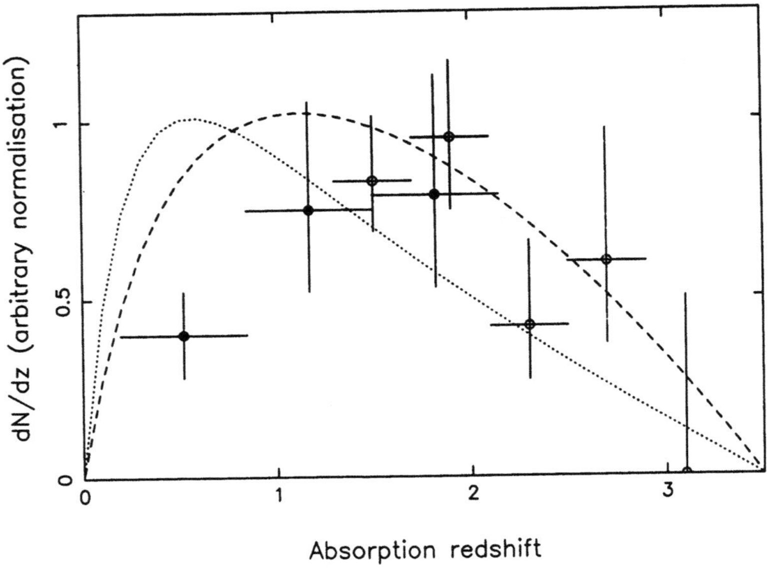

Figure 1. Solid circles: MgII data, open circles CIV data (adapted from SSB). The dotted and dashed curves show, for $q_0 = 0.5$ and 0.0 respectively, how the lensing cross-section varies with redshift for a typical quasar at a redshift of 3.5.

relative to the quasar of less than 5000 km s^{-1} should be excluded. In addition, care must be taken in the selection of absorption lines by equivalent width. Large systems often break-up into multiple components in higher resolution data, so we combine lines whose velocity separation is less than 2000 km s^{-1}. This should be done before selection of the complete subsample as a function of equivalent width. Another problem is that different data samples have different normalisations, so that combining them can lead to spurious indications of evolution.

Because of the methods we use to overcome these problems we arrive at slightly different conclusions than previous authors; for a more detailed discussion see Thomas & Webster (1989).

2 Tests and analysis of the data

2.1 EVOLUTION OF NUMBER DENSITY WITH REDSHIFT

We take as our null hypothesis that the absorption lines are randomly drawn from some probability function, $p(z)$, within the redshift window, $[z_{min}, z_{max}]$, in which they could have been detected. We test power-law models, $p \propto (1+z)^\gamma$ where, for constant comoving number density and $q_0 = 0$, $\gamma = 1$, or with $q_0 = 0.5$, $\gamma = 0.5$.

Fig. 2 summarises the results of our analysis of the SBS data-set which is the largest homogeneous sample, containing 60 absorption lines in 55 quasar spectra. In agreement with SBS we find we find evidence to reject no-evolution models when applied to the sample as a whole ($W_0 > 0.3$Å), although with a confidence of only 8%. However all this apparent evolution is due to the high-equivalent-width (hew, $W_0 > 0.9$Å) systems which are best fit by $\gamma < -3$; in contrast the low-equivalent-width (lew, 0.3Å $< W_0 <$ 0.9Å) line distribution is well described by no-evolution models.

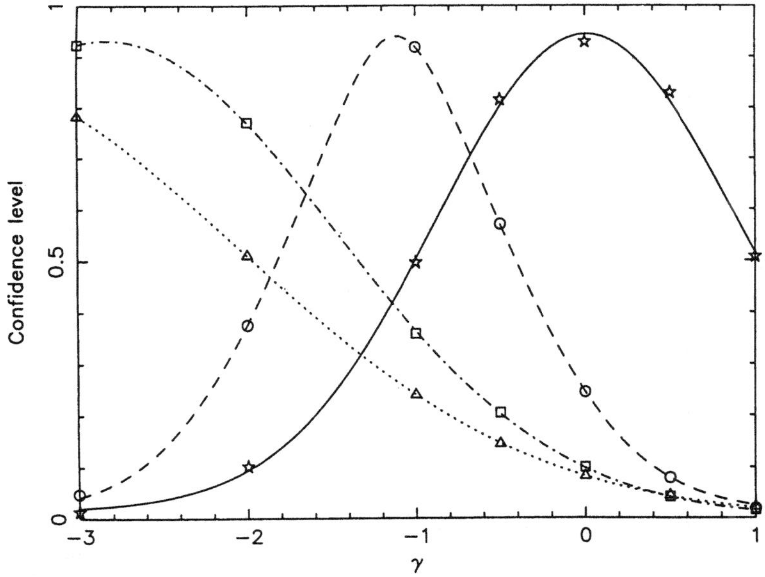

Figure 2. Confidence levels as a function of γ for various subsamples of the C IV data. The dashed line is for $W_0 \geq 0.3$Å , the dot-dashed line $W_0 \geq 0.6$Å , the dotted line $W_0 \geq 0.9$Å and the solid line 0.9Å $\geq W_0 \geq 0.3$Å .

A similar result is found for the magnesium systems. Here we have analysed the data sets of SSB1 and SSB2 which appear to have similar normalisations and which comprise a total of 38 absorption lines in 99 quasars. Once again the hew lines ($W_0 > 1.2$Å) show evidence for evolution (with a best fit $\gamma > 3$), although no evolution cannot be ruled out at the 10% level. Once the hew lines have been removed the remaining lew systems (0.6Å $< W_0 < 1.2$Å) are again well described by models with no evolution.

2.2 GRAVITATIONAL LENSING

The above result support a picture in which quasars are being magnified by massive halos associated with the hew absorption lines. The variation of lensing cross-section with redshift could mimic the apparent evolution. We would not expect the majority of lew systems to participate in the lensing since there is on average more than one such system along the line of sight to each quasar.

One useful measure of the amount of lensing we might expect to see is given by the lens-weighted path length, $s(z_1, z_2) = \int_{z_1}^{z_2} p(z)\,dz / \int_0^{z_q} p(z)\,dz$, where z_q is the quasar redshift. The mean values are 0.10 and 0.31 for the carbon and magnesium systems, respectively. Assuming a maximal contribution of 1 absorption line per quasar this gives 31 Mg II and 6 C IV systems associated with lensing halos. Thus the carbon data are unlikely to be seriously contaminated though the magnesium data could be.

How can we test the hypothesis that the sample of absorption line systems is biased by gravitational lensing? The test described in §2.1 is of no use because $p(z)$ is now a function of the magnification which is unknown. Instead we consider how the *detected* lines are distributed within $[z_{min}, z_{max}]$. Our null hypothesis is now that the set of numbers $\{s(z_{min}, z_a)/s(z_{min}, z_{max})\}$, where z_a are the absorption redshifts, is uniformly distributed over $[0,1]$. This test complements the previous one. It makes no use of null detections and therefore has less power to discriminate evolution in number density. On the other hand it is independent of system parameters and can be applied to a more general range of models.

The C IV results can be summarised as follows. The lew lines are well-fit by a large range of models in which the comoving number density varies as a power law (including no-evolution); they are poorly fit by lensing models. Conversely, the hew lines are adequately fit by lensing models but not by number density evolution. However, given that we are only looking for absorption over a small fraction of the available path length, nearly all the quasars would have to be lensed into the sample in order for this explanation to be viable.

For the magnesium systems the results are less clear. There is little difference between lensing and no-lensing models. The distribution of lew lines is adequately fit by all models, whereas the hew systems would seem to prefer models with evolution. However none of the models is rejected at the 5% level.

3 Discussion

This paper started with the premise that metal absorption line systems reside in massive halos, and, in particular, that the hew systems are associated with halos large enough to cause lensing. We have also assumed that the probability of finding an absorption line in such a halo is close to unity, which seems reasonable since the hew lines are often seen to break up into several lew lines when observed at high resolution. Is there any other evidence to support this view?

Of ten multiply imaged quasars in the literature which are claimed to be lensed, galaxies are seen in four cases. In two of these Mg II absorption associated with the galaxy is redward of the Lyman α emission line, but in neither case is it observed. In all four examples, C IV associated with the galaxies would fall in the Lyman α forest, and thus be difficult to detect. So, contrary to our assumption, not every lens is associated with Mg II absorption, and the covering factor of Mg II must be less than one.

Bergeron (1988) has found galaxies at the appropriate redshift in at least 10 out of the 13 cases of quasars with Mg II absorption which she observed. Using an argument similar to Webster *et al.* (1988), one can estimate the number of expected associations between a galaxy and a quasar as

$$N_{ass} \sim 3.6 \left(\frac{R}{12''}\right)^2 \left(\frac{N_g}{10^{3.9}}\right) \left(\frac{N_q}{13}\right)$$

where R is the radius searched for galaxies, N_g is the number of galaxies per square degree to the search limit, and N_q is the number of quasars in the sample. The estimate of N_g is taken from Tyson's (1988) R differential number counts, suitably corrected for the change of colour, where a limiting magnitude $m_r \sim 22.3$ for the galaxies is assumed. There are two possible explanations for this result. The first and perhaps most plausible is that Mg II lines are caused by gas clouds around the galaxy. In this case, the quasars with absorption lines at $z \lesssim 0.55$ as selected by Bergeron are a biased subsample of a larger sample of quasars. Since only a small fraction of quasars show Mg II absorption then a correction factor, $f_{bias} \sim 2.5$, could be applied to the counts, giving $N_{ass} \sim 9$ as observed. However not all the path length is searched for absorption lines: in our data a total of 38 absorption lines with $W_0 \gtrsim 0.6$ Å are observed in 99 quasars but only a third of the mean path length has been searched. Therefore each quasar will actually have ~ 1 absorption line and $f_{bias} \sim 1$. So, if the estimate for the galaxy number counts is correct, then the majority of bright quasars must be biased into the sample by gravitational lensing. We are obtaining more data with controls in order to clarify this issue.

4 Conclusions

We have looked for evolution and bias due to gravitational lensing in the redshift distribution of Mg II and C IV absorption lines in quasar spectra. In neither case is there evidence for evolution in the comoving number density of the weak absorption lines, nor is gravitational lensing to be expected since there are many such lines per quasar.

Strong evolution is apparent in the hew C IV systems (no evolution can be excluded at the 5% level). Alternatively a gravitational-lensing model fits the data equally well, however nearly every quasar would need to be lensed into the sample for this explanation to be viable.

For Mg II, it is not necessary to invoke evolution to account for the distribution of hew lines. Indeed, over the redshift range corresponding to the observations gravitational lensing would have little effect on the expected distribution. Neither type of model fits the data particularly well but the disagreement is not significant. Independent evidence suggests that lensing may be occurring — we are gathering more data to address this question.

We have investigated at some length the evidence for a lensing bias in the data as a way of reproducing the apparent evolution in the hew absorption line systems. An equally plausible explanation is that evolution in the clustering properties of lew systems, which appear as hew systems when close together in redshift, may account for this effect. This is discussed at greater length in an extended version of this paper which has been accepted for publication in the Astrophysical Journal (*viz.* Thomas *et al.* 1989).

Acknowledgements

We would like to thank the NSERC for support.

References

Bergeron, J. 1988. In *Proc. IAU Symp. 130*, eds. J. Audouze, M.-C. Pelletan and A. Szalay, Kluwer, Dordrecht, p.343.

Sargent, W. L. W., Boksenberg, A. and Steidel, C. C., 1988. *Astrophys. J. Suppl.*, **68**, 539. (SBS)

Sargent, W. L. W., Steidel, C. C. and Boksenberg, A., 1988a. *Astrophys. J.*, **334**, 22. (SSB1)

Sargent, W. L. W., Steidel, C. C. and Boksenberg, A., 1988b. *Astrophys. J. Suppl.*, , submitted. (SSB2)

Steidel, C. C., Sargent, W. L. W. and Boksenberg, A., 1988. *Astrophys. J. Lett.*, **333**, L5. (SSB)

Tyson, J. A., 1988. *Astron. J.*, **96**, 1.

Thomas, P. A. and Webster, R. L., 1989. *Astrophys. J.*, , in press.

Webster, R. L., Hewett, P. C., Harding, M. E. and Wegner, G. A., 1988. *Nature*, **336**, 358.

Preliminary Analysis of High-Resolution Spectra for UM 673 A & B *

A. Smette[1], J. Surdej[2,4], P. A. Shaver[1],
C. B. Foltz[3], F. H. Chaffee, Jr[3] and P. Magain[2,4]

[1] European Southern Observatory

[2] Institut d'Astrophysique, Université de Liège

[3] Multiple Mirror Telescope Observatory,
Smithsonian Institution and the University of Arizona

[4] also, Chercheur Qualifié au Fonds National de la Recherche Scientifique (Belgium)

Introduction. The most striking feature which appears shortward of the Lyman-α emission line in the spectra of distant quasars is a forest of narrow absorption lines due to hydrogen clouds located along our line-of-sight (Sargent et al. 1980, hereafter SYBT). Various models have been invoked to describe their physical state (SYBT, Ikeuchi & Ostriker 1986, Rees 1986, Tytler 1987, Lake 1988), which depend strongly on their adopted size. The fact that the light beams of a gravitationally lensed quasar follow close but different paths in the Universe provides us with a nearly unique way to determine this important parameter.

Comparing high-resolution (1 Å) spectra of the two images of 2345+007 (separated by 7.3"), in a spectral range of 900 Å between 3450 Å and 4350 Å , which includes 400 Å in the Lyman-α forest, Foltz et al. (1984) found 2 or 3 Lyman-α lines which are not common or are of substantially different strengths. This fact enabled them to determine that the radii of the Lyman-α clouds is roughly 5-25 kpc, depending on the redshift of the unknown lens.

Here we present preliminary results of the analysis of high-resolution spectra obtained for UM 673 A & B (cf. Surdej et al. 1987) at the MMT – more details will be given in Smette et al. (in preparation). In comparison with Foltz et al., our data are characterized by a better signal-to-noise ratio and they cover a much larger spectral range (1700 Å between 3200 and 4900 Å , with 1400 Å in the Lyman-α forest), at about the same resolution (1.5 Å). An advantage of UM 673 is that the lens has been identified (z = 0.49), but, on the other hand, the separation between the A and B images is more than three times smaller (2.2") than that for 2345+007.

The contamination of the B image by the A image is certainly less than 10 %, and probably more like 5 % in number of counts per second in all pixels, thanks to a good seeing.

Both spectra are remarkably similar, so that, except for a scaling factor, it has been possible to define the same continuum for both of them. Line lists for both spectra have

* Based on observations collected at the Multiple Mirror Telescope Observatory

been determined by visual inspection. Characteristics of the lines have been obtained by fitting them with gaussian profiles.

In order to pursue a detailed study of the Lyman-α forest, care had to be taken to eliminate all the lines produced in three metallic systems, found at redshifts z = 2.3564, z = 1.9404 and z = 1.9444 (Surdej et al. 1988, Sargent et al. 1988).

Results. Defining an anti-coincidence as a line detected in one spectrum but not in the other and a coincidence as a line found in both spectra, we can compute the ratio f between the number of anti-coincidences and the sum of the number of anti-coincidences and the number of coincidences. We find this ratio to be very small ($f \approx .02$), and this provides us with a conservative lower limit for the radius of a spherical Lyman-α cloud of about 20 kpc ($H_0 = 50 kms^{-1}Mpc^{-1}, q_0 = 0$), using the results of a Monte-Carlo simulation.

We find a strong correlation between the equivalent widths of the lines detected in both spectra. Such a result indicates that the lines-of-sight to both images of UM 673 actually cross the same clouds. The model of Tytler (1987) in which the clouds are small ($< 1pc$) is therefore rejected, since, in this model, there would be no relationship between the equivalent widths of lines in the two spectra.

Once the characteristic size of the Lyman-α clouds will have been accurately determined, it should be possible to set constraints on the redshift of an invisible lens from the value of f obtained using the spectra of QSO lensed images (if a sufficient number of Lyman-α absorption lines are available). For the particular case of 2345+007, we have, using the lower limit of 20 kpc for the cloud radii, a good argument to reject low values (z < 1) for the redshift of the lens, which may explain why the deflector has not yet been seen.

References

Foltz, C. B., Weymann, R. J., Röser H.-J., Chaffee F., H., Jr.: 1984, *Astrophys. J. (Letters)* **281**, L1

Ikeuchi, S., Ostriker, J.P.: 1986, *Astrophys. J.* **301**, 522

Lake, G.: 1988, *Astrophys. J.* **327**, 99

Rees, M.J.: 1986, *Monthly Notices Roy. Astron. Soc.* **218**, 25

Sargent, W.L.W., Young, P., Boksenberg, M., Tytler, D.: 1980, *Astrophys. J.* **42**, 41

Sargent, W.L.W., Boksenberg, A., Steidel, C.C.: 1988, *Astrophys. J. Suppl.* **68**, 539

Smette, A., Surdej, J., Shaver, P. A., Foltz, C. B., Chaffee, F. H., Jr, Magain, P.: 1989, in preparation

Surdej, J., Magain P., Swings, J.P., Borgeest, U., Courvoisier, T.J.-L., Kayser, R., Kellermann, K.I., Kühr, H., Refsdal, S.: 1987, *Nature* **329**, 695

Surdej, J., Magain P., Swings, J.P., Borgeest, U., Courvoisier, T.J.-L., Kayser, R., Kellermann, K.I., Kühr, H., Refsdal, S.: 1988, *Astron. Astrophys.* **198**, 49

Tytler, D.: 1987, *Astrophys. J.* **321**, 49

BIDIMENSIONAL SPECTROGRAPHY OF THE "CLOVER LEAF" H1413+117 AT SUB-ARCSEC. SPATIAL RESOLUTION.

M.C. Angonin[(1)], C. Vanderriest[(1)], J. Surdej[(2)].

(1) Observatoire de Paris-Meudon 5 place Janssen, F 92195 Meudon.
(2) Institut d'Astrophysique, Université de Liège Avenue de Cointe, B 4200 Ougrée-Liège.

Introduction:

The B.A.L. type quasar H1413+117 was found to be a new case of gravitational mirage (Magain et al., 1988). It is composed of 4 images with nearly equal brightness and mean separation ≤1", hence its nickname "clover leaf ". This configuration did not allow until now to obtain independant spectra of the 4 components.

Observations:

H1413+117 was observed in March 1989 at the Canada- France-Hawaii telescope with the integral field spectrograph SILFID (Vanderriest et al., 1987). Its principle is as follows: a fibre optics device transforms a hexagonal field in the focal plane of the telescope into a pseudo "slit" in front of the spectrograph. After the dispersion, the exactly inverse geometrical transformation could be done in order to reconstruct crude pictures of the object at any chosen wavelength. Furthermore we can select *a posteriori* the fibres corresponding exactly to a given part of the entrance field and build an integrated spectrum. Fig. 1 summarizes the data handling.

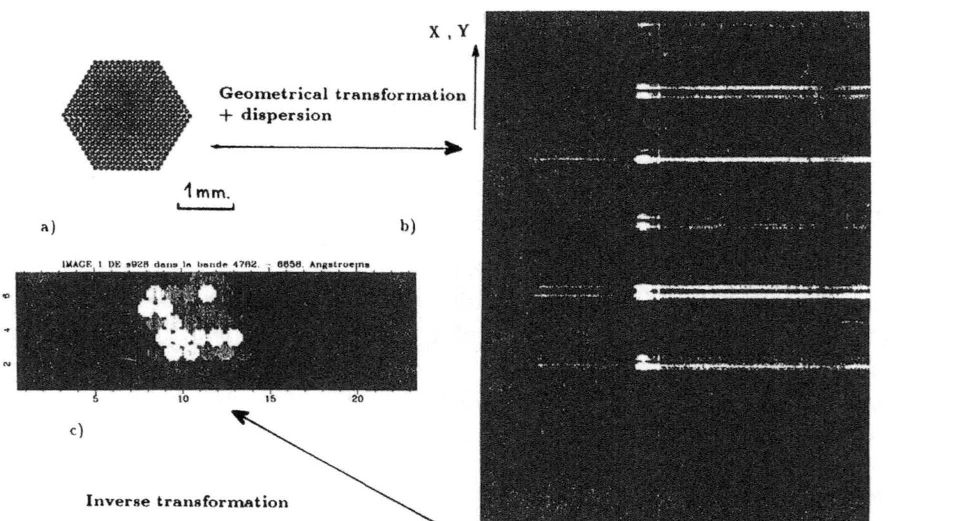

Fig.1: -a) Entrance side of the fibre optics device.
-b) Resulting bidimensional spectra.
-c) Reconstruction of the focal image, showing how the fibres have been lighted.

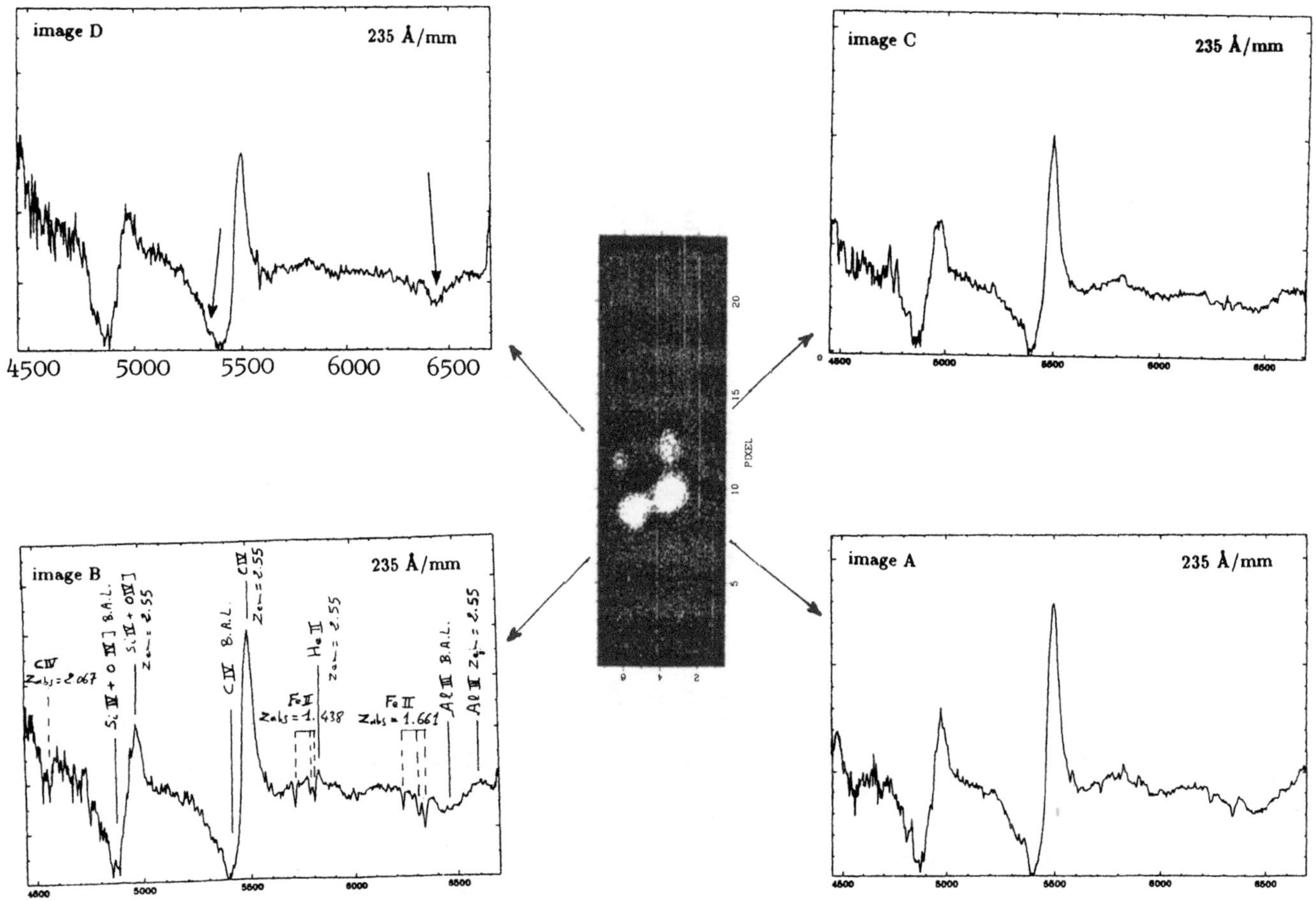

Fig.2: Extracting individual spectra of the 4 images of the "clover leaf". Sum of two 30 mn exposures at 235 Å/mm.

Our spectra were obtained during a period of very good seeing (F.W.H.M. $\simeq$ 0.5"- 0.6"). Each fibre subtended 0.33" on the sky and thus the images were slightly undersampled. Fig. 2 displays the spectra extracted from the sum of two 30 mn exposures at 235 Å/mm. To build them, we took great care of avoiding mutual contamination between the images.

Results:

Just by looking at Fig.2, several remarks can be made:

-1) The overall shapes of the 4 spectra are remarkably similar, confirming the gravitational lens hypothesis.

-2) The narrow absorption line systems at z = 1.438 and z = 1.661 previously detected by Magain et al. are seen most clearly in image B and less conspicuously in image A, but they seem absent in images C and D.

-3) With respect to the other 3 images, significant differences in the *emission lines / continuum* intensity ratio is detected for image D. This could be easily explained by a micro-lensing effect on D. This interpretation is supported by photometric evidences of its variability (Kayser et al., 1989).

It is more difficult to understand the other differences observed in the intensity of the C IV and Al III Broad Absorption Lines for image D (arrows on fig.2). They could suggest either an intrinsic variability of the source ionizing the B.A.L. Region (see e.g. Barlow et al., 1989) combined with the time delay and/or a highly inhomogeneous structure of this B.A.L.R..

Conclusion:

Bidimensional spectrography helped us to analyse the "clover leaf " and gave further evidence of gravitational lensing. A model with 2 lensing galaxies at the observed redshifts 1.438 and 1.661 could fit both the optical data and radio maps (Kayser et al., 1989).

However, other observations in similar seeing conditions seem necessary in order to disentangle the various effects suspected in the spectrum of image D.

Being at the same time a B.A.L. Quasar and a gravitational mirage, H1413+117 could be an important tool for better understanding the B.A.L. phenomenon.

References:

- Barlow T., Junkkarinen V. & Burbidge M., 1989, Astrophys. J., in press.

- Kayser R., Surdej J., Condon J.J., Hazard C., Kellermann K.I., Magain P., Remy M. & Smette A., 1989, Astrophys. J., in press.

- Magain P., Surdej J., Swings J.P., Borgeest U., Kayser R., Kühr H., Refsdal S. & Remy M., 1988, Nature, **334**, 325.

- Vanderriest C. & Lemonnier J.P., 1987, "Instrumentation in Astronomy", proc. IXth Santa Cruz workshop, 304.

The MIT Search Program for Gravitational Lenses

Bernard F. Burke
Massachusetts Institute of Technology
Cambridge, Massachusetts 02139

1. HISTORICAL BACKGROUND

The search program described here started with radio observations made with various instruments of the National Radio Astronomy Observatory (NRAO), combined with observations of the National Optical Astronomy Observatory (NOAO) [1], supplemented by key work at the Mount Palomar Observatory. The stimulus came from the discovery by Walsh, Carswell, and Weymann [2] that the radio source 0957+561 was an example of multiple imaging of a quasar by the lensing effect of the gravitational field of an intervening galaxy (more accurately described as a gravitational mirage, the French designation). This came about as a consequence of combined radio and optical observations. The radio source 0957+561 had been catalogued by the Jodrell Bank group, and in the subsequent precise positional measurements at the NRAO and the work of optical identification at the Steward Observatory, its properties as a gravitational lens were established. Interestingly, the relevant theory had been carried out in some detail, and the work of S. Refsdal [3] can be cited particularly as an outstanding pioneer treatment, carried out fifteen years before the actual discovery. Time delay effects had also been correctly treated by Cooke and Kantowski [4], and once the first discoveries were in hand, the approximate expectations could be calculated, with Turner, Ostriker and Gott [5] giving useful orders-of-magnitude expectations. A simple estimate showed that the most probable configuration for multiple imaging is for the lensing galaxy to be situated at a redshift of approximately one-half, with the object quasar at a redshift of one or greater; 0957+561 fits this prescription (z_Q=1.4, z_G=0.36).

The first three examples of lensing, 0957+561, 1115+080, and 2345+007, were each discovered among samples of several hundred objects, and it was evident that the phenomenon was sufficiently common to merit a systematic search effort. We had used the Very Large Array (VLA) of the NRAO to produce high-quality radio maps of 0957+561 [6], and our MIT group, consisting of C. Lawrence (now at CalTech), C. Bennett (now at Goddard Space Center), A. Garcia Barreto (now at Toluca University), G. Langston (now at NRAO), and J. Hewitt, had just completed the MG survey, carried out with the 300-ft transit telescope of the NRAO, at a frequency of 5 GHz [7]. The primary list contained 6000 sources along the band $-0.^{\circ}5 \leq \delta \leq 19.^{\circ}5$, $|b| > 10^{\circ}$ and the relatively high observing frequency favored the presence of quasars, which are usually easier to interpret when lensing occurs, compared to radio galaxies. We were joined by E. Turner of Princeton in a program that would consist of first mapping the MG sources with the VLA, looking for compact multiple structures, and then examining them optically to verify (or disprove) the existence of lensing. The first VLA maps were made in May 1981, with the 11th and final day of observation being in April 1986. The first example of lensing from this data base was 2016+112 [8], found on the same day that the Berkeley group discovered 1635+267 [9]. The sample sizes from which the lens examples came were of the order of a few hundred, and were consistent with the hypothesis that about one quasar in every three hundred would be multiply imaged.

2. SEARCH METHODOLOGY

The VLA mapping program, carried out at 5 GHz, was prepared by computerized methods, with little human intervention. The MG survey, consisting of a high-reliability (5 sigma) list of 6000 sources, supplemented by a secondary list of 4000 sources with lesser confidence was used as input. The first four observing days were set up using a preliminary list, since the final MG catalog was not ready. Two of these sessions were carried out with the VLA in the A configuration (resolution 0.3 arc-seconds at 5 GHz), while the other two used the B configuration (resolution one arc-second). For the remaining seven observing days the A configuration was used and the final MG catalog was used as the source list. The observations cover virtually all sources with flux greater than 100 mJy at 5 GHz, and a total of 4000 sources were mapped.

Data reduction was carried out by using the standard AIPS software routines of the NRAO, and special procedures were constructed to minimize human intervention. The initial MG positions, determined by the 300-ft telescope with a 2.4 arc-minute beam, frequently placed the source off-center in the first (uncleaned) VLA map, so in the first iteration the map offset of each source is determined. A second, clean map is then made, with self-calibration being used for those sources brighter than 10 mJy. The ratios of peak flux to rms noise in the maps were generally 17 db or better. A total of 1786 sources were used in the first large optical identification effort.

The first step in choosing candidates for optical study was to categorize each source. Two promising categories of maps were recognized: the "A" list consisted of maps showing two or more unresolved components, while the "B" list contained those examples of multiple source structure that showed resolution of one or more components. Within each list, the sources were ordered in priority: non-colinearity of components, with angular separation less than 10 arc-sec gave the highest priority, linear arrangements and doubles with less than 10 arc-sec size fell in the second priority group (many of these would certainly be double-lobed radio galaxies), while all multiples with size greater than 10 arc-sec formed the lowest priority group. Thus, a total of six categories were recognized, with emphasis on the upper two classes in each group. The first selection yielded 99 sources in categories A1 and B1, 311 in categories A2 and B2, and 133 sources in the lowest categories, A3 and B3, a total of 443 candidates. A total of 577 sources did not meet the classification criteria for one reason or another, and 766 sources were unresolved. Thus, a total of 1786 mapped sources were classified; details of the procedures have been given by J. Hewitt [10].

The optical identification work proceeded in two stages. Firstly, deep continuum CCD images were taken, looking for optical structure. The optical structure did not have to match the radio structure, but if a given optical object showed evidence of non-stellar appearance, it was selected for the second stage of optical study, in which spectra were taken. The principal optical telescope used in this work was the 4-meter Kitt Peak telescope of the NOAO, supplemented by observations with the 5-meter Palomar telescope of Cal Tech. Approximately one radio source in ten from the MG-VLA survey was chosen (mostly from lists A1, A2, B1, and B2) for optical study. Of these, roughly one in seven showed evidence of continuum optical structure. About forty of these sources have undergone spectroscopic study, of which four have been established as definite examples of lensing, a fifth is a strong candidate, and four more excellent prospects are awaiting verification.

3. SURVEY RESULTS

2016+112

This was the first clear example of lensing in our survey and was discovered at an early stage in the work [8]. The modelling of the source and lens has been studied in some detail by Narasimha et al [11], but the VLBI studies of Heflin [12] have raised questions that require further study. The radio source consists of three compact objects, and was immediately recognized as a prime candidate even before the optical observations were undertaken. Optically, coinciding with the upper two radio sources, there are two quasar-like objects at a redshift of z=3.27, distinguished by the extraordinary sharpness of their emission lines. A faint galaxy (z approximately 1) coincides with the southernmost radio source, and a second faint galaxy appears within the triangle. Two faint emission patches of Lyman-alpha emission at z=3.27 appear in the NW quadrant, well-separated from the quasar images [13]. A third quasar image appears to be located close by the southernmost galaxy [14].

MG1131+0456 and MG1654+1346

These examples of the "Einstein Ring" phenomenon were recognized from the radio data alone as being very strong candidates for lensing. The perfect alignment of a point source with a spherically symmetrical lens would yield a ring, at the "Einstein radius". If the alignment is not perfect, the primary image would fall just outside the ring and the secondary image would lie just inside, on the opposite side. MG1131+0456, the first instance to be discovered [15], was such a clear morphological example of ring formation, with supplementary primary images of a compact object just outside the ring on one side and with the secondary images just inside the ring on the other side, that even without optical data a strong case for lensing could be made. The modelling work of Kochanek et al [16] confirmed the case beyond question.

The second example, MG1654+1346 [17], is a different class of ring. In this case, the ring is accompanied by a clear quasar image off to one side, and the modelling work makes a persuasive case for the ring being the image of an extended lobe excited by the parent quasar. There should be other examples of this phenomenon that have not yet been recognized, since the radio lobes that accompany active radio galaxies and quasars cover a larger fraction of sky than the nuclei, so there is a greater probability that they will be lined up with a foreground lensing galaxy.

MG0414+0534

This triple radio source was identified as a prime prospect from its radio morphology initially. The optical appearance is also triple, and it is safe to categorize this as a secure example of lensing [18]. The strongest component is clearly double, making this, in fact, a quadruple image, in virtually the same morphological class as the famous "triple" (actually quadruple) lens system 1115+080. All three images in MG0414+0534 have continuous spectra that suggest that the initial object is an elliptical galaxy at a redshift in the range 0.8-1.0. The foreground lensing galaxy has not yet been identified, but the radio morphology resembles 1115+080 so strongly that it almost certainly must be an edge-on spiral galaxy of the type originally postulated by Young [19] for this configuration.

0023+171

This candidate, one of the earlier examples found in the survey [20], still requires further study. Optically, there are two quasar-like objects with similar emission-line spectra, but the complex radio structure and conspicuous lack of a foreground imaging galaxy raise questions about the interpretation. It may, indeed, be an example of lensing, but the possibility that one is seeing a close radio galaxy-quasar pair cannot be ruled out yet. Optical observers are encouraged to continue working on this object.

Other candidates

The sources 1042+178, 2018+102, and 2113+100 were cited by Hewitt as promising cases [10]. Events have conspired against obtaining conclusive optical data, but there is both optical and radio structure in all three cases. The radio structure are all suggestive of a high probability of lensing.

Comments on the results

One is immediately struck by the unusual character of each example of lensing found by the survey. 2016+112 is in a configuration that is hard to explain by models, the two "Einstein Rings" are examples of a new class of source whose possibility had been recognized but which had not been a major target of any study, while MG0414, establishes another class (of two objects) in which a fortuitous alignment is required; it appears likely that an edge-on galaxy, as in the case of 1115+080, is responsible for the lensing.

The simplest example of lensing that might be expected, a pair of quasar-like images, seems to be an infrequent occurrence among the radio-selected objects. Part of the answer lies in one of our selection criteria: the selected objects generally had radio structures of at least 2 arc-sec in size, since structures more compact than that would be hard to resolve optically. The statistical study of Turner, Ostriker and Gott [5] predicted that there should be a population of multiply-imaged quasars with a most probable separation of approximately 0.5 arc-sec, but these would not have been selected in the work carried out so far.

A second outstanding question, yet to be resolved properly, is the odd-image problem. Even though one still is unable to find complete examples of imaging with the required odd number of images (2016+112 may be the only example so far), there is a general belief that the last image is almost always extremely faint, because of the strong central mass concentration expected in most of the foreground lens galaxies.

Thirdly, one is struck by the fact that only one of the radio examples (0023+178) is lacking an obvious candidate for the lensing mass. In contrast, a substantial number of optical candidates fail to show a foreground galaxy. The selection criteria are different, and this may be an explanation; it is also possible that the candidates with no visible lensing objects in the foreground may be chance pairs of quasars, and not lenses at all. The alternative possibility, that one is seeing lensing by concentrations of dark matter, is exciting but less convincing.

4. COSMOLOGICAL IMPLICATIONS

The preliminary VLA sample examined so far is reasonably well-defined, although the study of all 4000 sources, when complete, will give a more definitive statistical picture. Nevertheless, several conclusions can be drawn. The first is statistical: the samples examined so far contain 1786 sources, including quasars and radio galaxies. Both classes of sources are candidates for lensing, although the quasar examples are easier to recognize. We have discovered at least four, and possibly eight, examples of lensing, giving a probability of between one in 225 and one in 450 of finding a "gravitational mirage" in the VLA sample. The search is certainly incomplete, so these probabilities are lower limits. Since quasars are easier to recognize, and about half of the radio sources in the sample are quasars, the probability of finding a lensed quasar is doubled. Thus, we conclude that between one in 100 and one in 200 quasars in the VLA sample are likely to be lensed.

The present results, tentative as they are, can be put to direct use for cosmology. Hewitt made extremely conservative assumptions for the purpose of placing an upper limit to Ω_L, the fraction of mass density in the universe attributed to lenses. She characterized the instrumental survey limits as a fraction of image separation Δ_α and flux ratio s_2/s_1, and mapped these into phenomenological representations, using point-source lenses and isothermal sphere lenses as the examples. In these two classes of model space, with source position offset β and Einstein ring radius α_o as the parameters, the domains covered by the observations were clearly defined. Analysis then proceeded by using standard random-variable techniques, in order to predict the results under various assumptions of cosmological model and lens characteristics. The observational results could then be compared directly to establish the limits on Ω_L .

Under the extremely conservative assumption that all multiple sources in the sample range were lensed, Hewitt concluded that Ω_L cannot exceed 0.9 for point masses in the range 10^{11} to 10^{12} solar masses. It should be emphasized that these upper limits are probably an order of magnitude or even two orders of magnitude on the conservative side. It appears most unlikely that Ω_L exceeds 0.1 for compact objects of a galactic mass.

5. ACKNOWLEDGEMENTS

The survey work has depended on the collaboration of many individuals. The recent work has been particularly dependent on the contributions of J.N. Hewitt, S.R. Conner, M.B. Heflin, M.R. Griffith, J. Lehar, E.L. Turner, C.L. Lawrence, D.P. Schneider, M. Schmidt, and J.E. Gunn.

REFERENCES

1. This work was supported by grants from the National Science Foundation. The National Radio Astronomy Observatory is operated by Associated Universities, Inc., under contract with the National Science Foundation. The National Optical Astronomy Observatories are operated by the Association of Universities for Research in Astronomy, Inc., under contract with the National Science Foundation.

2. Walsh, D., Carswell, R.F., and Weymann, R.J. 1979, Nature, 279 381.

3. Refsdal, S. 1964, M.N.R.A.S. 128 295.

4. Cooke, J.H. and Kantowski, R. 1975, Ap. J. 195 L11.

5. Turner, E.L., Ostriker, J.P., and Gott, R.J. 1984, Ap. J. 284 1.

6. Greenfield, P.E., Roberts, D.H., and Burke, B.F. 1985, Ap. J. 293 370.

7. Bennett, C.L., Lawrence, C.R., Burke, B.F., Hewitt, J.N., and Mahoney, J. 1986, Ap. J. Supp 61 1.

8. Lawrence, C.R., Schneider, D.P., Schmidt, M., Bennett, C.L., Hewitt, J.N., Burke, B.F., Turner, E.L., and Gunn, J.E. 1984, Science, 223 46.

9. Djorgovski, S., and Sprinrad, H. 1985, Ap. J. 282 L1.

10. Hewitt, J.N. 1986, Ph.D. Thesis (MIT).

11. Narasimha, D. Subramanian, K., and Chitre, S.M. 1987, Ap.J. 315, 434; Narasimha, D. and Chitre, S.M. 1989, Ap. J. 97 327.

12. Heflin, M.B. This symposium.

13. Schneider, D.P., Gunn, J.E., Turner, E.L. Lawrence, C.R., Hewitt, J.N., Schmidt, M., and Burke, B.F. 1986, A.J., 91 991.

14. Langston, G., Carilli, C., Conner, S.R., Heflin, M., Lehar, J., Lawrence, C., Dhawan, V., and Burke, B.F. 1989, Gravitational Lenses (Moran, J.M., Hewitt, J.N., and Lo, K.Y., Eds.). Springer-Verlag (Berlin) p. 100.

15. Hewitt, J.N., Turner, E.L., Schneider, D.P., Burke, B.F., Langston, G.I., and Lawrence, C.R. 1989, Nature 333 537.

16. Kochanek, C.S., Blandford, R.D., Lawrence, C.R., and Narayan, R. 1988, M.N.R.A.S.

17. Langston, G.I., Schneider, D.P., Conner, S.R., Carilli, C.L., Lehar, J., Burke, B.F., Turner, E.L., Gunn, J.E., Hewitt, J.N., and Schmidt, M. 1989, A.J. 97 1283.

18. Hewitt, J.N., Burke, B.F., Turner, E.L., Schneider, D.P., Lawrence, C.R., Langston, G.I., and Brody, J.P. 1988, Results of the VLA Gravitational Lens Survey (Moran, J.M., Hewitt, J.N., Lo, K.Y., Eds.). Springer-Verlag (Berlin) p. 147-158.

19. Young, P., Gunn, J.E., Kristian, J., Oke, J.B., and Westphal, J.A. 1981, Ap. J. 244 736.

20. Hewitt, J.N., Turner, E.L, Lawrence, C.R., Schneider, D.P., Gunn, J.E., Bennett, C.L., Burke, B.F., Mahoney, J.H., Langston, G.I., Schmidt, M., Oke, J.B., and Hoessel, J.G. 1987, Ap. J. 321 706.

First Epoch VLBI Observations of the Gravitational Lens System 2016+112

M. B. Heflin[1], M. V. Gorenstein[2], C. R. Lawrence[3], B. F. Burke[4], and I. I. Shapiro[5]

[1]Massachusetts Institute of Technology, 26-349, Cambridge, MA 02139
[2]Waters Chromatography Division, Millipore Corporation, Milford, MA
[3]California Institute of Technology, Pasadena, CA
[4]Massachusetts Institute of Technology, Cambridge, MA
[5]Harvard-Smithsonian Center for Astrophysics, Cambridge, MA

Abstract. VLBI data at 18 cm indicate that 2016+112 A and B have milliarcsecond structure. The A and B images can be adequately modeled with a linear magnification matrix, although a few features in each hybrid map cannot be explained by such a model. Three lens models by Narasimha, Subramanian, and Chitre (1987) and one by Narasimha and Chitre (1989) predict magnification matrices. The dark halo model by Narasimha and Chitre (1989) works best. Less than half of the VLA fluxes are seen in the respective VLBI hybrid maps, indicating some extended structures. The C component was too faint to be mapped. The VLBI flux ratio B/C is about 3.8.

Observations. On 1984 June 1 an 18 cm VLBI observation of the 2016+112 gravitational lens system was conducted. Mark-III mode A recording took place at Bonn (B), Green Bank (G), Fort Davis (F), Owens Valley (O), Haystack (K), and the VLA (Y). Correlation was performed at Haystack.

The VLBI array was pointed toward B and then A after the experiment via the process of fringe fitting at different delay and delay rates. Measured amplitudes and phases were obtained whenever A and B were resolved and detected with signal to noise ratios greater than eight. The C component was too weak to be detected during the 13 minute integration periods so it could not be mapped. Weak correlated flux density was detected in the C component region by coherently summing 403 minutes of data on the GY baseline using B as a phase reference. The C component region had a peak signal to noise ratio of 61 while the B component region had a peak signal to noise ratio of 233, implying a VLBI flux ratio B/C of roughly 3.8 at 18 cm.

Calibration and Editing. Amplitude calibration was performed using standard radiometry for each station. Five u-v crossing points were used to fine tune the antenna gains. Ten phases

Source and Method	r (mas)	θ (° E of N)	Total Flux (mJy)	Agreement Factor $(\chi^2/N)^{1/2}$
A				
Independent	9.3	– 59	23.3	3.995
Hybrid	8.5	– 66	22.2	3.568
Magnified	9.5	– 57	23.4	3.980
B				
Independent	7.8	– 42	23.0	4.884
Hybrid	6.5	– 49	27.3	2.908
Magnified	8.1	– 41	22.0	5.272

Table 1. Source parameters for different models.

were deleted because of interference or other problems at the VLA. The shortest baselines, KG and FO, undersampled the region of the u-v plane sensitive to extended structures. Seven points indicated extended structure but were not sufficient to constrain it. Hybrid maps made while including the seven points showed regularly spaced stripes at about the ten percent level. The seven points were excluded for final models and hybrid maps.

Independent Modeling. The first guess models were single elliptical gaussians. Data for both A and B contain features which cannot be explained by such one component models indicating that A and B have structure at the milliarcsecond level. The next models involved two elliptical gaussians. Preliminary two component models were published in the proceedings of IAU Symposium No. 129 (1988). Two component models for both A and B were optimized using a non-linear least squares fit to amplitude, closure phase, and residual difference phase data (Gorenstein *et al.* 1984). The best fit models for A and B were each composed of two unresolved circular gaussians. The two component models were used as input for hybrid mapping.

Hybrid Maps. Figure 1 contains the hybrid maps. The A map shows a central core with a companion to the northwest. The core and companion are separated by 8.5 milliarcseconds. The position angle is – 66°. Position angles are defined counterclockwise from north. The B map also shows a central core with a companion to the northwest. The core and companion are separated by 6.5 milliarcseconds. The position angle is – 49°. Notice that in each case the companion is fainter than the core but the ratios differ slightly. Also, A has an extension to the southeast which does not appear in B. Separations, orientations, total fluxes, and Caltech agreement factors for independent modeling, hybrid mapping, and magnified modeling are given in Table 1. Notice that in every case the separation is larger and the orientation is farther

Source	r (mas)	θ (° E of N)	Total Flux (mJy)
A	9.1 ± 0.5	-61 ± 5	23.0 ± 0.7
B	7.5 ± 0.9	-44 ± 4	24.1 ± 2.8

Table 2. Means and standard deviations of source parameters.

from north for the A model. Means and standard deviations are given in Table 2.

Magnified Modeling. Optical and radio evidence suggest that A and B are two gravitationally lensed images of a single source. If the source is not evolving on the time scale of the unknown time delay between the two images and the bending law is sufficiently smooth, A and B should be related by a linear magnification matrix (Gorenstein *et al.* 1984). After independent modeling and hybrid mapping confirmed the basic two component compact structure of A and B, modeling was performed with a magnification constraint. The magnification matrix has four independent parameters. Since the individual compact components were found to be unresolved, only two and a half constraints are derived from the data. The separation and orientation of the unresolved components yield two constraints. The absolute value of the magnification matrix determinant is constrained by the A/B flux ratio. Hence there are four unknowns and two and a half equations. Table 2 summarizes the magnification matrix constraints.

Table 3 compares the experimental and theoretical results. Theories I-III represent solutions I-III from the work of Narasimha, Subramanian, and Chitre (1987). Theory IV represents the model from Narasimha and Chitre (1989). The determinant of each matrix is shown along with the values permitted by the A/B flux ratio. The mean values of r and θ for A were used to calculate r and θ for B via the theoretical magnification matrices. None of the theoretical predictions agree exactly but theory IV is best. All parameters are within 7 σ of the experimental values.

Conclusions. VLBI data indicate 2016+112 A and B have milliarcsecond structure as shown in the hybrid maps. The A and B structures can be adequately modeled with a linear magnification matrix although the extension on A and the flux ratio of the unresolved components in each hybrid map are not explained by such a model. Three lens models by Narasimha, Subramanian, and Chitre (1987) and one by Narasimha and Chitre (1989) predict magnification matrices. Theoretical and experimental results are compared in Table 3. The dark halo model, theory IV, agrees better than the others. All parameters are within 7 σ of the

Theory or Experiment	Matrix Determinant	r for B. Assume 9.1 mas for A.	θ for B. Assume – 61° E of N for A.
Theory I	– 0.72	4.6	– 113
Theory II	– 0.56	3.9	– 91
Theory III	– 0.65	4.4	– 89
Theory IV	– 0.97	5.4	– 71
Experiment	± (0.95 ± 0.11)	7.5 ± 0.9	– 44 ± 4

Table 3. Comparison of theoretical and experimental results.

experimental values. Less than half of the VLA flux is seen in the VLBI hybrid maps indicating extended structure. Seven short baseline points also indicated extended structure. The C component was too faint to be mapped. The VLBI flux ratio B/C is approximately 3.8 at 18 cm. A more complete paper including the actual data has been submitted for publication in the *Astrophysical Journal.*

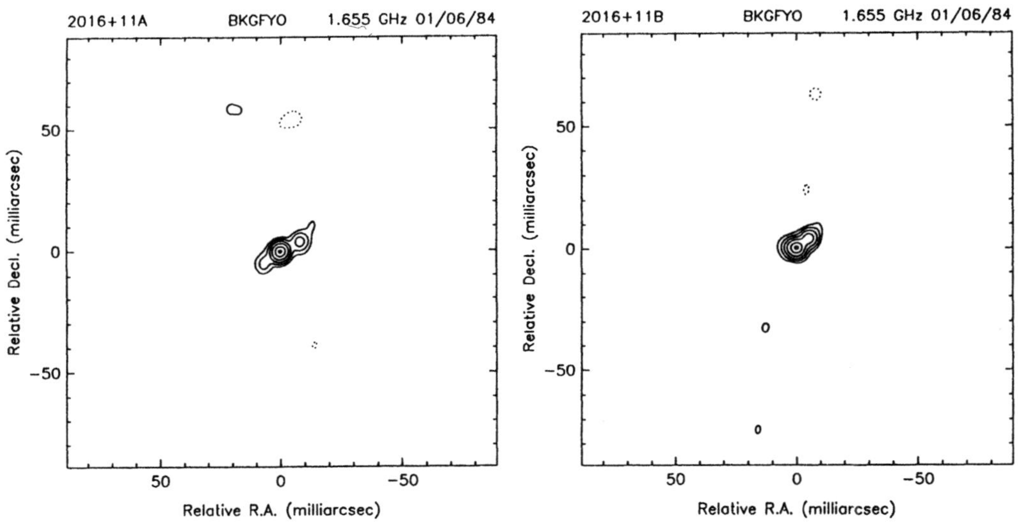

Figure 1. Hybrid maps of 2016+112 A and B. Contours are at -4, 4, 8, 16, 32, 64, and 95 percent of the peak. The beam FWHM is 4.9 milliarcseconds. North is up. East is to the left.

References.

Heflin, M. B., Gorenstein, M. V., Falco, E. E., Shapiro, I. I., Burke, B. F., Hewitt, J. N., Rogers, A. E. E., and Lawrence, C. R. 1988, *The Impact of VLBI on Astrophysics and Geophysics*, eds. M. J. Reid and J. M. Moran (Dordrecht: Reidel), p. 209.

Gorenstein, M.V., *et al.* 1984, *Ap. J.* **287**, 538.

Narasimha, D., Subramanian, K., and Chitre, S.M. 1987, *Ap. J.* **315**, 434.

Narasimha, D., and Chitre, S. M. 1989, *A. J.* **97**, 327.

X-RAY OBSERVATIONS OF GRAVITATIONAL LENSES

E. Kellogg, E. Falco, W. Forman, C. Jones and P. Slane
Harvard-Smithsonian Center for Astrophysics
Cambridge, Massachusetts

ABSTRACT

A preliminary status report is given on studies using the *Einstein* X-ray observations of distant clusters of galaxies that are also candidates for gravitational lenses. The studies will determine the location and surface brightness distribution of the X-ray emission from clusters associated with selected gravitational lenses. The X-ray emission comes from hot gas that traces out the total gravitational potential in the cluster, so its distribution is approximately the same as the mass distribution causing gravitational lensing. Core radii and X-ray virial masses can be computed for several of the brighter *Einstein* sources, and preliminary results are presented on A2218. Preliminary status is also reported on a study of the optical data from 0024+16. A provisional value of 1800 to 2200 km s^{-1} for the equivalent velocity dispersion is obtained. The ultimate objective is to extract the mass of the gravitational lens, and perhaps more detailed information on the distribution of matter as warranted. A survey of the *Einstein* archive shows that the clusters A520, A1704, 3C295, A2397, A1722, SC5029-247, A3186 and A370 have enough X-ray counts observed to warrant more detailed optical observations of arcs for comparison. Mass estimates for these clusters can therefore be obtained from three independent sources: the length scale(core radius) that characterizes the density dropoff of the X-ray emitting hot gas away from its center, the velocity dispersion of the galaxies moving in the cluster potential, and gravitational bending of light by the total cluster mass. This study will allow us to compare these three techniques and ultimately improve our knowledge of cluster masses.

INTRODUCTION

Gravitational lensing behavior in distant clusters of galaxies was discovered by Lynds and Petrosian (1986) - Soucail et al (1987). Faint arcs and rings in deep blue CCD exposures were found to be "mirage" images caused by bending of light from blue galaxies at very high redshift by the gravitational effect of clusters at smaller redshifts. The redshift of an arc in A370 was first measured by Soucail et al, 1988, confirming its nature as a gravitational lens image, and the first detection of the faint blue mini-arcs in A370 was reported by Fort et al, 1988. The information contained in these optical images of the arcs and rings in clusters of galaxies permits an estimate of the mass contained in the gravitational lens producing the arcs and rings.

This paper is a preliminary report on studies using the *Einstein* X-ray data on some of these distant clusters. We determine the location and surface brightness distribution of the X-ray emission from clusters associated with selected gravitational lenses. Since the X-ray emission comes from hot gas that traces out the total gravitational potential in the cluster, its distribution should be approximately the same as the mass distribution that gives rise to gravitational lensing effects. While most of the very distant clusters in the *Einstein* data are too weak to permit studying the brightness distribution with high resolution, we can obtain core radii and compute X-ray virial masses for several of the brighter sources.

The optical data on 0024+16 are also being analyzed to extract the mass of the gravitational lens. In many cases, optical observations have been made of individual galaxy velocities in a given cluster. From these, a velocity dispersion can be calculated, and a virial mass derived.

Mass estimates for these clusters can be obtained from three independent sources: the length scale(core radius) that characterizes the density dropoff of the X-ray emitting hot gas away from its center, the velocities of the galaxies moving in the cluster potential, and the bending of light by the total cluster mass. This study will allow us to compare these mass estimates for each cluster. If they are all consistent with a single value of the mass for a given cluster, we will have a strong statement on the total mass, including dark matter, contained in the cluster. If the estimates do not agree, further study of the differences will be indicated.

These studies will be of immediate interest in the planning of ROSAT (and ultimately AXAF) pointed observations of gravitational lenses. Detailed studies of these cluster X-ray sources will also be useful to optical astronomers investigating gravitational lensing. Centroid positions for the extended X-ray emission, which represent an improvement over those available in the *Einstein* production data processing, are derived and peculiarities in the X-ray brightness distribution are noted. This information will be a useful guide to planning and interpreting the optical observations.

Since the discovery of cluster X-ray sources in the Uhuru data (Gursky et al. 1971, Kellogg et al. 1972, Forman et al. 1972, Kellogg et al. 1973), when the brightest few clusters were observed, a large number of cluster X-ray sources have been detected and some of these have been studied in detail (Jones et al. 1979, Forman and Jones 1982, Jones and Forman 1984). Spectral analysis (Kellogg, Baldwin and Koch 1975) shows that the emission comes from hot gas with a significant concentration of iron series elements (Mitchell et al. 1976, Serlemitsos et al. 1977), although the abundances are $\sim$ 2-4 times below solar (Mitchell and Mushotsky 1980, Mushotsky 1980).

The gas can be used as an improved tracer of dark matter, because the gas is in hydrostatic equilibrium with the total mass, both dark and luminous. Thus, the gas distributes itself in the gravitational potential well formed by all the matter in the lens, including dark matter (David et al. 1989). Previously, galaxies have been used as test particles to trace the cluster's potential well. Hot gas atoms are much better tracers than galaxies; since their collision times are much shorter, the associated velocity distribution is much closer to thermal equilibrium.

The *Einstein* observatory obtained the first two-dimensional images of X-ray emitting clusters of galaxies at $z > 0.2$ (White, Silk and Henry 1981, Henry and Lavery 1984). A systematic study of the morphology of these distant cluster X-ray sources has recently begun (Kellogg et al. 1989). The investigation involves a search for evolution effects, by comparing the class of clusters with $z \leq 0.2$ and the class with $0.2 < z \leq 1$. These sources are very difficult to study because they are very faint. Therefore, fewer counts have been accumulated in their images than for the more intense sources at lower z. As a result, not only are the statistical uncertainties greater, but the background subtraction has been a problem. The combination of these two effects has discouraged detailed investigation of large redshift clusters in the *Einstein* data. In this study the most recent, corrected background estimates are used. In this study, a careful optimization of the removal of the telescope and detector point response is being done, to obtain the best possible interpretation of the observations. Further an intensive effort is going on at CfA to extract better quality data from the Monitor Proportional Counter (MPC) for faint sources, which may give X-ray spectra out to higher energies for some of these clusters.

During the past year, there have been significant optical observations of gravitational lens behavior in clusters of galaxies. At the Toulouse Workshop on Gravitational Lenses, a list of 11 clusters that show this behavior was circulated, in addition to about 25 other objects that exhibit rings, or multiple magnified images. There is a feeling among observers that roughly 1/4 to 1/3 of all compact clusters exhibit some obvious form of mirage behavior. A recent paper by Nemiroff and Dekel (1989) presents quantitative arguments in support of this notion. Recent results by Tyson (1989) argue that every cluster will show such behavior if a sufficiently deep exposure is taken so that background objects with $z \geq 1$ are detected.

In the past several years, various investigators (Schramm and Kayser 1987, Kochanek et al.1989) have developed techniques to analyze extended gravitational images such as arcs or rings, with the ultimate goal of estimating the mass distribution of gravitational lenses and possibly of estimating cosmological parameters. Such estimates are affected by uncertainties due to various problems, such as the difficulty in measuring redshifts for very faint, extended sources (*viz.* the arcs and rings) and the possible presence of unseen inhomogeneities of the cosmological medium itself, along the line of sight to the clusters (Gorenstein, Falco and Shapiro 1988). These uncertainties currently give a latitude of factors of a few to the best estimates of the mass in clusters. The analysis software that is currently implemented at CfA was written to treat unresolved gravitational images; it is currently being extended by Falco to treat extended images. Our hope is to define clearly the uncertainties in known analysis techniques by studying the uniqueness of models for the mass distribution in clusters, and to apply them to as many different clusters as possible. Then, we may be able to develop statistical measures of the power of the techniques to achieve the important cosmological goals indicated above.

EINSTEIN OBSERVATIONS OF GRAVITATIONAL LENS CLUSTERS

To begin the investigation of X-ray properties of lens candidates, radial count rate distributions and detailed enlarged contour maps of a 20 arc min region about each X-ray source were plotted for approximately 30 distant clusters. Examples of these for some of the better known gravitational lens candidates are given in Figures 1 to 4.

The examples shown illustrate the varying quality of the data, which is a function of the total counts observed and the intensity of the X-ray source.

We have been using image processing and analysis systems developed at CfA to analyze the optical image of 0024+16 and to derive a description of the underlying gravitational lens. A preliminary examination of the *Einstein* IPC and HRI data indicates that point sources will not make a major contribution to the total emission, so that the IPC data will be usable for deriving the X-ray virial mass.

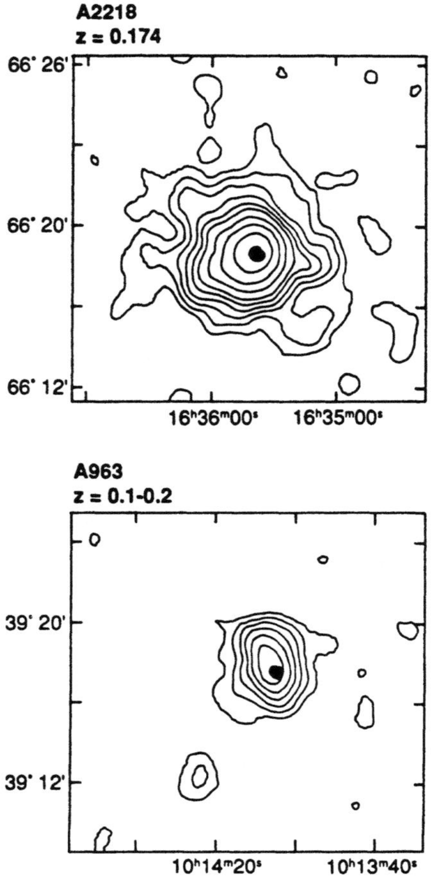

Figure 1: *Einstein* X-ray contour maps of A2218(203 X-ray counts) and A963(42 X-ray counts), both gravitational lenses in clusters of galaxies. IPC data. The filled circle is the optical position of the cluster lens. The IPC maps have a position accuracy of ±30 arc sec.

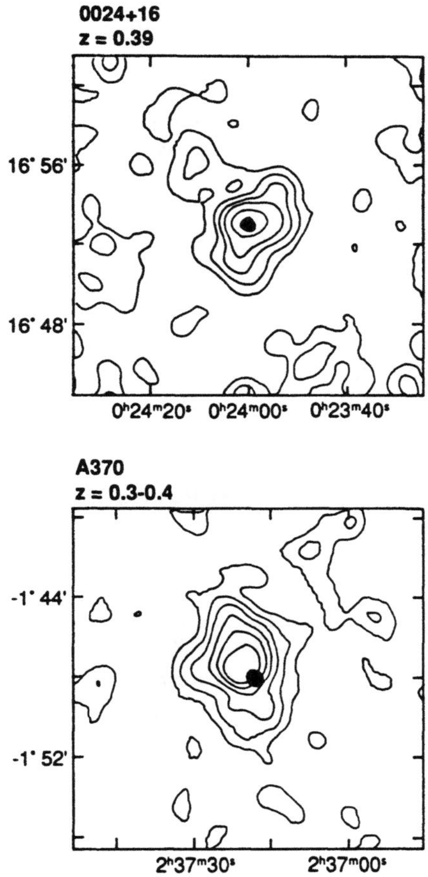

Figure 2: *Einstein* X-ray contour maps of Cl0024+16(120 X-ray counts) and A370(109 X-ray counts), both gravitational lenses in clusters of galaxies. IPC data. The filled circle is the optical position of the cluster lens. The IPC maps have a position accuracy of ±30 arc sec.

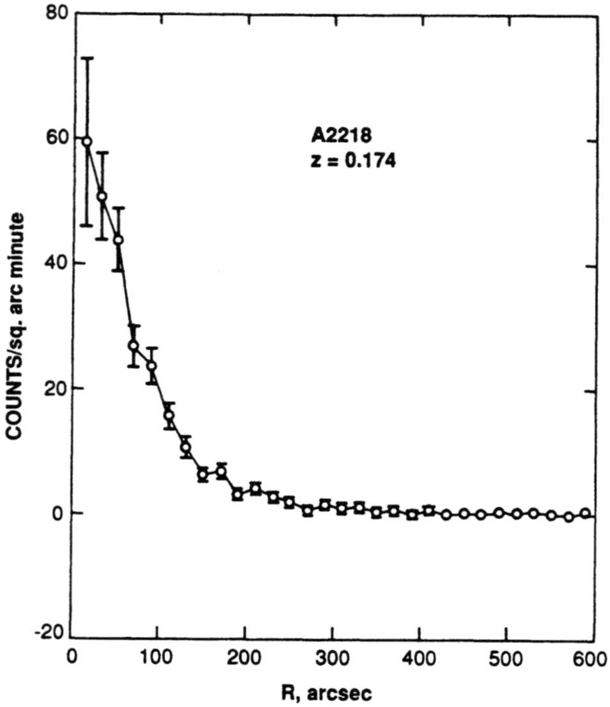

Figure 3: Radial X-ray count distribution for A2218. IPC data. Telescope response not removed.

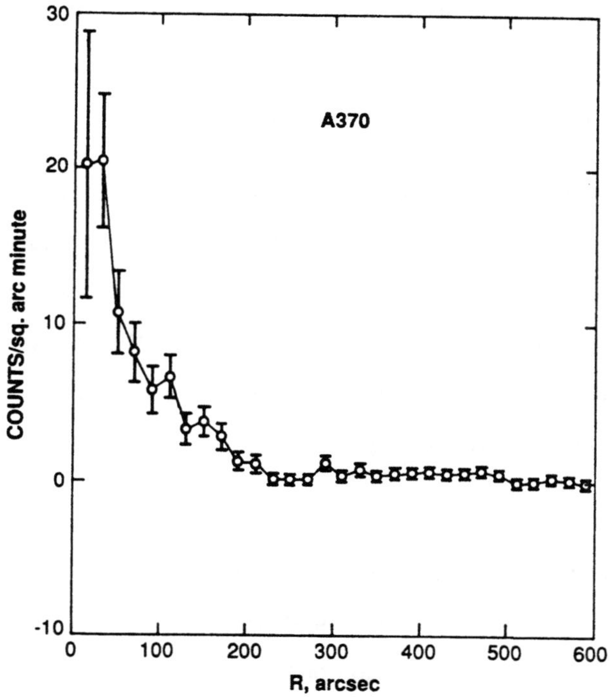

Figure 4: Radial X-ray count distribution for A370. IPC data. Telescope response not removed.

We have generated a list of 19 cluster X-ray sources that are good candidates for optical lensing searches. The list includes centroid positions of the X-ray emitting region, accurate to about 30 arc seconds (see Table 1).

The first column of Table 1 gives the name of the source. The second column gives the redshift if one has been measured and the third column gives a reference to the redshift measurement. The fourth column gives the total counts observed (after background subtraction) and can be used to evaluate the statistical precision of the data. The fifth and sixth columns give the position of the X-ray source, and the seventh column gives the position uncertainty.

The statistical quality of the data is a function primarily of the total number of counts observed, although it also depends on the intensity of the source relative to the background counting rate. The sources with the greatest number of counts are those that will permit the best determination of the X-ray core radius and virial mass, and so should be the ones most useful for obtaining optical data for comparative analysis of the mass determinations.

This list will be useful to optical observers who are searching for gravitational lens behavior in clusters, because it points to the location of the majority of the mass in these clusters. This position is not always in the same place as the concentration of brightest galaxies. Therefore, the location of the X-ray source is probably the best center about which to search for arcs and rings with deep CCD exposures.

CLUSTER MASS DETERMINATIONS

The goal of this study is to determine the mass contained in clusters of galaxies in three different ways. First, mass estimation from the *Einstein* X-ray data; second, analysis of the gravitational lensing images in the visible spectrum, modelling the mass distribution

Table 1: Distant Cluster X-ray Sources Detected by *Einstein* Observatory

Name	z	Ref.	X-ray Counts	R.A.(1950)	Dec	Position Error (arc min)
A520	0.203	1	~253	04 51 36.0	02 50 42.5	32.0
A1704	0.220	2	239	13 12 37.84	64 50 36.1	31.0
3C295	0.464	4	195	14 09 33.36	52 26 22.0	32.0
A2397	0.224	1	158	21 53 36.0	01 09 18.2	32.0
A1722	0.328	2	~137	13 18 35.92	70 20 18.4	35.0
SC2059-247	0.188	3	113	20 59 14.92	-24 43 47.9	32.0
A3186	-	-	~112	03 53 12.76	-74 11 06.7	32.0
A370	0.373	1	~109	02 37 21.85	-01 46 17.2	36.0
A913	0.3 - 0.4	-	92	09 59 40.66	20 45 40.6	36.0
A1246	0.216	1	62	11 21 22.5	21 45 39.7	35.0
A2111	0.229	1	60	15 37 44.67	34 35 02.3	30.0
A1557	0.210	2	54	12 30 15.48	63 09 33.5	36.0
3C 19	0.483	4	52	00 38 13.64	32 53 14.3	36.0
A2645	0.246	1	45	23 38 42.0	-09 17 58.9	35.0
PHL1093	0.270	-	40	01 37 23.39	01 16 39.7	36.0
A1655	0.234	2	38	12 56 37.0	65 38 19.0	36.0
A41	0.275	1	35	00 26 22.16	07 34 08.4	36.0
4C74.13	-	-	34	07 35 38.48	74 21 38.0	35.0
Cluster 249	-	-	26	03 46 40.05	-45 23 45.4	37.0

1 = Sarazin et al 1982.
2 =Huchra et al 1990.
3 = White et al 1981.
4 = Giacconi and Seward, 1979

required to explain the observed distorted images of background galaxies; third, use of available redshift data for galaxies in these clusters to find a corresponding virial mass.

X-ray Virial Mass Estimates

In cases where sufficient counts are detected, say ≥ 175, detailed fitting of brightness functions is justified.

Pulse height spectroscopy can be done with the *Einstein* IPC. There is also some potential for using the *Einstein* Monitor Proportional Counter for sources that do not have another source of comparable strength within the field of the 3/4 degree collimator. The MPC can give an estimate of the gas temperature to higher energy because it does not suffer from the 4 keV upper energy cutoff of the *Einstein* mirror; it has reasonable efficiency up to ≥ 15 keV. For some sources, it may be possible to obtain spectra with the Ginga satellite.

The IPC images, the spectral hardness in at least two energy bands for some sources, and the comparison of IPC and HRI data can be analyzed for a few sources to check for compact sources. Fits are computed to specific model brightness distributions as appropriate. The degree to which the analysis of cluster brightness distributions is pushed depends upon the number of photons detected.

The hot gas pervading a typical cluster may be used to trace the density profile of the cluster. Given this density distribution, the virial mass of the cluster may be calculated and compared with that determined by velocity dispersion measurements from cluster galaxies or by modeling of gravitational lens images associated with the cluster.

We may treat the cluster gas as an ideal gas in hydrostatic equilibrium in a (spherically symmetric) cluster potential. Thus, we have

$$P = \frac{kT}{\mu M_H}\rho$$

and

$$\frac{dP}{dr} = -\frac{GM(r)\rho}{r^2}$$

where $M(r)$ is the mass interior to the radius r. We may combine these to find

$$M(r) = -\frac{kTr}{G\mu M_H}\left[\frac{d\ln\rho}{d\ln r} + \frac{d\ln T}{d\ln r}\right]$$

or, using an isothermal approximation,

$$M(r) = -\frac{kTr}{G\mu m_p}\frac{d\ln\rho}{d\ln r}.$$

Typically, the X-ray surface brightness distribution may be represented by:

$$I(r) = I_0\left[1 + \left(\frac{r}{r_c}\right)^2\right]^{-3\beta+1/2}.$$

For an isothermal distribution, the density profile may then be deprojected:

$$\rho(r) = \rho_0\left[1 + \left(\frac{r}{r_c}\right)^2\right]^{-3\beta/2}.$$

Differentiating, we may substitute to find

$$M(r) = \frac{3\beta kT}{G\mu M_H}\frac{r}{[1 + \left(\frac{r_c}{r}\right)^2]}.$$

Thus, we need only determine r_c, β, and T from the X-ray data in order to derive the mass distribution of the cluster. The core radius r_c and β may be determined by fitting the X-ray data to trial expressions for $I(r)$ (properly convolved with the instrument point response function) as β and r_c are varied. A temperature value for the cluster may be determined if a sufficiently long IPC observation exists. For observations lacking such spectral information, typical values of $T \sim 5$ keV may be assumed without serious loss of accurcacy. Similarly, for sparse data sets, the average value $\beta \approx 2/3$ (Jones and Forman, 1984) may be used to facilitate determination of r_c. In this manner, the mass contained within the image-producing portion of the cluster-lens may be calculated (given the lens radius inferred from optical observations) and compared with calculations based upon the optical lensing models.

We have begun such an analysis for A2218. From fits to the *Einstein* data (Figure 3) we find an X-ray core radius of 0.16 – 0.36 Mpc for $H_0 = 50$, and $\beta = 0.55 - 0.75$. Using $kT = 6.7^{+0.5}_{-0.4}$ keV as determined by Ginga observations (McHardy et al, 1989), we obtain the curve in Figure 5 for the contained X-ray virial mass as a function of angular distance from the center of the cluster. The confidence limits in the curve plotted are obtained by plotting a curve for each combination of the extreme values of core radius, T and β obtained from the fits to A2218 data, and estimating the envelope of curves corresponding to all such values within the extremes.

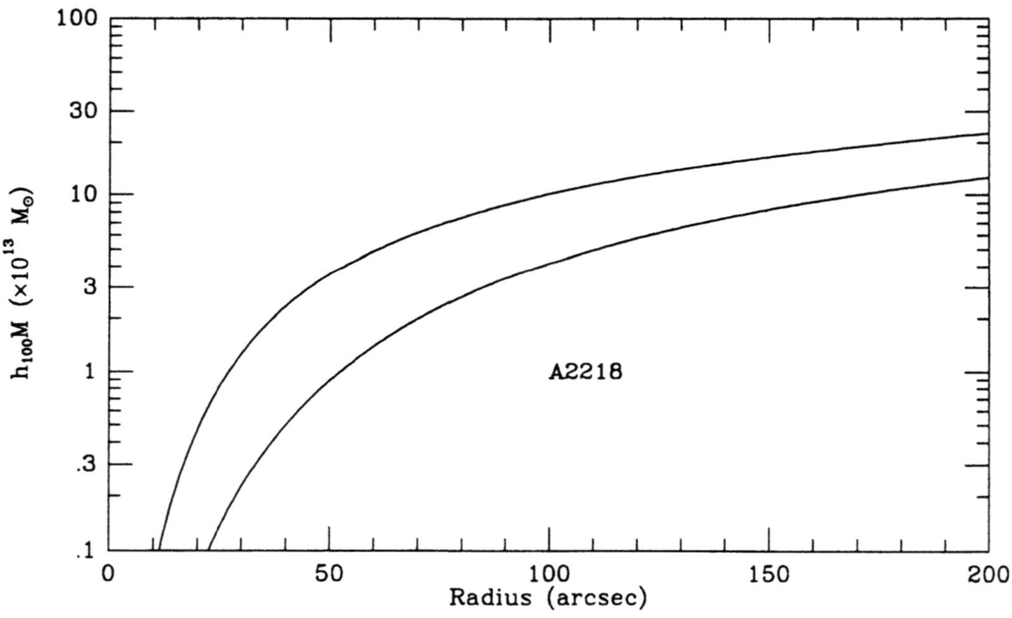

Figure 5: Contained mass of X-ray emitting hot gas versus angular radius for A2218. The two curves represent the approximate extrema of the confidence region, although a precise assignment of level of confidence has yet to be done.

Cluster Mass Estimates by Gravitational Lens Analysis of Optical Images

Giant luminous arcs are now a nearly common feature formed by compact clusters. Now that such features are expected, more are certain to be found. The potential database of candidates is very large. Therefore, the analysis of optical views of gravitational images is a promising technique for cosmological studies. One of the critical possible results is a determination of the total mass in clusters, including dark matter, because the gravitational deflection of light is sensitive to all matter, regardless of its emissivity. It is expected that most of the matter in clusters of galaxies is non-luminous. Therefore, using gravitational bending of light is a promising technique to obtain a new type of mass estimate, to be compared with optical and X-ray virial masses. Such new estimates, if sufficiently accurate and numerous, will provide new constraints, not only on the mass-to-light ratios in clusters, and therefore, on types of dark matter therein, but also on the mean matter density of the universe.

There are two regimes of interest for clusters acting as gravitational lenses. In the first regime, the surface mass density is too low to yield multiple images of backgound galaxies, and no large arcs can form. In such a case, the images of background galaxies will be elongated, preferentially in a tangential direction with respect to the cluster's center. Deep CCD images of the cluster field then may best be analyzed statistically (Tyson, Valdes and Wenk 1989) to determine the mass distribution of the cluster. In the second regime, the surface mass density suffices to yield multiple images of background galaxies, and large arcs can appear. The presence of large arcs is due to chance alignment of caustics of the lens and background galaxies. The larger the mass of the cluster, the larger the cross-section for such chance alignments. The shapes of large arcs (from 30 degrees to a full ring) provide very strong constraints on the mass distribution of the lensing cluster, because these shapes are very sensitive to the separation between the caustics and the background galaxies. Of course, the two regimes mentioned overlap, because of the granularity of the mass distribution in clusters, as evinced by luminous galaxies. Each case requires careful accounting of the individual galaxies that are observed.

What optical observations are needed once a candidate lensing cluster has been identified? Deep CCD imaging and spectrophotometry are necessary for two reasons: first, the shape of arcs must be mapped carefully to minimize geometrical uncertainties; second, a determination of the redshift of the galaxies that are stretched into arcs is critical to remove degeneracies in estimates of the mass distribution of lensing clusters. Sufficient interest has been generated in the lensing properties of clusters that we can safely expect observations of these sources to be constantly pushed to their limits.

Images of the cluster 0024+16 show many arcs that are attributed to gravitational deflection (Turner, 1988, 1989. Also see Koo, 1988). The arcs are thought to be produced by gravitational lensing due to a foreground cluster of red galaxies, at $z \cong 0.39$ (Dressler, Gunn, and Schneider 1985). We have started modeling efforts to account for the presence of the arcs. As a first approximation, we assumed the main mass component of the cluster to be a smooth surface mass distribution with a King-type profile, parametrized by a line-of-sight velocity dispersion and a core radius, and centered at the approximate center of brightness of the X-ray emission from the cluster. The values of the model parameters that account for the arc separations are in the range 1800 to 2200 km s^{-1} for the velocity dispersion, and 30 to 40 arcsec for the core radius. These are preliminary values that indicate simply that reasonable values of the parameters will eventually suffice to account for the properties of the arcs.

We plan to model the surface mass distribution of the cluster in greater detail, based on a more refined description of the distribution of X-ray brightness, and on accounting for the presence of individual galaxies that affect the shapes of the arcs.

Comparisons of Mass Estimates

The optical virial mass of a cluster is determined by measuring the velocities of a sufficiently large number of galaxies. The X-ray virial mass of a cluster is determined by calculating the volume mass density in the cluster that is necessary to produce the observed X-ray emission, and then integrating the density. The lensing mass of a cluster is determined by analyzing the properties of the optical images produced by the cluster. If the only matter in clusters were of the luminous variety, these three mass estimates would coincide. Since the X-ray and optical virial masses disagree by at least an order of magnitude with mass estimates obtained by accounting only for the luminous matter, we know that dark matter must be found in clusters. Thus, the third proposed type of estimate will add to the evidence for dark matter, and may yield more accurate values for the amounts in which it is present in clusters. Each of the individual techniques can be in error for its own reasons. Therefore, the three techniques used jointly can compensate for the individual errors of each technique.

The optical virial mass estimates are susceptible to errors caused by misclassifying individual galaxies as cluster members, when they are really background or foreground objects. In that case, their redshift is associated with the Hubble flow, not with their virial velocity in the cluster. Optical virial estimates are also distorted by substructure in the cluster; if there is subclustering, some of the galaxies within the cluster are also members of groups within the cluster and have their own peculiar velocity field distortions as a result.

X-ray virial masses are presently limited by the small number of X-ray photons detected, with resultant poor counting statistics. This makes measurements of the core radius, β, and temperature uncertain. As the number of photons detected increases with future X-ray astronomy observations from ROSAT, BBXRT, Astro-D, Spectrum X-γ, and AXAF, the statistical uncertainties will decrease and other sources of confusion and error will emerge. The X-ray virial mass determinations will also have problems from foreground and background source confusion. Substructure within the cluster will result in temperature, density, and velocity inhomogeneities that cause errors in measuring the total mass. Some of these errors can be mitigated by measuring the redshift of the gas that emits the X-rays, using advanced spectrometer detectors such as the CCD and the bolometer on AXAF. However, some confusion and inaccuracy will remain in the X-ray mass determinations.

SUMMARY AND DISCUSSION

This paper has given a preliminary report on studies using the *Einstein* X-ray data on some of these distant clusters. Examples of the X-ray radial brightness distribution have been given, and the core radius and β have been derived for A2218.

A preliminary value of the mass of the gravitational lens in 0024+16 has been obtained. The value, 1800 to 2200 km s^{-1}, is somewhat higher than the value of 1287 km s^{-1} obtained from optical redshift measurements on the galaxies (Dressler, Gunn, and Schneider 1985), although no formal error limit is quoted for their value. The work for deriving an X-ray virial mass is still under way, so no value is yet available. For A2218, we have derived a contained X-ray mass vs. radius, which can be compared with future determinations of mass from the optical velocity dispersion and the gravitational lens analysis. As this paper was being finalized, we learned that a preliminary value of (1050 +700 -200) km s^{-1} has been obtained (Birkinshaw, 1989), with suggestion that the correct value is closer to 1300.

Further work will focus on obtaining and comparing mass estimates from three independent sources in each cluster: the length scale(core radius) that characterizes the density dropoff of the X-ray emitting hot gas away from its center, the velocities of the galaxies moving in the cluster potential, and the bending of light by the total cluster mass. We urge the optical observers to concentrate their efforts on obtaining data for the clusters listed in Table 1, especially those at the top of the list for which well in excess of 100 counts were detected in X-rays.

REFERENCES

Birkinshaw, M. (1989) private communication.

David, L., Arnaud, K., Forman, W., Jones, C. (1989) submitted to Ap. J.

Dressler, A., Gunn, J., and Schneider, D. 1985 Ap. J. 294 70.

Forman, W. and Jones, C. (1982) Ann. Rev. Astron. Astrophys. 20 547.

Forman, W., Kellogg, E., Gursky, H., Tananbaum, H. and Giacconi, R. (1972) Ap. J. 178 309.

Fort, B., Prieur, J.L., Mathez, G., Mellier, Y., Soucail, G., 1988, Astron. Astrophys. 200, L5

Giacconi, R. and Seward, F. (1979) CFA/HEA 79-285, Smithsonian Astrophysical Observatory, Cambridge, Massachusetts.

Gorenstein, M. V., Falco, E. E., and Shapiro, I. I. (1988) Ap. J. 327 693.

Gursky, H., Kellogg, E., Murray, S., Leong, C., Tananbaum, H. and Giacconi, R. (1971) Ap. J.(Letters) 167 L81.

Helfand, D., Ku, and Abramopoulos, F. (1980) Highlights of Astronomy 5 747.

Huchra, J., Henry, J. P., Postman, M., and Geller, M. 1990 Ap. J. in press.

Henry, J. P. and Lavery, R. J. (1984) Ap. J. 280 1.

Hughes et al, 1987(8)

Jones, C., Mandel, E., Schwarz, J., Forman, W., Murray, S., and Harnden, F. (1979) Ap. J.(Letters) 234 L21.

Jones, C. and Forman, W. (1984) Ap. J. 276 38.

Kellogg, E., Gursky, H., Tananbaum, H. and Giacconi, R. (1972) Ap. J.(Letters) 174 L65.

Kellogg, E., Murray, S., Giacconi, R., Tananbaum, H. and Gursky, H. (1973) Ap. J.(Letters) 185 L13.

Kellogg, E., Baldwin, J. and Koch, D. (1975) Ap. J. 199 299.

Kochanek, C. S., Blandford, R. D., Lawrence, C. R., and Narayan, R. (1989) M.N.R.A.S. 238 43.

Koo, D.C. (1988) "Large-Scale Motions in the Universe: A Vatican Study Week. V.C. Rubin and G.V. Coyne, eds. Princeton U. Press. p. 253.

Lavery, R. J., and Henry, J. P. (1988) Ap. J. (Letters) 329 L21.

Lynds, R., and Petrosian, V. (1986) Bull. Am. Astr. Soc. 18 1014.

Mitchell, R. J., Culhane, J.L., Davison, P.J. and Ives, J.C. (1976) Mon. Not. Roy. Astron. Soc. 189 329.

Nemiroff, R. and Dekel, A. 1989 Ap. J. 344 51.

Sarazin, C., Rood, H., and Struble, M. 1982 Astron. and Ap. 108, L7.

Serlemitsos, P.J., Smith, B.W., Boldt, E.A., Holt, S.S., and Swank, H. (1977) Ap. J. (Letters) 211 L63.

McHardy, I., Stewart, G., Edge, A., Cooke, B., Yamashita, K. and Hatsukade, I. 1989 (preprint).

Mitchell, R.J. and Mushotsky, R. (1980) Ap. J. 236 730.

Mushotsky, R.F. (1980) in ***X-ray Astronomy***, ed. R. Giacconi and G Setti, Reidel, Dordrecht, p. 171.

Nemiroff, R.J. and Dekel, A. (1989) Ap. J. 344 51.

Sarazin, C. L., Rood, H. J. and Struble, M. F. (1982) Astron. and Ap. 108 L7.

Schramm, T. and Kayser, R. (1987) Astr. Ap. 174 361.

Soucail, G., Mellier, Y., Fort, B., Picat, J.P., 1987, Astron. Astrophys. 172, L14

Soucail, G., Mellier, Y., Fort, B., Mathez, G., and Cailloux, G. (1988) Astr. Ap. (Letters) 191 L19.

Turner (1988) Proc. 14th Texas Symposium on Relativistic Astrophysics.

Turner (1989) private communication.

Tyson (1989) Proc. Toulouse Workshop on Gravitational Lenses, to be published

Tyson, J. A., Valdes, F., and Wenk, R. A. (1989) preprint.

White, R.A., Sarazin, C.L., Quintana, H. and Jaffe, W.J. (1981) Ap. J. (Letters) 245 L1.

White, R. H., Sarazin, C. L., Quintana, H. and Jaffe, W. J. (1981) Ap. J. Lett. 245 L1.

White, S.D.M., Silk, J. and Henry,J.P. (1981) Ap. J. (Letters) 251 L65.

Consequences of Gravitational Lensing for Gamma Radiation from 3C273

L. Metcalfe[1] and B. McBreen[2]

[1] Astrophysics Division, ESTEC, Postbus 299, 2200 AG Noordwijk, The Netherlands.

[2] Physics Department, University College Dublin, Dublin 4, Ireland.

Abstract

The absorption of gamma radiation by photon-photon interactions in the core region of 3C273 has been reconsidered because 3C273 may be gravitationally lensed by a galaxy along the line of sight to the quasar. It is shown that the gamma radiation detected by COS-B can escape from the same compact core as the quasar's x-ray flux if the intensification of the gravitational lens is about 100.

Introduction

It has been suggested (Chitre et al. 1984) that the unusual brightness of the quasar 3C273 and the observations (Pearson et al. 1981) of superluminal motion in its core can be explained by postulating a faint lensing galaxy about halfway along the line of sight to the quasar. Deep CCD photometry (Tyson et al. 1982) of 3C273 reveals a fuzz around it with isophotes similar to those of an elliptical galaxy. The red-shift measurements (Wychoff et al. 1980) of the fuzz suggest that it is hosting the quasar. However, 3C273 is displaced from the centre of the fuzz and the absolute luminosity of the host galaxy is much higher than that of other galaxies hosting quasars. These facts may be explained by assuming the presence of a lensing galaxy of high mass-to-light ratio along the line-of-sight to 3C273. Gravitational intensification of the flux from 3C273 could have important consequences for the opacity of the quasar to gamma radiation. Gamma radiation in the energy range 50 MeV to 800 MeV was observed from the direction of 3C273 by the COS-B satellite (Bignami et al. 1981). It was pointed out (McBreen, 1979) that the quasar's x-ray flux is such that the gamma radiation should have been absorbed by photon-photon interactions to produce electron positron pairs if the gamma rays originated in the same compact source as the x-rays. This conclusion is not correct if 3C273 has been intensified by a factor of about 100 by a suitably placed galaxy acting as a gravitational lens.

Photon-Photon Absorption in 3C273

The gamma ray optical depth is given by (Bignami et al. 1981, Metcalfe and McBreen, 1986)

$$\tau = r\sigma n_X(2E_T)2.5E_T \qquad (1)$$

where r is the size of the x-ray emitting region, σ is the photon-photon interaction cross-section and n_X is the density of ambient x-ray photons, henceforth assumed constant, within the source and approximated by its value at $2E_T$, where E_T is the threshold energy for the interaction. The function σ has been approximated by a rectangular function of height 1.7×10^{-25} cm^2 and width $2.5E_T$. If ϕ is the flux of x-ray photons at $E = 2E_T$ measured at the Earth, then

$$n_X(2E_T) = \phi(2E_T)\, 4\pi D^2 \,/\, Ic4\pi r^2 \qquad (2)$$

where D is the distance to the source and I is the intensification of the source by an intervening gravitational lens. The Einstein Observatory provided results on the short term variability of the keV x-rays from many quasars including 3C273 (Zamorani et al. 1984). In particular, 3C273 was observed to vary by about 10% over a period of about twelve hours. The Ariel V experiment observed an 80% increase in the x-ray flux over a similar period (Marshall et al. 1981). If we exclude relativistic expansion effects and beaming of the radiation then the light travel time argument yields $r = 1.5 \times 10^{15}$ cm for the dimensions of the x-ray source.

Table I gives the gamma ray optical depth τ for two different intensifications of the source, I = 1 and 100, at several gamma ray energies and corresponding threshold x-ray energies. It has been assumed that 3C273 is at the cosmological redshift distance of 380 Mpc (z = 0.158 and H_0= 100 km s^{-1} Mpc^{-1}). The values of the x-ray flux are taken from the 2 to 60 keV measurements of Worral et al. (1979). If there is no gravitational lens and hence no intensification of the source, it is apparent from the large optical depths that the COS-B gamma rays cannot originate in the small x-ray source associated with the nucleus of 3C273. This conclusion is in agreement with the computations of McBreen (1979), Bassini and Dean (1981) and Bignami et al. (1981) and has lead to models where the gamma rays originate in a different region of the source from the x-rays (Jones, 1979; Protheroe and Kazanas, 1983; Morrison et al. 1984).

Table 1. The Optical depth for Gamma Radiation of Energy E_γ from 3C273.

E_γ(MeV)	E_T(keV)	τ_1(I=1)	τ_2(I=100)
50	20	20	0.2
200	5	30	0.3
500	2	60	0.6
800	1.3	72	0.72

It is apparent from Table I that gamma-rays can escape from the central core with little attenuation if 3C273 is gravitationally intensified by a factor of about 100. In the case of gravitational lensing the COS-B gamma-ray source can coincide with the variable Ariel V and Einstein Observatory source associated with the nucleus of the quasar. In this context the model of Kafatos (1980) is of particular interest because the gamma rays are produced by the Penrose pair production mechanism in the ergosphere of a massive rotating black-hole and radiated into a 40° fan beam around the equator.

This beaming is not sufficient to avoid the pair production problem even if it is assumed that the x-rays are beamed into the same angle. However, it could allow the escape of the COS-B gamma radiation with a smaller gain in the lensing model. Chitre et al. (1984), using different types of galaxies in suitable lens configurations, have shown that intensifications between 12 and 80 are readily obtained for 3C273. These lens configurations can account for the observed superluminal motion and the appearance of only one bright image of 3C273. It should be noted that very high intensifications occur if compact sources are located near the caustics or microcaustics of gravitational lenses (McBreen and Metcalfe, 1987,1988). The competing relativistic beam model (Begelman et al. 1984) can also account for the same observations, but in this case, if the gamma rays originate from the core region of 3C273, it is necessary to beam the x-rays and gamma rays into an opening angle of about 10° to avoid the pair-production difficulty (Königl, 1981).

While the various analyses of lensing statistics for objects at cosmological distances make it entirely reasonable to propose that 3C273 is gravitationally lensed,probability considerations would be irrelevant if future observations clearly reveal the lensing galaxy. In particular, spectral observations of the optical nebulosity may reveal the absorption line system of the fainter galaxy lensing the galaxy hosting 3C273. More detailed measurements of the gamma ray spectrum and its variability may be helpful in assessing the importance of the lensing and relativistic beaming models.

Conclusions

The photon-photon absorption of gamma radiation in the core region of the quasar 3C273 has been analysed because the quasar may be gravitationally lensed by a suitably placed galaxy of high mass-to-light ratio along the line of sight. It has been concluded that the gamma radiation observed by COS-B can escape from the core region with little attentuation if the gravitational intensification is about 100. The gravitational lensing scenario may be tested by searching for the absorption-line system expected if the intervening galaxy exists.

References

Bassani, L. and Dean, A.J. : 1981, Nature, 294, 332
Begelman, M.C. et al. : 1984, Rev. Mod. Phys., 56, 255.
Bignami, G.F. et al. : 1981, Astron. Astrophys., 93, 71.
Chitre, S.M. et al. : 1984, Astron. Astrophys., 139, 289.
Jones, T.W. : 1979, Astrophys. J., 233, 796.
Kafatos, M. : 1980, Astrophys. J., 236, 99.
Königl, A. : 1981, Astrophys. J., 243, 700.
McBreen, B. : 1979, Astron. Astrophys., 71, L19.
McBreen, B. and Metcalfe, L. 1987,Nature, 330, 348 : 1988, Nature, 332, 234.
Marshall, N. et al. : 1981, Monthly Notices Roy. Astron. Soc. 194, 987.
Metcalfe, L. and McBreen, B. : 1986, Astron. Astrophys. 166, L5.
Morrison, P. et al. : 1984, Astrophys. J., 280, 483.
Pearson, T.J. et al. : 1981, Nature, 290, 365.
Protheroe, R.J., Kazanas, D. : 1983, Astrophys. J., 265, 620.
Tyson, J.A. et al. : 1982, Astrophys. J., 257, L1.
Worral, D.M. et al. : 1979, Astrophys. J., 232, 683.
Wychoff, S. et al. : 1980, Astrophys. J., 242, L59.
Zamorani, G. et al. : 1984, Astrophys. J., 278 28.

SEARCHING FOR COSMIC STRINGS
(A LA RECHERCHE DES CORDES PERDUES)

Mark Hindmarsh
Department of Physics, The University
Newcastle NE1 7RU, U.K.

1. INTRODUCTION

The ideas of grand unification are unlikely ever to be tested in the laboratory. However, if the hot big bang is correct, the universe should have been "grand unified" at around 10^{-36} s after the initial singularity. As the universe cooled it would have gone through a phase transition, which may have left behind defects in the vacuum (Kibble 1976) which could have the topology of a point, a line, or a surface – monopoles, strings or domain walls. The implications for cosmology of all these objects are profound: the absence of monopoles and walls already tells us that certain classes of theory are incompatible with a non-inflationary universe (Guth 1981, Zel'dovich *et al* 1974). Strings have perhaps the most exciting possibilities. If they provide the density perturbations that seed structure formation (Zel'dovich 1980, Vilenkin 1981), then their mass per unit length μ should be about 10^{-6} in units of G (Turok and Brandenberger 1986). This corresponds to about 10^{21} kg m^{-1}, or in natural units ($h = 2\pi$, $c = 1$) about $(10^{16}\text{GeV})^2$, which is in the range of values for a grand unified phase transition. It is tempting to regard this as more than coincidence, and ask if direct observational evidence could be obtained. The primary interactions of strings are gravitational, although if they are superconducting (Witten 1985) they may become visible through electromagnetic or more exotic processes. If we restrict ourselves to the gravitational interactions, which are common to both ordinary and superconducting strings, then there three main avenues of observation: gravitational radiation, lensing, and distortions in the cosmic microwave background (see, for example, Vilenkin (1987) and references therein). In fact, the "standard model" of cosmic strings (see Vilenkin 1985 for a review) is in trouble as regards gravitational radiation, as millisecond pulsar timings (Rawley *et al* 1987) give an upper bound on the energy density in waves which is incompatible with $G\mu = 10^{-6}$ (Albrecht and Turok 1989). However this need only be the death knell of this particular version of the scenario, which required strings to form after inflation. There are other models which avoid the overproduction of waves with periods of a few years (Yokoyama 1989). It is also possible that the cosmic string simulations are not quite right in some of the details of the production of loops which makes a big difference to the amount of gravitational radiation.

However, if GUT scale strings are around today they will definitely show up either as lines of discontinuity in the apparent temperature of the CMB with $\Delta T/T \simeq 2.5 \times 10^{-5}$ (Kaiser and Stebbins 1984) or as lines of double images of distant galaxies and quasars. In this talk I shall outline an idea for searching for the lensing signature of cosmic strings in deep photographic material. I shall describe some salient points of the cosmic string scenario, including their lensing properties and probabilities. Finally I outline a possible search strategy and what such a search can achieve.

2. COSMIC STRINGS IN THE EARLY UNIVERSE

A cosmic string is a very thin and very massive tube of mass/energy. Many particle physics models contain such objects as solutions to the field equations, and if one such model is the correct one they would have been formed at a phase transition in the early universe, in a tangled network consisting of about 20% closed loops and 80% infinitely long strings. In an expanding universe the network of long string straightens out, mainly by chopping off closed loops, at just the right rate to keep its characteristic scale a constant fraction of the horizon size ct. The closed loops can chop further, until eventually they reach non-self-intersecting solutions which gradually disappear into gravitational radiation. This is the so-called "scaling solution". Many important properties of the system of strings are expressible in terms of a few dimensionless parameters and the horizon; for example, the number density distribution of loops of size L is, at time t in the matter era,

$$n(L,t)dL = \frac{\nu}{c^2t^2}\frac{dL}{(L+\Gamma G\mu ct)^2} \tag{1}$$

where Γ is a number of order 100 measuring the efficiency of radiation of gravity waves of a typical loop and ν is a normalisation detemined to be about 1 from numerical simulations (Albrecht and Turok 1989). For a complete list of references on the basic string model see Vilenkin (1985).

3. GRAVITATIONAL LENSING BY STRINGS

The metric around a straight string is well known to be conical. This means that there are two paths either side of an intervening string from a source to an observer, who will therefore see two images. In the observer's rest frame the splitting angle σ is (see Vilenkin 1987)

$$\sigma = 8\pi G\mu \sin\phi \left(1-\frac{x_s}{x_g}\right)\gamma^{-1}(1-\mathbf{n}\cdot\mathbf{v})^{-1} \tag{2}$$

where $\mathbf{v}$ is the velocity of the string according to the observer, $\mathbf{n}$ is the normal vector in the direction of the point on the string between the two images and x_s and x_g are the distances between the observer and the string and the source respectively. This formula should be approximately true for curved string in an expanding universe, provided the impact parameter is much smaller than the string curvature radius and the Hubble radius. For redshifts less than 1 we may write

$$\sigma \simeq 5.2\left(1-\frac{z_s}{z_g}\right)\mu_6 \sin\phi U(v)\ \text{arcsec} \tag{3}$$

where $\mu_6 = G\mu/10^{-6}$ and $U(v)$ is the velocity factor in (2). A straight string does not magnify or distort the images: however, if the source is resolvable and lies partially inside the wedge beyond the string that is lensed the observer will see only part of it, with a sharp edge cut off by the string. It has even been suggested that strings may be found by searching for these truncated images (Paczynski 1986). If the image is not resolved there will just be a difference in magnitude and possibly colour, if for example just the arms of a spiral lie in the wedge.

Gravitational lensing by loops of string is coniderably more complicated. If the angular size of the loop is large compared with the angular separation of the images then the part of the loop in front of the source behaves like a segment of almost straight moving string. If the angular size of the loop is much smaller than σ than in behaves as a point source with a Schwarzschild metric. However, for the intermediate case, the light ray sees the full time-dependent gravitational potential of the loop and no approximations can be made, but from general optical considerations we can expect multiple images, caustics and all the usual lens phenomena (Hogan and Narayan 1984).

4. THE AMOUNT OF STRING IN THE SKY

Firstly, we define a few of the dimensionless parameters that will be important. A is the average number of horizon lengths of string per horizon volume $(ct)^3$. Thus the energy density in infinite string is expressible as $\rho_\infty = A\mu/(ct)^2$. We have already come across the loop distribution normalisation ν, in equation (1). Finally, we denote the typical size of loops at birth relative to the horizon by ϵ. The simulations of Albrecht and Turok (1989) give $A \simeq 10$, $\nu \simeq 1$ and $\epsilon \simeq 0.1$, although there is some debate over whether the distribution of loop

sizes at birth really does scale with ct (Bennett and Bouchet 1989).

Using (1) we find that in the matter era the total length per unit volume of string in loops is

$$\mathcal{L}_L = \nu \ln(\epsilon/\Gamma G\mu)(ct)^{-2} \tag{4}$$

and the total number density is $\nu/\Gamma G\mu(ct)^3$. For low redshifts, the volume element is $dV = (ct_0)^3 27\pi z^2 dz/2$, while the angular distance $d_A(z)$ grows as $3ct_0z/2$ (Weinberg 1972). Thus we may write the projected number density of loops out to a redshift z as

$$N_L \simeq 3\nu\Gamma_{100}^{-1}\mu_6^{-1}z^3 \text{ deg}^{-2} \tag{5}$$

while the total angular lengths of string per square degree in loops (Θ_L) and in infinite string (Θ_∞) are

$$\begin{aligned} \Theta_L &\simeq 0.1\nu z^2 \text{ deg}^{-1} \\ \Theta_\infty &\simeq 0.1A_{10}z^2 \text{ deg}^{-1} \end{aligned} \tag{6}$$

where we have taken $\ln(\epsilon/\Gamma G\mu) = 10$. Thus, out to a redshift of 0.3, where the peak of the number distribution of galaxies with redshift lies for a limiting blue magnitude $J \simeq 24$ (Broadhurst *et al* 1988), there are about 800 degrees of string, distributed roughly equally between infinite string and closed loops. From the distribution (1) it is clear that most of the loops have length $L = \Gamma G\mu ct$, and so if β is the average ratio between the length of a loop and its RMS radius we find that the angular diameter of string loops at redshift z is

$$\lambda \simeq 3(\Gamma_{100}\mu_6\beta_{10}^{-1})z^{-1}\text{arcsec} \tag{7}$$

Finally, we need to know how straight we can expect the network of infinite string to be, projected against the sky. The infinite strings form Brownian walks with persistence length $\xi = \alpha ct$, as it scales with the growth of the horizon. The parameter α is about 0.3 in the matter era (Albrecht and Turok 1989). Including a factor of $2/\pi$ as an average over angles to the line of sight, the projected persistence length ς is given by $4\alpha/3\pi z$ or

$$\varsigma \simeq 7\alpha_{0.3} z^{-1} \,\mathrm{deg}\,. \tag{8}$$

5. LENSING PROBABILITES

Out of a few thousand observed quasars, of order ten are known to be gravitationally lensed, and the images typically have a separation of a few arc seconds. This enables us to estimate an upper bound on the amount of string with $\mu_6 \simeq 1$ in the universe. Roughly speaking, each bit of string can be thought of as lensing a strip $4\pi G\mu$ radians wide. Hence the lensing probability is about $4\pi G\mu(\Theta_\infty + \Theta_L)$ or $10^{-4}(\nu + A_{10})\mu_6 z^2$. Given that most of of the redshifts of the sources are greater than 1, and taking the fraction of quasars lensed to be about 10^{-3} we have that

$$\nu + A_{10} < O(10) \tag{9}$$

Most quasar lensing events can be attributed to intervening matter that is detected, but for the few cases where there are no observed lenses there might be a cosmic string. However, in order to resolve the question one would need to make high resolution, high sensitivity observations of the CMB, which will be a long time coming. Instead of looking for double quasars we could look for double galaxies. One such search using the CTI is under way (Hogan 1987). I shall briefly describe an alternative approach, using deep photographic material obtained at the 4m AAT, which is to look for the lines of double images produced by infinite strings. In a one hour exposure at the prime focus, we can expect, under good conditions, to reach a limiting magnitude of 24-25 using enhanced contrast techniques (Couch *et al* 1984). Coupled with a sky coverage of 1 deg^2 this is an efficient way to maximise survey volume. On a plate with a limiting blue magnitude $J_{lim} = 24.5$ we can expect to find 30,000 or so galaxies (Tyson 1988), with an $N(z)$ peaked at around 0.3 (Broadhurst *et al* 1988). Given that there about 400 degrees of infinite string out to $z = 0.3$, the probablity of finding a detectable string in a plate is about $0.01A_{10}$. This is rather a long shot, but with a big prize for hitting the target.

6. A SEARCH STRATEGY

When looking for lens candidates, one usually looks for close pairs of objects as nearly identical in magnitude and colour as possible. In our approach, every close pair is a lens candidate, provided its members are not too grossly different in magnitude. However, because on any image most galaxies are faint, most pairs will consist of two faint galaxies. Infinite string at a redshift of 0.3 has an angular persistence length of about $20\alpha_{0.3}$ deg: thus across a 1 deg^2 plate we can expect it to produce an approximately straight line of pairs with their position angles aligned. How can we search for aligned pairs? We are looking against a background of random pairs: if σ is the upper cut-off in the separation of objects in our search, then we will find approximately $N_p \simeq \frac{\pi}{2}(\bar{N}_J\sigma L_p)^2$ random pairs, where $\bar{N}_J$ is the number of galaxies at limiting magnitude J and L_p is the linear dimension of the plate. The number produced by the string, N_s, is approximately $\bar{N}_J\sigma L_p$ which, up to numerical factors, is the square root of N_p: hence, independent of μ_6, the signal is hopelessly lost in the noise. However, we can improve this state of affairs by a couple of tricks. Firstly, we only only search in strips across the plate

of width W. Secondly, we put the position angles of the pairs in the strip into B angle bins. Then the average number of random pairs in a bin is about $\bar{n} \simeq \bar{N}_J^2\sigma^2 L_p W/B$, and provided the strip and the bin are wide enough to accomodate the string and its pairs, we find that $N_s \simeq \bar{n}^{\frac{1}{2}}(BL_p/W)^{\frac{1}{2}}$. Thus by maximising B and minimising W we can improve our signal-to-noise ratio. There are practical limitations on this approach: W must not be so narrow as to miss out sections of curving string, so we require $W > L_p^2/\varsigma$ or $W/L_p > 20$; also, we cannot expect to determine the position angles of objects separated by a few arc seconds to better than about 20°. These factors act to limit the effective volume of the survey, for the further away a piece of string is the fewer pairs it produces per degree of length. For strips drawn at random we can estimate at what redshift a string merges into the noise. The N_p pairs on the plate are distributed amongst the B position angles with b_i pairs in each bin. When we draw a strip let it contain N pairs: the distribution of N is not quite binomial because the areas of the strips are not always the same, but on average $N \simeq N_p W/L_p$. The probability of getting n_i pairs in the ith bin on a strip with N pairs drawn from the N_p random pairs is

$$P(n_i|b_i,N) = \binom{n_i}{b_i}\binom{N_p - b_i}{N - n_i}\Big/\binom{N_p}{N} \qquad (10)$$

which, for $n_i \ll b_i$ and $N \ll N_p$, is given approximately by the binomial distribution

$$P(n_i|b_i,N) \simeq \binom{N}{n_i}\left(\frac{b_i}{N_p}\right)^{n_i}\left(1 - \frac{b_i}{N_p}\right)^{N-n_i} \qquad (11)$$

We can define a threshold number in a bin $\hat{n}_i(N)$ by asking that the probability of getting $\hat{n}_i$ or more pairs in the ith bin be sufficiently unlikely as not to occur by chance when N_{str} strips are selected. In order to ensure that every pair of points on the plate falls into a strip at least once we need $N_{str} > (W/L_p)^2$, so $\hat{n}_i$ is defined as the smallest such that

$$\sum_{n_i \geq \hat{n}_i}^{N} P(n_i|b_i,N) \leq P_{cut} \simeq N_{str}^{-1} \qquad (12)$$

The estimated depth of the search is the redshift at which N_s falls below $\hat{n}_i$. Using $b_i = N_p/B$, $\bar{N}_J = 30000$, $\mu_6 \sin\phi = 1$ and $N = (z/7)N_p$ this turns out to be about $z = 0.25$ if all pairs are resolved. However, with automated plate scanning machines it is hard to do better than about 3″ on AAT plates before merging becomes a problem, which limits the depth for $\mu_6 \sin\phi = 1$ to about $z = 0.15$ (Hindmarsh 1989). It may well be possible to find more efficient search algorithms (the advantage of this one is purely in the ease of estimation of the depth), but all share the problem of the resolution of automated plate scanning which limits the search to $\mu_6 > 1$.

ACKNOWLEDGEMENTS

This is a very brief summary of work in progress in collaboration with Richard Ellis and Philip Tuckey (Durham) and Yannick Mellier (Toulouse). I am especially grateful to the Astronomy group at Durham for extending their hospitality. This work is funded by the SERC.

REFERENCES

Albrecht A. and Turok N. (1989), *Phys. Rev.* **D40**, 973.

Bennett D. and Bouchet F. (1989), in *"The Formation and Evolution of Cosmic Strings"* Gibbons G. and Vachaspati T. eds., (CUP Cambridge).

Broadhurst T. J., Ellis R. E. and Shanks T. (1989), *Mon. Not, R. Astron. Soc.* **235**, 827.

Couch W. J. *et al* (1984), *Mon. Not. R. Astron. Soc.* **209**, 307.

Guth A. (1981), *Phys. Rev.* **D23**, 347.

Hindmarsh M. (1989), in *"The Formation and Evolution of Cosmic Strings"*, Gibbons G. and Vachaspati T. eds., (CUP Cambridge).

Hogan C. (1987), in *"Cosmology and Particle Physics"*, Hinchliffe I. ed., (World Scientific, Singapore) , 116.

Hogan C. and Narayan R. (1984), *Mon. Not. R. Astron. Soc.* **211**, 575.

Kaiser N. and Stebbins A. (1984), *Nature* **310**, 391.

Kibble T.W.B. (1976), *J. Phys.* **A9**, 1387.

Paczynski B. (1986), *Nature* **319**, 567.

Rawley L. A.,Taylor J. H., Davis M. M., Allan D. W. (1987), *Science* **238**, 761.

Turok N. and Brandenberger R. (1986), *Phys. Rev.* **D33**, 2175.

Tyson J. A. (1988), *Astron. J.* **96**, 1.

Vilenkin A. (1981), *Phys. Rev. Lett.* **46**, 1169, 1496(E).

Vilenkin A. (1985), *Phys. Rep.* **121**, 263.

Vilenkin A., (1987) in *"300 Years of Gravitation"*, Hawking S. and Israel W. eds., (CUP Cambridge).

Weinberg S. (1972), *"Gravitation and Cosmology"* (Wiley, New York).

Witten E. (1985), *Nucl. Phys.* **B249**, 557.

Yokoyama J. (1989), *Phys. Rev. Lett.* **63**, 712.

Zel'dovich Ya. B. (1980), *Mon. Not. R. Astron. Soc.* **192**, 663.

Zel'dovich Ya. B., Kobzarev I. Yu., Okun' L.B. (1974), *Sov. Phys. JETP* **40**, 1.

III

Microlensing

Microlensing

Peter Schneider
Max-Planck-Institut für Astrophysik
Karl-Schwarzschild-Str. 1
D-8046 Garching, FRG
BITNET: PVS at DGAIPP1S

Abstract. This review summarizes the main aspects of microlensing (ML), i.e. the effects of small compact objects on a background source due to gravitational light deflection. After mentioning the basic ML models, the methods to detect ML are discussed. The main effect of ML is magnification, both of multiple images of quasars and of isolated sources. Directly observable is a variation of the magnification in time which causes lens-induced variability. This provides a potential diagnosis for the sources, but is difficult to be separated from intrinsic variability for single image sources. The magnification of sources can lead to the amplification bias, i.e. it can affect the number of sources in flux-limited samples. Consequences of this effect are mentioned, and its possible detections, the most direct of which is the recently found association of high-redshift quasars with (foreground) galaxies, are briefly discussed. The main ML model which has been investigated in detail in the literature (random star field) is considered, and some results are presented. Since ML has been (and is expected to continue to be) a very active field of research, I can only scratch the surface for most topics, and refer the interested reader to the relevant papers. Finally, some remarks about inhomogeneous universes are added.

1. Introduction

The term microlensing (ML) stems from the fact that the angular separation of images caused by a gravitational lens of mass M is $\theta_0 \sim 3 \times 10^{-6}\sqrt{M/M_\odot}$ arcseconds, which corresponds to about a microarcsecond for stellar mass lenses. This scale is much too small to be resolvable by telescopes; therefore, the only observable effect ML has is its influence on the magnification of distant sources. In order for a source to be magnified by ML, it has to be smaller than $\xi_0 \sim 10^{-2}\sqrt{M/M_\odot}$ parsecs, otherwise the effects of ML are averaged out. This basically restricts the relevance of ML to active galactic nuclei (and possibly supernovae). The above quoted values for θ_0 and ξ_0 depend on the distances to the source and the lens, and have been estimated for a 'typical' situation with $z_d \sim 0.5$, $z_s \sim 2$.

Microlensing models. The simplest model for a (micro-)lens is a point mass lens (Einstein 1936, Liebes 1964, Refsdal 1964), which produces always two images. However, the two images are of comparable brightness only if the relative misalignment of source and lens is smaller than about θ_0. In order that a spherical mass can be considered as a 'point mass' in its lensing behaviour, its radius has to be much smaller than ξ_0. This is nearly always the case for 'compact' objects; e.g., even a red giant can be considered as a point mass. If a star is not isolated, but if there are other masses close to it, the simple point mass model has to be replaced by a more complicated lens. To lowest order of a perturbation series, one can consider the effects of the other masses as a tidal effect, i.e. a quadrupole term is superposed onto the point mass gravitational potential (Chang and Refsdal 1979, 1984, Kovner 1987). This lens model, which has two essential parameters (convergence and shear), produces either no, two, or four images, depending on the parameters and the relative source position. The next more complicated model is that of two point masses (Schneider and Weiss 1986, Erdl 1989), where three or five images are obtained. Although more complicated models could be considered in detail, the number of parameters soon becomes very large, and one better considers a model with a statistical distribution of microlenses. This then leads to the lens model of a random star field, which will be described in somewhat more detail below.

How to detect microlensing. Since ML is observable only due to its effects on the apparent brightness of sources, the methods to detect ML are based on a comparison of different flux measurements of one source. If these measurements are separated in time, so that the light curve of a source is measured, one can detect variations of the flux caused by ML. However, it is very difficult to separate intrinsic variability of a source from ML effects; the only hope for this method is to observe features in the light curve which are characteristic for ML events (e.g. Wambsganss, this volume). If the flux measurements are separated in space, so that the source is observed from two different locations, the relative positions of source, lens, and observer are slightly different for the two measurements. ML could then be identified if the flux measured at both telescopes is different (parallax method, Grieger *et al.* 1988). The separation of the two telescopes must be of the order of ξ_0, except for 2237+0305 where the lens is very close to us; there, a baseline of 10 AU would be sufficient. If the flux measurements are separated in angle, i.e. if a galaxy-mass lens produces multiple images of a source, and the light curves of the images can be obtained, intrinsic variability should show up in all images, separated by their respective time-delays. Thus, if variations are observed in one image which are not matched by variations in the other images within a time corresponding to the time-delays, one has a clear signature of ML. As reported by Corrigan (this volume), this method has lead to the probable discovery of ML in 2237+0305.

2. Effects and applications of microlensing

ML of multiply imaged sources. The original motivation of Chang and Refsdal (1979) to consider ML was the possibility that the magnification of the images of a multiply imaged quasar cannot be obtained by the lens model of the galaxy responsible for the image splitting, but that stars close to the light bundles affect the flux of the images. Therefore, the optical flux ratios of the images is not a 'good' constraint on models for the deflector (in contrast, flux ratios obtained from VLBI measurements are not affected by ML, and should be used for modelling). Furthermore, ML can induce variability of images which is not easily separated from intrinsic variability of the source; thus, ML makes it more difficult to obtain the time-delay from observations.

Since ML is very sensitive to the size of the source, source components of different sizes are affected differently. In particular, the continuum emitting region in quasars is believed to be smaller than ξ_0, whereas the broad line region (BLR) is larger. Nevertheless, ML should have an effect on the broad emission lines, in particular if the BLR has substantial substructure. A preliminary investigation (Nemiroff 1988) shows the influence of a single point mass lens on the broad lines. A more detailed study, using a random star field, will be done; it is possible that the comparison of the line profiles in multiply imaged quasars can rule out some specific models for the BLR, e.g. systematic rotation.

ML induced variability. If the light curve of a source (or one of its images) has been monitored and identified as being due to ML, it can be compared with theoretical 'light curves' (see Wambsganss, this volume, and references therein). Due to the statistical nature of the simulations, no exact match of the observed light curve can be expected, but some 'general features' can be brought to agreement between observations and simulations, such as the characteristic rise- or decline time-scale, the frequency of rapid variability, the peak magnification etc. These features then allow to constrain lensing parameters like typical mass of the microlenses, relative transverse velocity of source and lens, size of the source, and density of the star field. Light curves obtained from two spatially separated telescopes would be of great importance for this analysis (Grieger *et al.* 1988).

This last reference also pointed out a method to obtain information about the brightness structure of sources which undergo high magnification events. Theoretically speaking, these occur when a source crosses a caustic. If one assumes that this caustic is isolated (which is justified if the optical depth of the star field is sufficiently small), the universal behaviour of the lens mapping close to caustics can then be used to extract from the observed light curve the one-dimensional brightness profile of the source, integrated along the direction parallel to the caustic. Although the observational requirements for applying this method are considerable, one should point out that this method provides a resolution several orders of magnitude higher than existing telescopes. Thus, if a ML candidate source is identified, the effort to obtain a high-quality light curve may well be rewarded.

Except for quasars with resolvable multiple images, no variability due to ML has been identified yet, since separating it from intrinsic variability is extremely difficult, mainly because our understanding of the physics of AGNs is not sufficient to rule out certain intrinsic variability behaviour. For some extremely variable sources, a ML origin for their variability has been suggested in the literature, e.g. 3C345 (Grieger *et al.* 1988), 1156+295 (Schneider and Weiss 1987), AO0235+164 (Stickel *et al.* 1988a, Kayser 1989) and 0846+51W1 (Nottale 1986). Ostriker and Vietri (1985) have suggested that (at least some) BL Lac objects can be understood as gravitationally microlensed quasars, where the optical continuum is highly magnified, but not (due to its larger size) the BLR. Thus, the continuum would outshine the broad lines. One would then expect that in those sources, the surrounding galaxy of the BL Lac is indeed the lensing galaxy, and not the host galaxy. In fact, for the last two objects listed above, which are BL Lac objects, and an additional one, a foreground galaxy has been identified recently (Stickel *et al.* 1988a,b, 1989). It should of course be mentioned that this hypothesis has its difficulties, since BL Lac differ from quasars in many respects. If ML could indeed explain the most violent variability, some constraints for AGN models would be considerably changed.

Lensing of supernovae. The problem to identify magnified quasars is due to the fact that their intrinsic luminosity is unknown. But suppose we could identify a type of sources for which the spread in intrinsic luminosity is small. Then, if we measure the apparent brightness and distance (redshift) of such a source, and if the luminosity we obtain by using the standard luminosity-distance relation is well outside the range for this type of source, one might conclude this source to be brightened by lensing. The best candidate for this purpose are supernovae (SNs), since the spread in their luminosity is much smaller than that for quasars, their size is sufficiently small to be affected by ML, and they are sufficiently luminous to be observable at high redshift. In fact, the recent discovery of a $z \sim 0.31$ SN (Norgaard-Nielsen *et al.* 1989) is encouraging for surveys of high-z SN. The statistical expectations to find lensed SNs were considered by Linder *et al.* (1988); we found that roughly 1% of all SNs discovered up to a limiting magnitude of $R = 22$ will be magnified by more than a factor of 10, if the universe contains about 20% of its closure density in form of compact objects with mass $M \gtrsim 10^{-2} M_\odot$. Determining the fraction of lensed SN would thus allow to measure the amount of cosmic matter contained in compact objects — perhaps the only known method to investigate this question.

3. Amplification bias

Allowing for the possibility that sources can be magnified, it is not possible to relate the observed flux S uniquely to the luminosity L; if a source is magnified by a factor μ, one has $L = SD_L^2/\mu$, where D_L is the luminosity distance to the source. Since μ is unknown for an individual source, its luminosity can not be determined. The *amplification bias* is the effect that the number of sources in a flux-limited sample does not necessarily yield direct information about the luminosity function of sources: let $\Phi(L)$ be the density of sources of a given kind at a fixed redshift which have a luminosity in excess of L. If these sources are not magnified, a flux limited sample with threshold S would contain $n_0(S) \propto \Phi(SD_L^2)$ objects. Allowing for magnification with a probability density $p(\mu)$, the source counts are $n_{obs}(S) \propto \int d\mu\, p(\mu)\, \Phi(SD_L^2/\mu)$. Taking the factors of proportionality into account, one obtains

$$n_{obs}(S) = \frac{1}{\langle\mu\rangle} \int d\mu\, p(\mu)\, n_0(S/\mu), \tag{1}$$

where $\langle\mu\rangle$ is the mean solid angle distortion in the sample region. Thus, the amplification bias changes the observed source counts relative to the case that no magnification occurs. If the observed source counts are interpreted as $n_0(S)$, the luminosity function derived from it will differ from the true luminosity function, unless $p(\mu) = \delta(\mu - \langle\mu\rangle)$.

Quasar luminosity function. The quasar space density $\Phi(L, z)$ as a function of luminosity and redshift is one of the few landmarks of the universe at high redshift; in addition, the dependence of Φ on redshift constrains the luminosity evolution of individual quasars and thus corresponding models on fueling the central engine. If the source counts are not affected by amplification bias, $\Phi(L, z)$ can (at least in principle) be obtained directly from observations. On the other hand, if $n_0(S,z) \neq n_{obs}(S,z)$, determining the luminosity function Φ requires a knowledge of the probability distribution $p(\mu)$. Although the suggestion that all quasars are 'just' gravitationally lensed nuclei of Seyfert galaxies (Barnothy and Barnothy 1968) can be disregarded, the possibility that the apparent redshift evolution of Φ is in part due to amplification bias (Turner 1980) has found considerable attention in the literature (e.g. Canizares 1982, Peacock 1982, Vietri 1985, Ostriker and Vietri 1986, Schneider 1987a,b). The results of these studies can be summarized as follows:

If the luminosity function is flatter than L^{-2}, amplification bias is unimportant. Since for faint sources, the observed source counts are flatter than S^{-2}, one can conclude that they are basically identical to $n_0(S)$. The high-S end of the source counts *can* be affected by lensing, depending on the (unknown) slope of $\Phi(L,z)$ and on $p(\mu)$. Considering point sources only, one finds $p(\mu) \propto \mu^{-3}$ for large μ. If the high-S part of the source counts are strongly affected by the amplification bias, the luminosity function must be steeper than L^{-2}, and the source counts must have a slope close to S^{-2}. However, the continuum emitting region of quasars has a size which is comparable to ξ_0, and therefore the point source approximation is not justified. Considering sources of finite size, the probability density $p(\mu)$ depends on the mass spectrum of the microlenses. It is thus possible to obtain any slope of $n_{obs}(S) \propto S^{-\alpha}$ for large S with $\alpha \geq 2$ by choosing the appropriate mass function, provided $\Phi(L)$ is steeper than $L^{-\alpha}$ for large L. In particular, for a Salpeter mass function, one obtains $\alpha = 2.6$, close to the observed slope.

Quasar-galaxy associations. Gravitational lensing cannot account for the apparent overdensity (e.g., Arp 1987) of quasars around low-redshift galaxies (Canizares 1981, Vietri and Ostriker 1983, Schneider 1986, 1987c,d, Linder and Schneider 1989), due to the fact that for faint quasars, the source counts are too flat to allow the amplification bias to operate, and the sky density of bright quasars is too low for a statistically significant overdensity to be discovered, unless the surroundings of $\sim 10^5$ galaxies are searched for quasars. The recent discoveries of an overdensity of galaxies around high-redshift quasars (Tyson 1986, Fugmann 1988, 1989, Webster *et al.* 1988) can probably be explained by the lens action of matter associated with the excess galaxies on the quasars, provided the galaxies are indeed in the foreground - which in fact is the case for those galaxies for which a redshift has been measured (R. Webster, this volume, and private communication). I will concentrate here on the Webster *et al.* sample, since it has well-defined selection criteria. In a sample of 296 quasars with redshift ≥ 0.5, they have found 11 galaxies within 6" of the quasar, whereas from the density of galaxies on their plates, a random superposition would yield 2.6 associations. The flux threshold for the quasar sample is $B_0 = 18.7$. The quasar source counts can be parameterized as

$$\frac{dn}{dS} = \text{const.} \begin{cases} (S/S_b)^{-1.8} & \text{for } S < S_b \\ (S/S_b)^{-3.6} & \text{for } S > S_b \end{cases}, \tag{2}$$

where S_b is the flux at the break, corresponding to about $B_b = 19.2$. Using the assumption that the overall source counts are not seriously affected by lensing, one can identify (2) with $n_0(S)$ and use it in (1). Schneider (1989a) computed a probability density by assuming that the matter in the foreground galaxies consists of microlenses, whereas Narayan (1989) considered a smooth mass distribution in the lens; both considerations roughly yield the same overdensity. Kovner (1989) has shown that one can obtain an upper limit to the fractional overdensity, which no longer depends on the specific lens model, but only on $n_0(S)$ and the mean magnification $\langle\mu\rangle$. For an isothermal sphere with velocity dispersion $300\sigma_{300}$ km/s, one obtains $\langle\mu\rangle = (1-\alpha_0^2/\theta^2)^{-2}$, where $\alpha_0 \approx 2.6\sigma_{300}^2(D_{ds}/D_s)$ arcseconds, and $\theta = 6$".

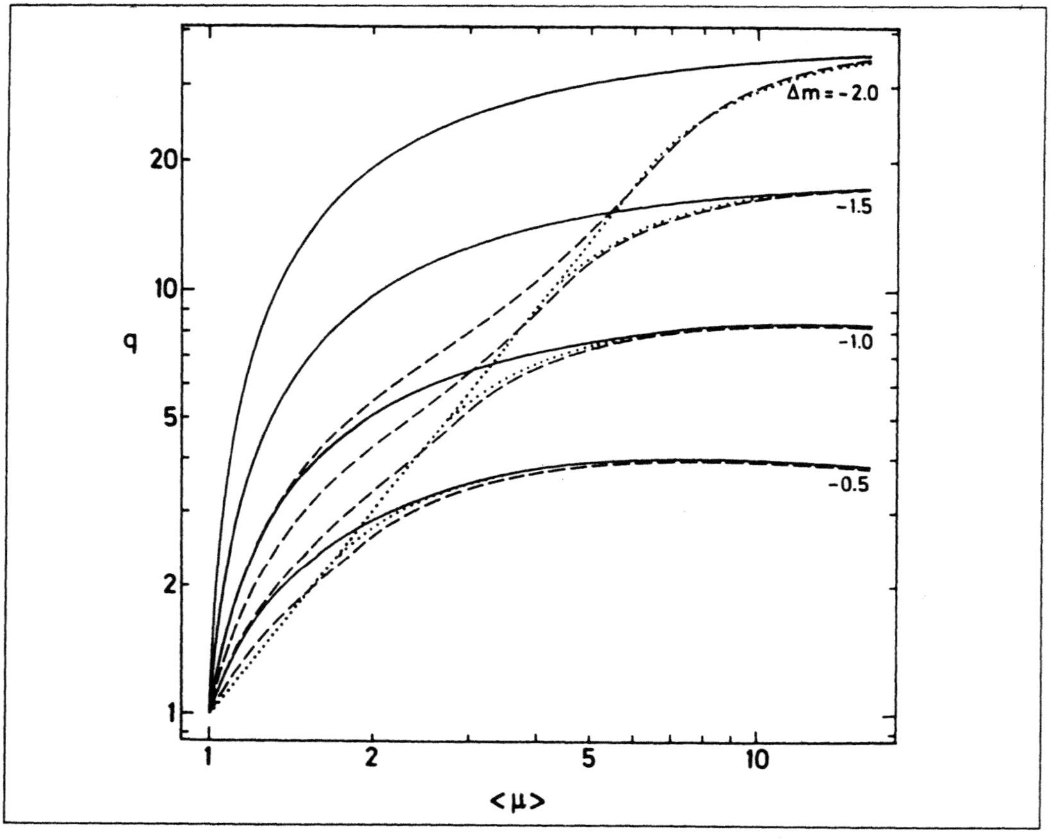

Fig. 1. The fractional overdensity q of galaxies around quasars, as a function of the mean magnification $\langle\mu\rangle$ within the solid angle which has been investigated for excess galaxies; as explained in the text, $\langle\mu\rangle$ depends on the maximum angular separation $\theta = 6$" and the velocity dispersion σ of the galaxies. Four families of curves have been drawn, for different values of the difference $\Delta m = B_0 - B_b$ between the flux threshold B_0 of the sample and the break B_b of the source counts (2). The dotted curves result from assuming the matter in the galaxies to be smoothly distributed (Narayan 1989), whereas the dashed curves represents a simplified version of the microlensing model of Schneider (1989a). The solid curves is the upper limit of the fractional overdensity as obtained by Kovner (1989). For small $|\Delta m|$, all three curves agree very well, i.e. the constraint of flux conservation near the break basically determines q. For larger $|\Delta m|$, the upper limit is considerably larger than the two specific model curves, in particular for reasonable values of $\langle\mu\rangle$. For large values of $\langle\mu\rangle$, all three curves agree. Note that for $\Delta m = -0.5$, $q \lesssim 3$, independent of $\langle\mu\rangle$.

The results are summarized in Fig.1, from which one sees that the overdensity of a factor 4.4 claimed by Webster *et al.* cannot be obtained by lensing under the listed assumptions. However, the galaxy excess derived from lensing is sufficiently close to the observed one, so that by relaxing the assumptions, one would bring observations and theory into agreement. Possibilities are: (i) the overdensity is smaller than 4.4, for which there are some indications (Webster, this volume); (ii) the difference $\Delta m = B_0 - B_b$ is smaller than -0.5; (iii) $n_0(S)$ cannot be identified with (2), but the overall source counts are affected by the amplification bias, so that the intrinsic luminosity function is steeper than suggested by (2), and, therefore, the overdensity becomes larger. One should bear in mind, however, that the galaxy excess can also be similar to that seen for low-redshift quasars (where the galaxies are at the same redshift as the quasar and just form physical associations), until the redshift of the galaxies has been determined.

Hints for amplification bias. Besides the just mentioned overdensity of galaxies around high-redshift quasars, there are three additional hints from observations that the amplification bias is in fact affecting the source counts of quasars (or some subgroup of AGNs):

Stocke *et al.* (1987) have found that in a flux-limited sample of about 200 X-ray selected AGNs, a subsample of 10 have a foreground galaxy nearby. The objects in this subsample have a significantly higher mean redshift than the rest of the sample. They have interpreted this difference as being due to (micro)lensing of the AGNs by matter associated with the foreground galaxies, an interpretation shared by Rix and Hogan (1988, 1989). However, this result should be considered as preliminary, since the statistical significance vanishes if two out of the ten objects are removed from the subsample.

The surprising result that three high-redshift BL Lac objects have a superposed galaxy on top of them (Stickel *et al.* 1988a,b, 1989) has been interpreted as being due to ML; as discussed in Sect. 2, the microlenses in the galaxies may also be (partly) responsible for the variability of these objects (Ostriker and Vietri 1985).

As discussed by J. Surdej (this volume), apparently luminous quasars have a larger chance to be multiply imaged than apparently less luminous ones. This observational result is predicted qualitatively by theoretical statistical lensing studies (e.g. Turner *et al.* 1984) and is due to the fact that image splitting is related to a magnification of the source. Recently, Bartelmann (1989) has undertaken a study in which he reconsidered the multiple image statistics by taking into account the possibility that the individual images can be affected by ML. Preliminary results include: the fraction of multiply imaged quasars in a flux limited sample decreases with increasing fraction of matter contained in microlenses (for given velocity dispersion of the galaxies), and decreases with increasing radius of the galaxies. This latter result is easily understood, since the larger the galaxies, the more ML are outside that region of galaxies where multiple images can occur; they can thus magnify sources which are not multiply imaged. In addition, the mean flux ratio of the multiple images increases with the fraction of mass in compact objects.

4. The random star field

As mentioned in Sect.1, the 'natural' model for ML is the random star field, i.e. randomly distributed point masses. This model is realistic insofar as one typically deals with a two-scale problem: the natural scale associated with a point mass is ξ_0, which is many orders of magnitude smaller than the size of a galaxy. Therefore, a gradient in the matter distribution of a galaxy does not show up on scales relevant for ML. The model neglects, however, correlations in the distribution of stars. A random star field is characterized by its optical depth κ_* (i.e. the fraction of its area covered by the circles around each star with radius ξ_0 - the so-called Einstein circles), the dimensionless density κ_c of smoothly distributed matter, and the external shear γ_g produced by the galaxy as a whole. For many purposes, the size of the star field can be taken to be infinite; otherwise, the radius of the field (or, equivalently, the total number N_* of stars) has to be specified. In addition, a mass spectrum of the stars is sometimes considered.

Deflection probability. Using a Markov method, Katz *et al.* (1986) have derived the probability distribution for a light ray traversing a random star field to be deflected by an angle α. The natural scale for the deflection is $\alpha_* = 4GM/(c^2\langle d\rangle)$, where $\langle d\rangle$ is the mean separation of stars. For $\alpha \ll \alpha_*$, the probability distribution is a Gaussian, as expected from the central limit theorem, whereas for $\alpha \gg \alpha_*$, the deflection probability declines as α^{-4}. This latter result can be obtained by adding up the probabilities for large deflection at individual stars and is thus not due to collective action of all stars in the field, as for $\alpha \ll \alpha_*$. The ("thermodynamic") limit $N_* \to \infty$ does not exist; it exhibits an 'infrared' divergence due to the long range of gravity. Alternatively, one can consider the probability that two rays with separation $\Delta\theta$ have a difference in the deflection angle of $\Delta\alpha$. For this correlated deflection probability, the limit $N_* \to \infty$ does exist. The main motivation for considering correlated deflections is that the effect of very distant stars does not show up in the difference of the deflections.

Shear and magnification. A small light bundle traversing a star field is tidally distorted by the shear from the galaxy as a whole, and the random component of the shear caused by the point masses. Nityananda and Ostriker (1984) have calculated the probability density for the random shear γ_*, $p(\gamma_*, \phi) = (2\pi)^{-1}\kappa_*(\kappa_*^2 + \gamma_*^2)^{-3/2}$; here, ϕ is the (random) direction of the shear caused by the stars. Note that the limit $N_* \to \infty$ exists and has been taken to obtain the above result. The total shear is then $\gamma = \sqrt{(\gamma_* \cos\phi + \gamma_g)^2 + (\gamma_* \sin\phi)^2}$, and the magnification along this light bundle is $\mu = \left[(1-\kappa_c)^2 - \gamma^2\right]^{-1}$. One can thus obtain the probability that a light bundle is magnified by μ, $p(\mu) = 2\pi \int d\gamma_*\, \gamma_* \int d\phi\, p(\gamma_*, \phi)\, \delta(\mu - \mu(\gamma_*, \phi))$. The probability for a point source behind a random star field to be magnified by more than μ_s behaves asymptotically as $\tau(\kappa_*, \kappa_c, \gamma_g)/\mu_s^2$; the function $\tau(\kappa_*, \kappa_c, \gamma_g)$ has been derived in Schneider (1987e). All these results do not depend on the mass spectrum of the lenses!

Position dependent magnification. In order to make predictions about the magnification of sources behind a random star field as a function of time, which can then be compared with observed light curves, one considers the source to change its relative position in time and computes the magnification as a function of relative position. There are two fundamentally different methods to do this: Paczynski (1986) has solved the lens equation for a random star field with N_* stars. This is a highly non-trivial task, as for each source position, the lens equation has about N_* solutions. The advantage of this method is that it can (and in fact, has to) deal with point sources, that information about all the images is available, and that a velocity dispersion of the stars can be taken into account. The disadvantage mainly is that it is complicated and that computing time is very large, so that $N_* \lesssim 10^3$.

The second method is ray shooting (or inverse ray tracing), as described by Wambsganss (this volume). The advantage of this method is that it can (and has to) be applied to extended sources, that a two-dimensional magnification pattern is obtained simultaneously, and that this method is basically simple and foolproof and can be applied to very general lensing situations. Disadvantages are the finite resolution, and the need for a static star field (which is not too bad if the systematic velocity of the star field as a whole is much larger than the internal velocity dispersion). The number of stars that have to be taken into account depends sensitively on the optical depth of the star field, and the maximum values for N_* this method can handle depends on the details of its implementation: $N_* \lesssim 200$ for a brute force method (Kayser *et al.* 1986), $N_* \lesssim 10^4$ for including series expansion of the deflection angle (Schneider and Weiss 1987), and $N_* \lesssim 10^6$ by including a hierachical tree code (Wambsganss *et al.* 1989, 1990). The only analytic result that has been obtained for a random star field with extended sources is the magnification fluctuation $\langle\mu^2\rangle/\langle\mu\rangle^2$ (Deguchi and Watson 1987b); this fluctuation increases with decreasing source size, but has a minimum for optical depth very close to one: at this point, the caustic network is so dense that a small source is sufficient to average out all magnification structure. This is in agreement with results of the simulations mentioned above.

5. The clumpy universe

Although this topic does not directly belong to ML, I thought that a few remarks should be included in these proceedings, since an inhomogeneous universe lies at the heart of gravitational lensing. Unfortunately, no exact solution of Einstein's equation in known which can describe an inhomogeneous universe; the Swiss-Cheese model, which is an exact solution, describes an unrealistic matter distribution (for further discussions on the Swiss-Cheese model, see Schneider 1989b). Standard perturbation theory does not really help, since it assumes that the density perturbations $\Delta\rho/\rho$ are small; however, the density contrast of galaxies or the solar system is large. So the standard approach, which is always used in lensing and which appears to be rather plausible, is to assume that the overall geometry of the universe is described by the Friedmann-Lemaitre metric, and local perturbations are treated with Newtonian gravity. This approach seems to be justified in gravitational lensing, since the deflection is a very local phenomenon. On the other hand, General Relativity is a nonlinear field theory, and such 'plausible' assumptions can really go wrong.

Recently, however, Futamase (1988) has developed an approximation scheme which can describe situations in which the perturbations of the metric are small, but where the density contrast is arbitrarily large. In particular, Futamase and Sasaki (1989) have shown that the metric of a clumpy universe is well approximated by

$$ds^2 = (1 + 2\phi_N)dt^2 - R^2(t)(1 - 2\phi_N)d\sigma^2, \tag{3}$$

where ϕ_N is the Newtonian potential, calculated from the Poisson equation, where the source term is given by the density contrast $\Delta\rho = \rho - \bar{\rho}$, $R(t)$ is the cosmological scale factor, which obeys the Friedmann equation for the mean density $\bar{\rho}$, and $d\sigma^2$ is the three-dimensional line element of a space of constant curvature. The approximation (3) is obtained by neglecting higher order terms in the small parameters, which are the ratio of the size of coherent matter inhomogeneities and the Hubble length, and the value of the Newtonian potential ϕ_N; e.g., (3) fails if one tries to describe a binary pulsar in the universe. One should still be cautious, since the expansion in small parameters does not automatically lead to a good approximation for the metric, since the corresponding field equations are non-linear. However, the work by Futamase and Sasaki has, for the first time, yielded a justification for the use of (3), which has been implicitly assumed in most papers on lensing.

If in further work it turns out that the metric (3) is indeed a good approximation for a clumpy universe, we are justified to study the light propagation in such a universe by using the multiple lens plane equation, which has been done in several papers. For references and more discussion on these topics, see Schneider (1989b).

References.

Arp, H.: 1987, *Quasars, Redshifts, and Controversies* (Interstellar Media, Berkeley).
Barnothy, J., Barnothy, M.F.: 1968, *Science*, **162**, 348.
Bartelmann, M.: 1989, in: Börner *et al.*1989.
Börner, G., Buchert, T., Schneider, P.: 1989 (editors): *Cosmology and Gravitational Lensing*, MPA-P3, Garching 1989.
Canizares, C.R.: 1981, *Nature*, **291**, 620.
Canizares, C.R.: 1982, *Ap. J.*, **263**, 508.
Chang, K., Refsdal, S.: 1979, *Nature*, **282**, 561.
Chang, K., Refsdal, S.: 1984, *Astron. Ap.*, **132**, 168.
Deguchi, S., Watson, W.D.: 1987, *Phys. Rev. Lett.* **59**, 2814.
Einstein, A.: 1936, *Science*, **84**, 506.
Erdl, H.: 1989, in: Börner *et al.*1989.
Fugmann, W.: 1988, *Astron. Ap.*, **204**, 73.
Fugmann, W.: 1989, *Astron. Ap.*, **222**, 45.
Futamase, T.: 1988, *Phys. Rev. Lett.* **A2**, 2175.
Futamase, T., Sasaki, M.: 1989, *preprint*
Grieger, B., Kayser, R., Refsdal, S.: 1988, *Astron. Ap.*, **194**, 54.
Katz, N., Balbus, S., Paczynski, B.: 1986, *Ap. J.*, **306**, 2.
Kayser, R.: 1989, *Astron. Ap.*, **206**, L8.
Kayser, R., Refsdal, S., Stabell, R.: 1986, *Astron. Ap.*, **166**, 36.
Kovner, I.: 1987, *Ap. J.*, **312**, 22.
Kovner, I.: 1989, *Ap. J.*, **341**, L1.
Liebes, S.: 1964, *Phys. Rev.* **133B**, 835.
Linder, E.V., Schneider, P.: 1988, *Astron. Ap.*, **204**, L8.
Moran, J.M., Hewitt, J.N., Lo, K.Y. 1989 (editors): *Gravitational Lenses* (Springer: Berlin).

Narayan, R.: 1989, *Ap. J.*, **339**, L53.
Nemiroff, R.J.: 1988, *Ap. J.*, **335**, 593.
Nityananda, R, Ostriker, J.P.: 1984, *J. Astrophys. Astr.* **5**, 235.
Norgaard-Nielsen, H.U. *et al.*: 1989, *Nature*, **339**, 523.
Nottale, L.: 1986, *Astron. Ap.*, **157**, 383.
Ostriker, J.P., Vietri, M.: 1985, *Nature*, **318**, 446.
Ostriker, J.P., Vietri, M.: 1986, *Ap. J.*, **300**, 68.
Paczynski, B.: 1986, *Ap. J.*, **301**, 503.
Peacock, J.A.: 1982, *M.N.R.A.S.*, **199**, 987.
Refsdal, S.: 1964, *M.N.R.A.S.*, **128**, 295.
Rix, H.W., Hogan, C.J.: 1988, *Ap. J.*, **332**, 108.
Rix, H.W., Hogan, C.J.: 1989, in: Moran *et al.*, p.206.
Schneider, P.: 1986, *Ap. J. (Letters)*, **300**, L31.
Schneider, P.: 1987a, *Astron. Ap.*, **183**, 189.
Schneider, P.: 1987b, *Ap. J. (Letters)*, **316**, L7.
Schneider, P.: 1987c, *Astron. Ap.*, **179**, 71.
Schneider, P.: 1987d, *Astron. Ap.*, **179**, 80.
Schneider, P.: 1987e, *Ap. J.*, **319**, 9.
Schneider, P.: 1989a, *Astron. Ap.*, **221**, 221.
Schneider, P.: 1989b, in: Börner *et al.*1989
Schneider, P., Weiss, A.: 1986, *Astron. Ap.*, **164**, 237.
Schneider, P., Weiss, A.: 1987, *Astron. Ap.*, **171**, 49.
Stickel, M. Fried, J.W., Kühr, H.: 1988a, *Astron. Ap.*, **198**, L13.
Stickel, M. Fried, J.W., Kühr, H.: 1988b, *Astron. Ap.*, **206**, L30.
Stickel, M. Fried, J.W., Kühr, H.: 1989, *Astr. Ap.* (in press).
Stocke, J.T., Schneider, P., Morris, S.L., Gioia, I.M., Maccacaro, T., Schild, R.E.: 1987, *Ap. J. (Letters)*, **315**, L11.
Turner, E.L.: 1980, *Ap. J.*, **242**, L135.
Turner, E.L., Ostriker, J.P., Gott, J.R.: 1984, *Ap. J.*, **284**, 1.
Tyson, J.A.: 1986, *Astron. J.*, **92**, 691.
Vietri, M.:1985, *Ap. J.*, **293**, 343.
Vietri, M., Ostriker, J.P.: 1983, *Ap. J.*, **276**, 488.
Wambsganss, J., Paczynski, B., Katz, N.: 1989, in: Moran *et al.*, p.209.
Wambsganss, J., Paczynski, B., Katz, N.: 1990, *Ap.J.* (in press).
Webster, R.L., Hewett, P.C., Harding, M.E., Wegner, G.A.: 1988, *Nature*, **336**, 358.

This article was processed by the author using the TEX Macropackage from Springer-Verlag.

Microlensing calculations with a hierarchical tree code: New Results

Joachim Wambsganss
Max-Planck-Institut für Astrophysik
8046 Garching, West-Germany
(bitnet: JKW at DGAIPP1S)

1. Introduction

Microlensing deals with the effects of individual compact objects inside the lensing galaxy on the (macro-) image of a lensed background quasar (for reviews see, e.g., Schneider 1990 or Watson 1989). Due to the enormous amount of computing time, microlensing calculations have been performed only for a comparatively small number of lensing stars and for moderate values of the (normalized) surface mass density $\sigma = \Sigma/\Sigma_{cr}$, (see e.g. Schneider and Weiss 1987 (SW); Kayser, Refsdal and Stabell 1986 (KRS); Paczynski 1986; Young 1981) where the critical density is given by $\Sigma_{cr} = (c^2/4\pi G)(D_S/D_L D_{LS})$ (here c is the velocity of light, G is the gravitational constant and D_L, D_S, D_{LS} are the angular diameter distances observer-lens, observer-source, lens-source, resp.). For QSO 2237+0305 this value is $\Sigma_{cr} \simeq 2.5\text{g cm}^{-2} \simeq 1.2 * 10^4 M_\odot \text{pc}^{-2}$.

With a new code based on the hierarchical tree method (Barnes and Hut 1986) the time consuming part of the computations, the calculation of the deflection angles of each star, can be reduced in an efficient and elegant way (Wambsganss *et al.*1989, 1990), so that a much larger number of stars can be included, allowing, e.g., for a mass spectrum of the lensing stars or surface mass densities closer to the critical one. Here we present new results of microlensing calculations obtained with this hierarchical tree code in four different regimes:

- surface mass densities $\sigma \to 1$
- microlensing at macrocaustics
- effects of a mass spectrum with different lower cutoff masses (for constant σ)
- macroimages for $\sigma \to 1$.

We used the inverse ray shooting method (described, e.g., in SW, KRS) and obtained two-dimensional distributions of the magnification factors for extended sources. The side lengths of these maps are $20\xi_0$, where $\xi_0 = \sqrt{(4GM_\odot/c^2)(D_{LS}D_L/D_S)}$ is the Einstein radius of a $1M_\odot$ star. The resolution is 1024^2 pixels for figs. 1a,b and 500^2 pixels for figs. 1c-f and 2a-c, so that one pixel length corresponds to about 0.02 ξ_0 or 0.04 ξ_0. The lightcurves are calculated along the three marked tracks for sources with Gaussian profiles of halfwidths 1 (thin line), 4 (medium) and 16 (thick) pixellengths. In the case of QSO 2237+0305 the smallest effective source size is of order 10^{-4} pc.

The time unit is $t = \xi_0(D_S/D_{LS})(1+z_L)/v_{600}$, where z_L is the lens redshift and v_{600} is the lens velocity (relative to observer/source) in units of 600 km/sec. For QSO 2237+0305 one time unit is about 9 years.

2. Microlensing for surface mass densities $\sigma \to 1.0$

In figs. 1a,b,c we show the magnification patterns for $\sigma = 0.8, 0.9$ and 0.95. The numbers of individual lenses used are 41820, 184723, 172616, respectively. For the latter case just $1M_\odot$ stars where used, whereas for the first two cases a Salpeter mass function ($f(m)dm = m^{-2.35}dm$) with upper cutoff at 1 $M_\odot$ and lower cutoff at 0.1 $M_\odot$ was used. The lightcurves along the marked tracks are displayed in figures 3a,b and c. The results of raising σ up to 0.95 can be summarized as follows:

- total number of (micro-) caustics increases rapidly, overlapping of caustics gets more and more important
- "clustering tendency" of caustics (already known from lower values of σ, see e.g. SW) increases dramatically
- time scale of fixed variations (e.g. $\Delta m \geq 0.5 mag$) grows
- lightcurves get smoother and flatter (for fixed source size); they do *not* show the U-shaped double peaks, that are well known for *low* surface mass densities
- average and peak amplitudes of variations decrease (for fixed source size)
- the typical slopes for our smallest sources decrease from about 2 mag within 3 normalized units ($\sigma = 0.8$) to about 1 mag within 8 time units ($\sigma = 0.95$)
- individual caustics can not be resolved any more in the lightcurve for our source sizes

3. Microlensing at Macrocaustics

We modelled a realistic macrocaustic of an isothermal sphere with a finite core and put part of the matter into stars. Without stars one gets a smooth decrease in magnification inversely proportional to the square root of the distance d between source and (macro-) caustic: $A \propto d^{-0.5}$ (see figs. 2a,3d). Figs. 2b,c show the "grainy" situation (corresponding lightcurves in figs. 3e,f) for different values of the scaling parameter α_0/β_0, the ratio of the deflection angles of the isothermal sphere and a single star:

$$\alpha_0/\beta_0 \approx 10^6 \sigma_{200}^2 \frac{1}{\sqrt{M/M_\odot}} \sqrt{\frac{r_{LS} r_L}{r_S}},$$

where σ_{200} is the velocity dispersion of the isothermal sphere in units of 200 km/sec and r_{LS}, r_L and r_S are the values of the angular diameter distances D_{LS}, D_L, D_S in units of c/H_0 (H_0 – Hubble constant). Microlensing at a macrocaustic has three main effects:

- the (micro-) lenses produce small "holes" in the magnification pattern, that are very much compressed in the direction perpendicular to the macrocaustic. They can be as deep as a few magnitudes, depending on the mass of the lensing stars and on the source size. This happens on a very small scale and results in dramatic changes in the lightcurve in small time intervals.
- the global effect of microlensing on a macrocaustic is a softening of the magnification peak, compared to a situation with smooth matter distribution. But for finite source sizes this decrease in magnification is moderate, we still get a very high maximum.
- the lightcurves show a very irregular behaviour with very steep slopes, which for our smallest sources reach values of up to 4 mag in a small fraction of one time unit.

4. Constant Surface Mass Density – Different Masses

We used $\sigma = 0.2$ as an example to illustrate the effects due to a mass function with different lower cutoffs on the magnification patterns and on the lightcurves. In one case we used lenses of constant mass $M = 1M_\odot$ (figs. 1d and 3g), in the other cases we used Salpeter mass functions with upper cutoffs at $1M_\odot$ and lower cutoffs at 0.1 and $0.001M_\odot$ (figs. 1e,f and 3h,i), corresponding to average masses of 0.22 and $0.0034M_\odot$, respectively.

The magnification patterns (figs. 1d,e,f) look very similar in terms of colors. One can understand the features as scaled due to the average masses of the lenses with some additional features due to a few higher mass objects. The lightcurves (figs. 3g,h,i), however, look *completely* different (because time does not scale according to the masses)! In the case with $M = 1M_\odot$ there are rare, clearly distinguishable magnification events on a constant background. In the cases with mass spectra one gets more and more small peaks which result in a steady flickering in the case with the lowest cutoff. Therefore it is *very* important to use realistic mass distributions for the microlensing objects in order to get reliable results for special microlens models!

5. Macroimages for $\sigma \to 1$

Macroimages of a circular background source with a radius $r = 2\xi_0$ are shown in figs. 2d,e,f for three different values of the surface mass density: $\sigma = 0.5$, 0.8 and 0.95. They consist of many microimages. All micro-images larger than one pixel are displayed. The angular scale of the total field is expressed in terms of Einstein angles $\beta_0 = \xi_0/D_L$ for a 1 $M_\odot$ star. Typical values for an angle that covers about 95% of the light of a macroimage reach from 8 β_0 (for $\sigma = 0.5$) up to 100 β_0 ($\sigma = 0.95$). For QSO 2237+0305 these numbers correspond to $2 * 10^{-5}$ and $8 * 10^{-4}$ arcsec. Although the angular scale is far below current optical observation techniques, these macroimages may be resolvable in VLBI observations; unfortunately, however, QSO 2237+0305 is radio quiet.

It is interesting to see the effect of increasing surface mass density on the shape of the macroimages. For the smallest value of σ the macroimage looks quite compact and circular, i.e. most of the (bright) microimages are close together. Increasing σ does not only cause many more microimages, but they are no longer concentrated so tightly! The overall impression of the macroimage varies strongly (even for fixed σ, just due to different positions!): linear, arc-like, ring-like, consisting of different-sized and -shaped parts!

References

Barnes, J., Hut, P. : 1986, *Nature*, **324**, 446.

Kayser, R., Refsdal, S., Stabell, R.: 1986, *Astr. Ap.*, **166**, 36 (KRS).

Paczynski, B.: 1986 *Astroph. J.*, **301**, 503.

Schneider, P.: 1990, this volume

Schneider, P., Weiss, A.: 1987, *Astr. Ap.*, **171**, 49 (SW).

Wambsganss, J., Paczynski, B., and Katz, N. in: *Lecture Notes in Physics,* Vol. **330**: "Gravitational Lenses", p. 209 (1989) (Springer-Verlag, Berlin, Eds.: J. M. Moran, J. N. Hewitt, K. Y. Lo).

Wambsganss, J., Paczynski, B., and Katz, N.: 1990, April 1, *Astroph. J.*, (in press).

Watson, W. D. in: *Lecture Notes in Physics,* Vol. **330**: "Gravitational Lenses", p. 195 (1989) (Springer-Verlag, Berlin, Eds.: J. M. Moran, J. N. Hewitt, K. Y. Lo).

Young, P.: 1981, *Astroph. J.*, **244**, 756.

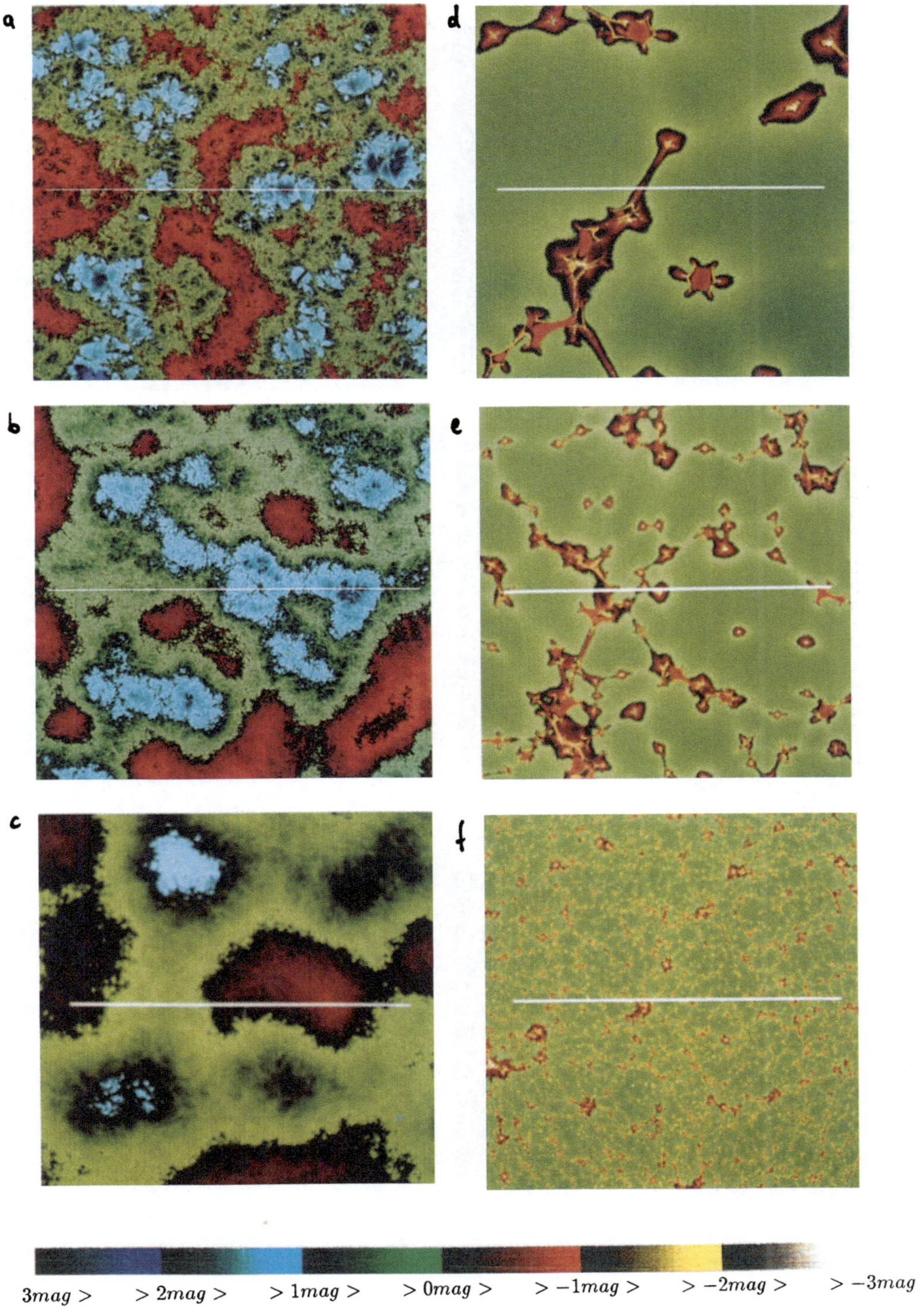

Fig. 1: Left: Magnification patterns for surface mass densities close to one: $\sigma = 0.8$ (a), 0.9 (b), 0.95 (c); right: magnification patterns for $\sigma = 0.2$ with Salpeter function: $M/M_\odot = 1$ (d), $0.1 < M/M_\odot < 1$ (e), $0.001 < M/M_\odot < 1$ (f); each color corresponds to a magnification range of one magnitude, according to the scale at the bottom.

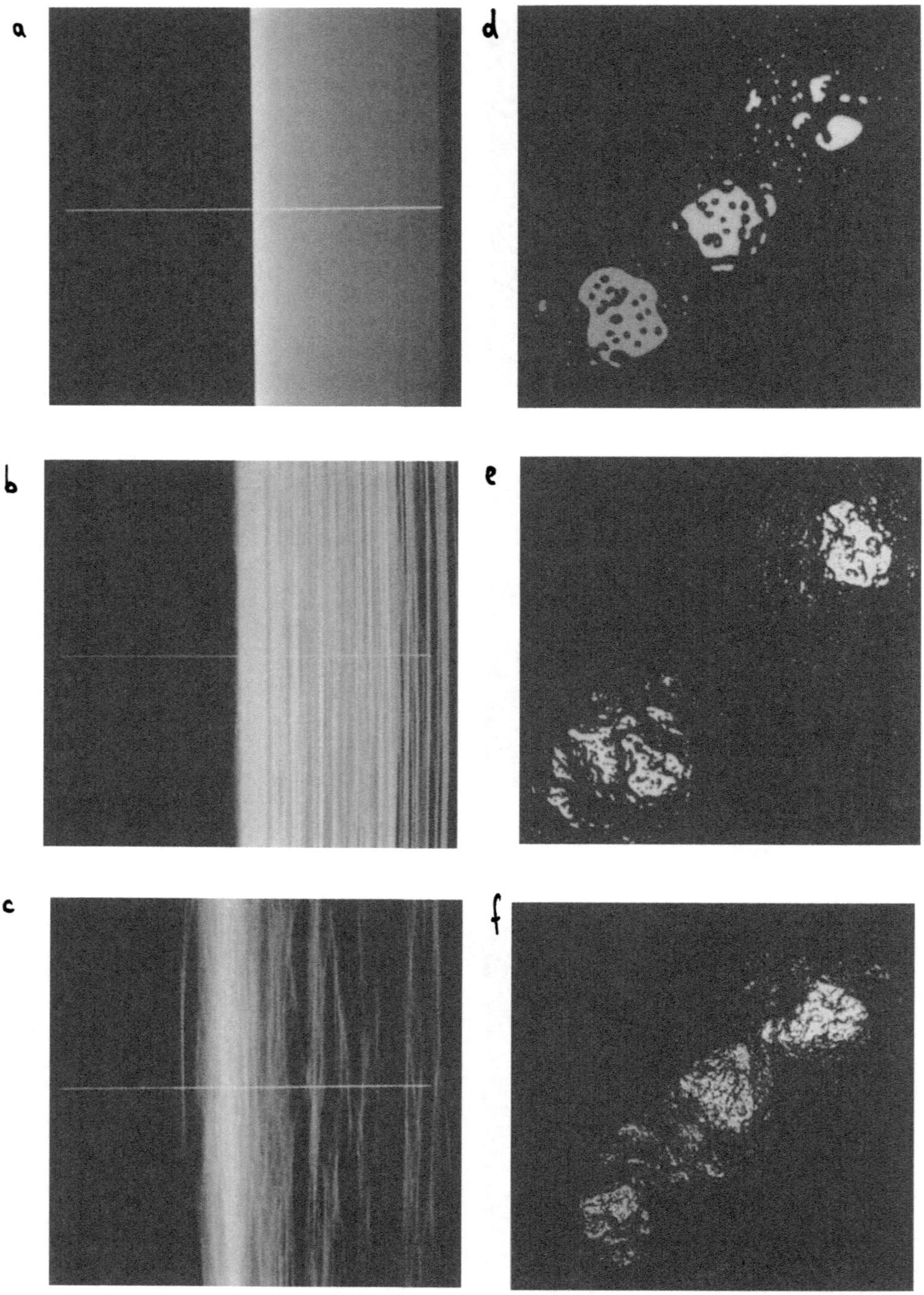

Fig. 2: Left: Magnification patterns for microlensing at macrocaustics (magnification according to a gray scale that covers four magnitudes: black > -3 mag > gray > -7 mag > white): without stars (pure macrocaustic) (a), with $\sigma_* = 0.01$ and $\alpha_0/\beta_0 = 10^6$ (b), with $\sigma_* = 0.1$ and $\alpha_0/\beta_0 = 10^5$ (c);
right: macroimages for $\sigma = 0.5$ (d), 0.8 (e), 0.95 (f) (in each figure the macroimages of three different source positions along a diagonal from lower left to upper right are shown with three different gray scales).

Light Curves:

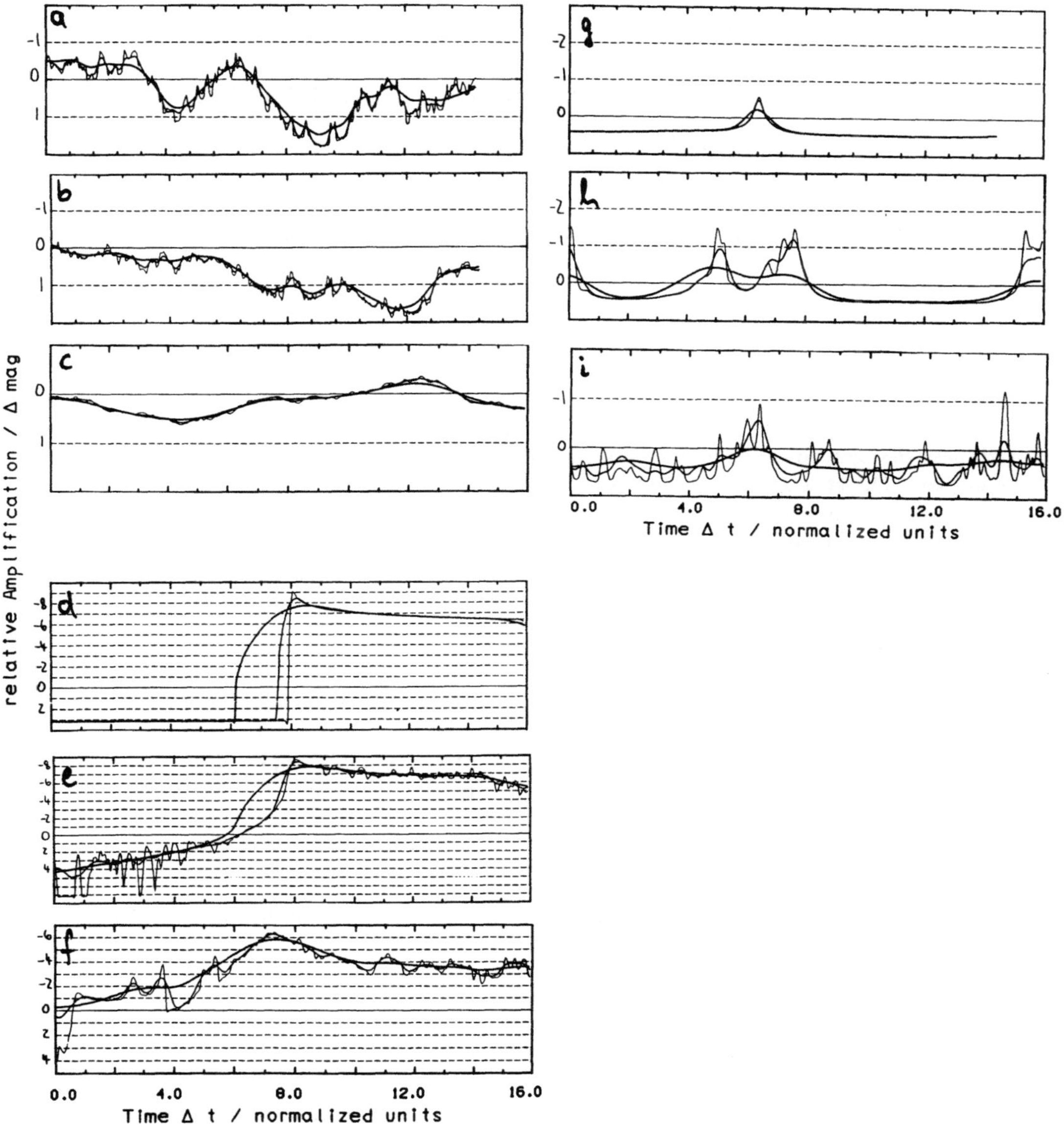

Fig.3: Lightcurves for sources with Gaussian profiles of halfwidths 1, 4, 16 pixels along the horizontal tracks marked in the magnification patterns in figs. 1a-f and 2a-c are shown; dashed horizontal lines indicate (de-) magnifications of full magnitudes relative to the average magnification (solid line):
- for surface mass densities close to one: $\sigma = 0.8$ (a), 0.9 (b), 0.95 (c);
- for microlensing at macrocaustics: no stars (d), $\sigma_* = 0.01, \alpha_0/\beta_0 = 10^6$ (e), $\sigma_* = 0.1, \alpha_0/\beta_0 = 10^5$ (f);
- for $\sigma = 0.2$ and Salpeter mass function: $M/M_\odot = 1$ (g), $0.1 < M/M_\odot < 1$ (h), $0.001 < M/M_\odot < 1$ (i).

Statistical investigations of the amplification near gravitational lens caustics

H.J. Witt
Hamburger Sternwarte, Gojenbergsweg 112, D–2050 Hamburg 80

Abstract

If stars in a lensing galaxy get close to the light path to a quasar due to relative motion, large brightness changes, so called high amplification events (HAEs), may occur in the light curve of the quasar. The strength of the high amplification events, which depends individually on the caustic crossing point of the observer (the source), can be calculated by using the flux factor K. I present statistical distributions of the flux factor for different samples of stars. It is shown that low mass stars have the main influence on the statistical distribution of the flux factor.

1 Introduction

Since Irwin et al. [2] could manifest microlensing effects by observational data of the quadruple quasar system 2237+0305, it is necessary to have more accurate theoretical techniques to investigate the light curves of microlensed quasars. I introduce a new approach to get statistical information about the strength of HAEs in a light curve.

2 Lens equation and the flux factor

I investigate different samples of stars which are as usually described by the normalized lens equation [6,3,4]:

$$\zeta = z + \begin{pmatrix} \gamma & 0 \\ 0 & -\gamma \end{pmatrix} z + \sum_{i=1}^{n} \varepsilon_i \frac{z_i - z}{|z_i - z|^2} \tag{1}$$

where γ is called the shear and the masses of the stars are given in units of the solar mass $\varepsilon_i = M_i/M_\odot$.

If for instance a radially symmetric source crosses a caustic as shown in Fig. 1 and two new (sub)images appear (or disappear), the maximum amplification depends on 3 quantities [5]:

- the radius of the source R_s
- the form factor f which depends on the brightness profile of the source
- the flux factor K which depends individually on the crossing point at the caustic.

The maximum amplification of the two (sub)images is given by (compare the light curve in Fig. 1).

$$A_{\max} = \frac{Kf}{\sqrt{R_s}} \tag{2}$$

The flux factor K was first derived by Chang [1]. We simplified her expression [5]. The result

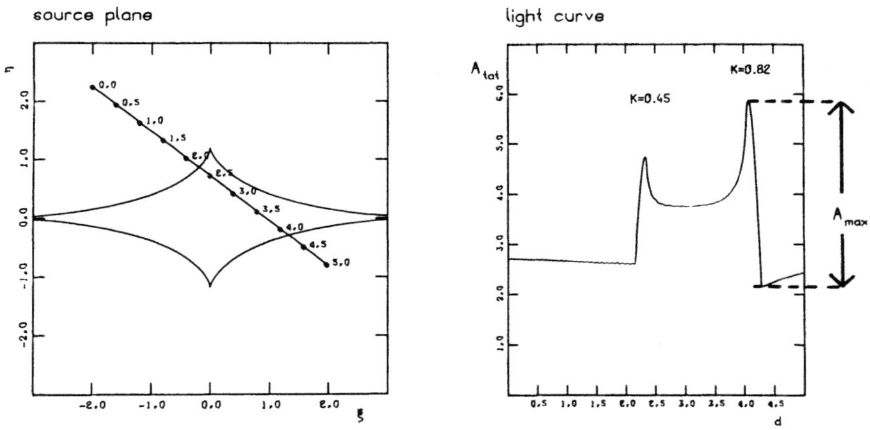

Figure 1: On the left side the caustic of a star with shear $\gamma = 0.8$ is shown. On the right side the light curve of a source with $R_s = 0.1$ and constant surface brightness ($f = 1.39$) moving along the track indicated on the left side is shown.

is given by the following formula:

$$K = \sqrt{\frac{2}{|\boldsymbol{T}_\zeta|}} \tag{3}$$

where

$$\boldsymbol{T}_\zeta = J\boldsymbol{T}_z \qquad \boldsymbol{T}_z = \begin{pmatrix} -\frac{\partial |J|}{\partial y} \\ \frac{\partial |J|}{\partial x} \end{pmatrix} \tag{4}$$

$\boldsymbol{T}_\zeta$ is the tangential vector of the caustic and $\boldsymbol{T}_z$ is the tangential vector of the critical curve. J is the Jacobian of the normalized lens equation and $|J|$ denotes the determinant of the Jacobian. The equation (3) holds if the radius of the source is small compared with

- the radius of curvature of the caustic,
- the distance to other caustics, and
- the distance to cusps.

If the source crosses a cusp, the formula is not valid because the tangential vector $\boldsymbol{T}_\zeta$ vanishes at a cusp.

3 Distributions of the flux factor K

To obtain more information from the light curve of a quasar, it is important to know in which way the flux factor fluctuates for special samples of stars. Therefore I compute the distributions of the flux factor K for samples of stars with the same masses and different densities σ. Furthermore I investigate the influence of low mass stars in the distribution of the flux factor.

I now introduce the probability density of the flux factor, which is given by the following formula:

$$p(K) = \frac{1}{\lambda} \sum_m \frac{d\lambda_m}{dK} \tag{5}$$

where λ is the total length of the caustic of the sample. If the observer crosses a caustic the probability to get a flux factor between K_1 and K_2 is given by

$$P(K_1 \leq K \leq K_2) = \int_{K_1}^{K_2} p(K)dK \tag{6}$$

Furthermore I define the mean K–value and the mean square K–value.

$$\langle K \rangle = \int_0^\infty Kp(K)dK = \frac{1}{\lambda}\int_\lambda K d\lambda \tag{7}$$

$$\langle K^2 \rangle = \int_0^\infty K^2 p(K)dK = \frac{1}{\lambda}\int_\lambda K^2 d\lambda \tag{8}$$

I have evaluated the K–value along the caustics in order to obtain the K–value distribution. Fig. 2 shows the critical curves (at the top) and the caustics (in the middle) of a sample of 400 stars with the same mass of $\varepsilon_i = 1$. The stars are randomly distributed in a circle with radius 1. The normalized coordinates of the stars are obtained by scaling the positions with the factor $\sqrt{\sum_i \varepsilon_i/|\sigma|}$.

If all stars have the same mass I find empirically the following relation between $\langle K \rangle$, $\langle K^2 \rangle$ and the density of stars σ:

$$\langle K \rangle \propto \sigma^{-1/4} \tag{9}$$

$$\langle K^2 \rangle \propto \sigma^{-1/2} \tag{10}$$

In Fig. 3 I use a sample of 10 stars with 10 solar masses. In the next Figures 4 – 6 I add more and more low mass stars with 0.1 solar masses to the sample. The density σ is kept constant. It can be seen from Fig. 4 – 6 that the influence of the low mass stars in the K–distribution is very strong. In Fig. 6 the low mass stars clearly dominate the K–value distribution. This is demonstrated in the last figure. I here let the low mass stars fixed and take the massive stars away and the K–value distribution remains nearly unchanged. It is interesting to note that the low mass stars in Fig. 6 contain only 30% of the total mass of the stars.

For low mass stars the K–value distribution is pretty local (see Fig. 7). Thus we are able to obtain good estimates of the masses of low mass stars in a galaxy out of a few HAEs in a light curve of a microlensed quasar.

References

[1] Chang,K.: (1984) *Astron. Astrophys.* **130**, 157

[2] Irwin,M.J., Webster,R.L., Hewett,P.C., Corrigan,R.T., Jedrzejewski,R.I.: (1989) submitted to *Astrophys. J.*

[3] Kayser,R., Refsdal,S., Stabell,R.: (1986) *Astron. Astrophys.* **166**, 36

[4] Kayser,R., Weiss,A., Refsdal,S., Schneider,P.: (1989) *Astron. Astrophys.* **214**, 4

[5] Kayser,R., Witt,H.J.: (1989) *Astron. Astrophys.* **221**, 1

[6] Paczyński,B.: (1986) *Astrophys. J.* **301**, 503

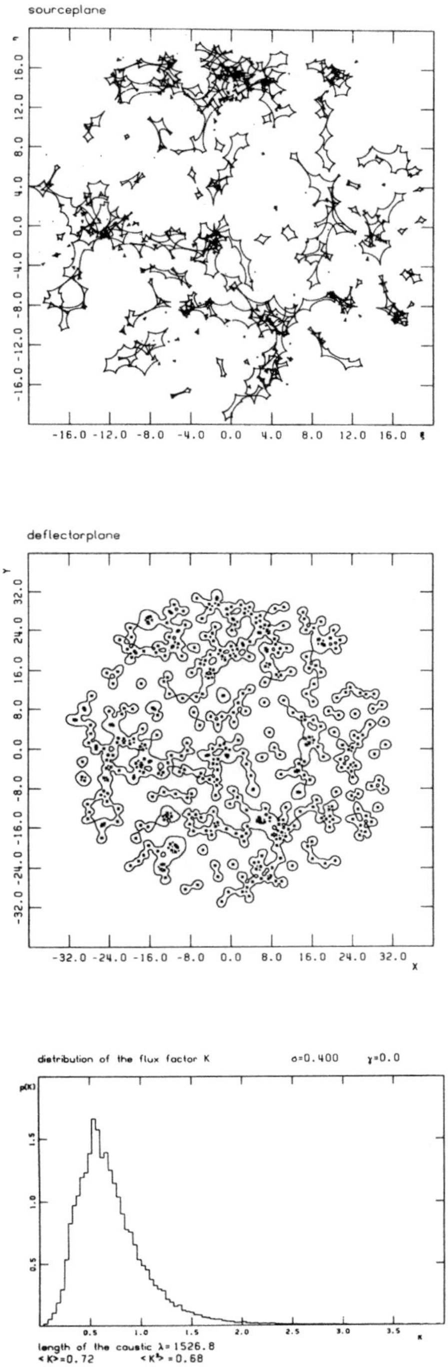

Figure 2: K–value distribution of the caustics. The sample contains 400 stars of 1 solar mass.

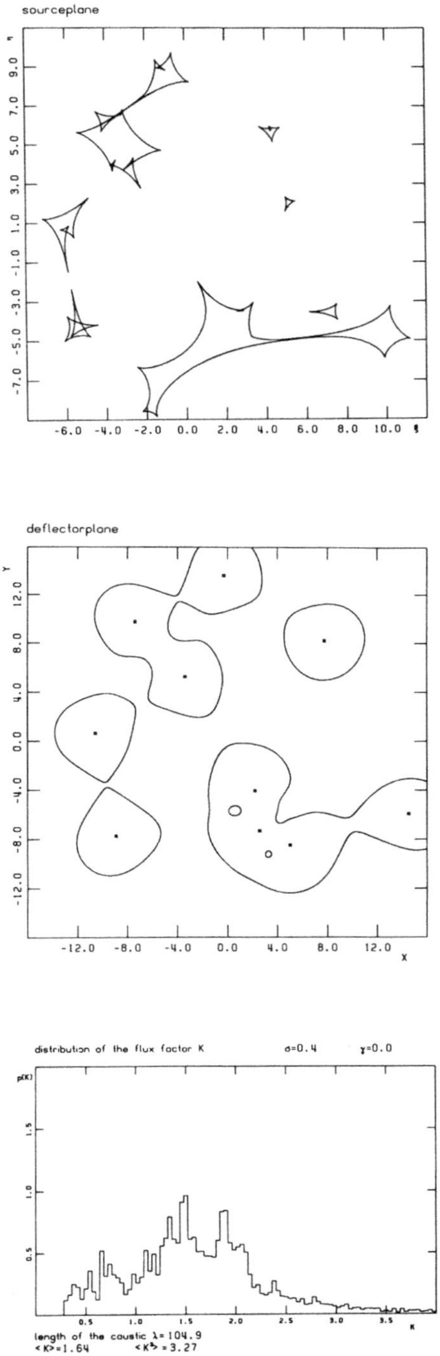

Figure 3: K–value distribution of the caustics. The sample contains 10 stars of 10 solar masses.

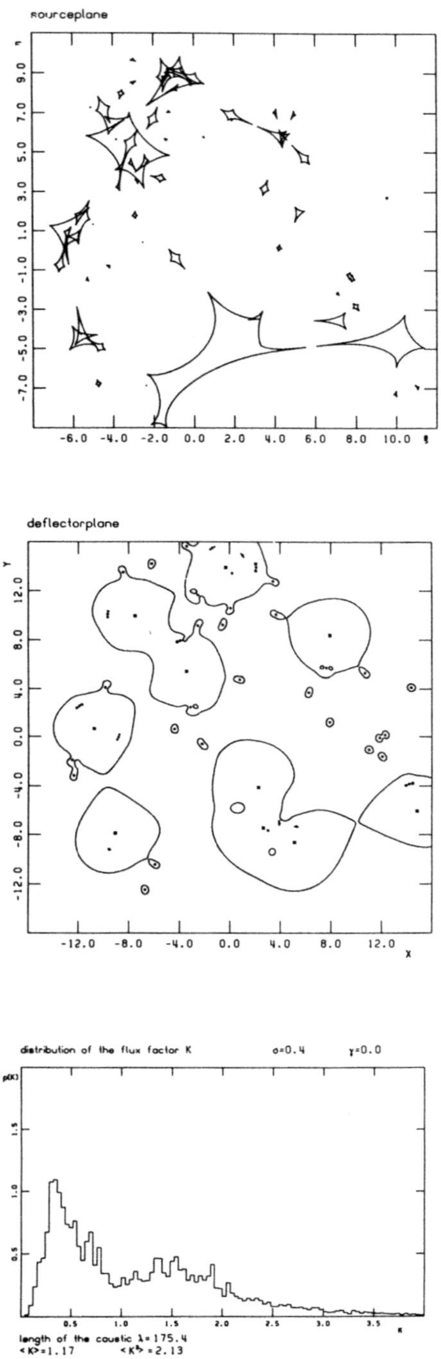

Figure 4: K–value distribution of the caustics. The sample contains 10 stars of 10 solar masses and 40 stars of 0.1 solar masses.

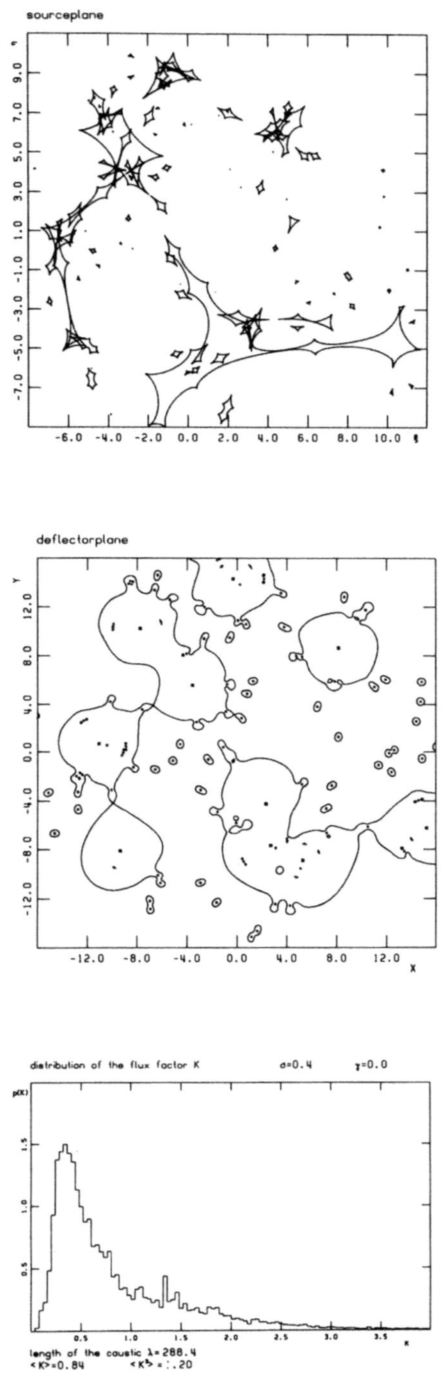

Figure 5: K–value distribution of the caustics. The sample contains 10 stars of 10 solar masses and 100 stars of 0.1 solar masses.

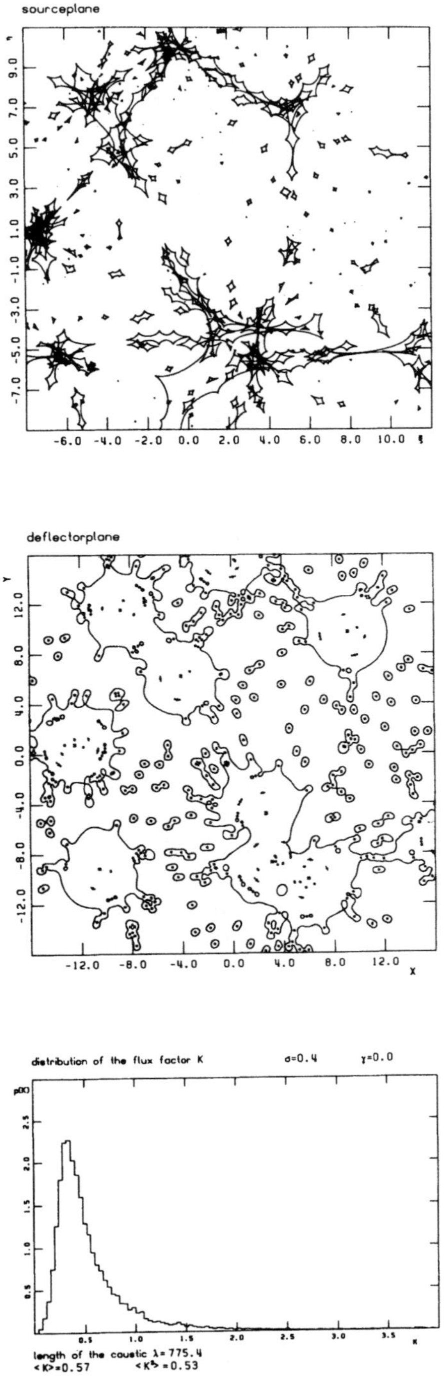

Figure 6: K–value distribution of the caustics. The sample contains 10 stars of 10 solar masses and 400 stars of 0.1 solar masses.

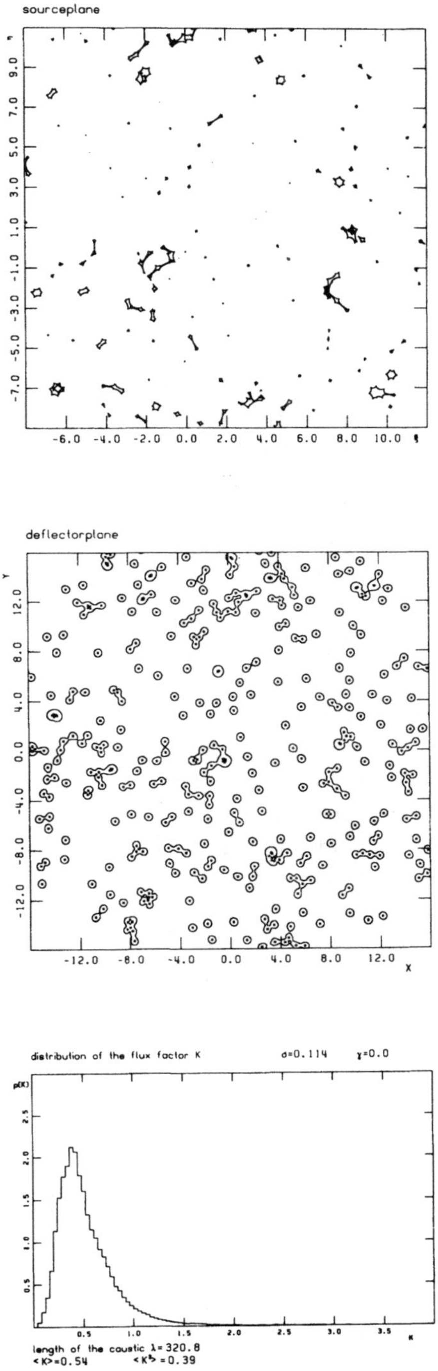

Figure 7: K–value distribution of the caustics. The sample contains 400 stars of 0.1 solar masses.

The deconvolution of the quasar structure from the lightcurve of a high amplification event with the regularisation method

B. Grieger
Hamburger Sternwarte, Gojenbergsweg 112, D-2050 Hamburg 80, FRG

Abstract

Gravitational microlensing—lensing by compact objects like stars in a galaxy in the lightpath to a quasar—can cause large brightness changes, so called high amplification events (HAEs). We present a method to calculate the one-dimensional intrinsic luminosity profile—i. e. the surface luminosity integrated in one direction—of a quasar from the lightcurve of a HAE using the non-classical inversion technique of regularisation. The results of test calculations on simulated lightcurves and first applications on real measurements are shown.

1 Introduction

Amplification due to gravitational microlensing was investigated by Chang and Refsdal (1979, 1984), and various simulations were presented by Kayser, Refsdal, and Stabell (1986), Paczyński (1986), and Schneider and Weiss (1987). It was shown by Grieger, Kayser, and Refsdal (1988), that a one-dimensional projection of the surface brightness distribution of the source can be reconstructed from the light curve of a high amplification event. The idea of this method is illustrated in Fig. 1. The classical, direkt reconstruction can lead to drastic noise amplification, which is a general feature of deconvolution problems. In section 2 we discribe the non-classical approach to deconvolution problems—as it was discussed by e. g. Turchin *et al.* (1971), Craig and Brown (1986)—and present the results of some test calculations. In section 3 we present the results of the deconvolution of a light curve of ON+231 and we briefly discuss the observational situation.

2 Theory

We consider a compact source which is crossed by an anti-caustic orientated parallel to the η-axis. Let $P(\xi)$ be the one-dimensional parallel to the anti-caustic integrated luminosity profile of the source. When the anti-caustic is at $\xi = \xi_c$, the flux from the appearing micro-images is given by

$$F(\xi_c) = \int_{\xi_{\min}}^{\xi_c} \frac{P(\xi)}{\sqrt{\xi_c - \xi}} d\xi \tag{1}$$

If P is represented by a polygonian $(\xi_j, P_j)_{j=1\ldots m}$, we can calculate a $n \times m$-matrix $\mathcal{K}$, so that the points of the lightcurve $(t_i, F_i)_{i=1\ldots n}$ of a HAE are given by

$$F = \mathcal{K}P. \tag{2}$$

The classical approach to deconvolution problems is the solution of the linear equation system (2). In our case the matrix $\mathcal{K}$ is simple enough to obtain an explicit solution, see Grieger *et al.* (1988).

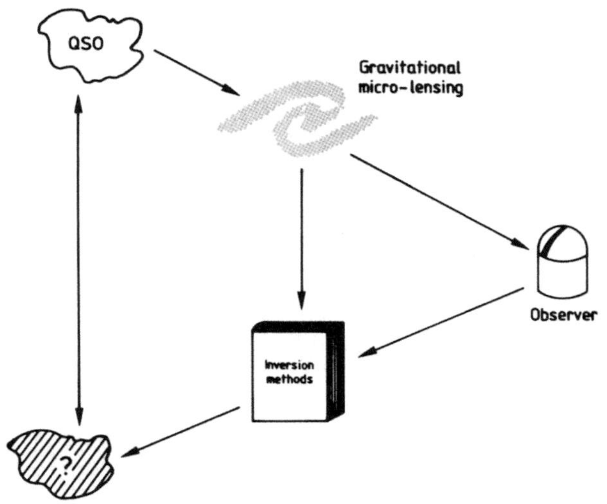

Figure 1: When a compact source like the optical core of a quasar crosses a gravitational lens caustic, the surface brightness of the source is convoluted with the amplification near the caustic to produce an observable brightness variation. Because the amplification is always $\propto d^{-1/2}$, where d is the distance from the caustic, we can use standard deconvolution methods to obtain information on the structure of the source.

Because the convolution smoothes out structures of high frequency, the deconvolution leads to an amplification of such structures. Therefore an error noise of only a few percent can be drastically amplified, so that the signal to noise ratio of the restored source function becomes much less than unity. This is demonstrated in Fig. 2, which shows a comparison between the solutions of the classical and the non-classical inversion method.

To stabilize the solution, we have to take into account the inevitable measuring errors of the data. The deconvolution problem reads completely

$$F = \mathcal{K}P + \delta, \tag{3}$$

where $(\delta_i)_{i=1\dots n}$ are the individual errors of the measured data F_i. Let $(\sigma_i)_{i=1\dots n}$ be the accuracy of the measurements F_i, then every source function P that satisfies

$$(F - \mathcal{K}P)^2 \leq \sigma^2 \tag{4}$$

is compatible with the measured data. To obtain a unique solution, we indroduce a function $\mathcal{L}$, which attaches a "badness" $\mathcal{L}(P)$ to a source function P, and select a solution by the condition

$$(F - \mathcal{K}P)^2 + \lambda\mathcal{L}(P) = \min, \tag{5}$$

where λ is the so-called *smoothing parameter.*

The *regularisation method* uses as $\mathcal{L}$ the norm of the second deriative, i. e.

$$\mathcal{L}(P) = \int_{x_{\min}}^{x_{\max}} (P''(x))^2 dx. \tag{6}$$

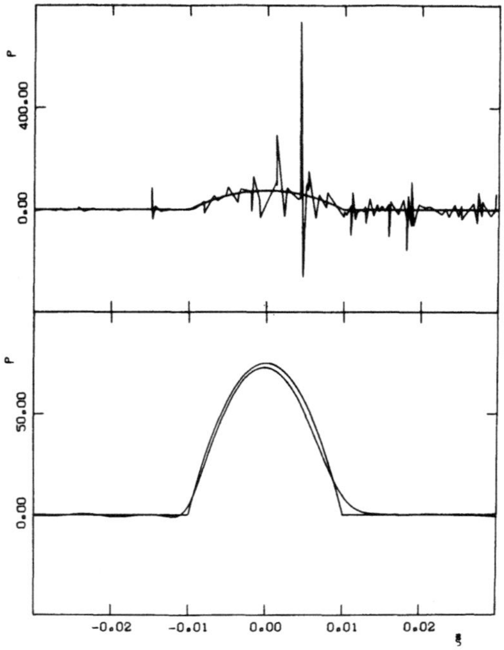

Figure 2: Comparism of a classical and a non-classical deconvolution of the Abelian integral equation out of 120 data points with an error of 5%. At the top the very noisy direkt reconstruction together with the true source function, at the bottom the solution stabilized with the regularisation method, also together with the true source function. Note the different scaling.

Inserting this (in a discrete form) in Eq. (5) leads to the linear equation system

$$(\mathcal{K}^T\mathcal{K} + \lambda\mathcal{H})P = \mathcal{K}^T F, \tag{7}$$

where $\mathcal{H}$ is a $m \times m$-matrix, the *smoothing matrix.* From Eq. (7) we obtain the "smoothest" solution P which is compatible with the measured data.

For test calculations we use a radial symmetric source model with radius $R = 0.01$ normalized units and a a surface brightness $I(r) \propto \sqrt{1 - (r/R)^2}$, which has the advantage of leading to an analytical solution for the light curve. We take the brightness values of this light curve at random selected points and add a normal distributed random error. Some results of the deconvolution of such lightcurves with the described method are shown in Fig. (3).

3 Observations

Observations of the quadruple quasar system 2237+0305 by Irwin *et al.* in August/September 1988 show a significant brightness change of the component A (see the contribution of R. T. Corrigan in this volume). Because the light-travel time differences for the images in the 2237+0305 system are predicted to be about a day or less by reliable theory, the

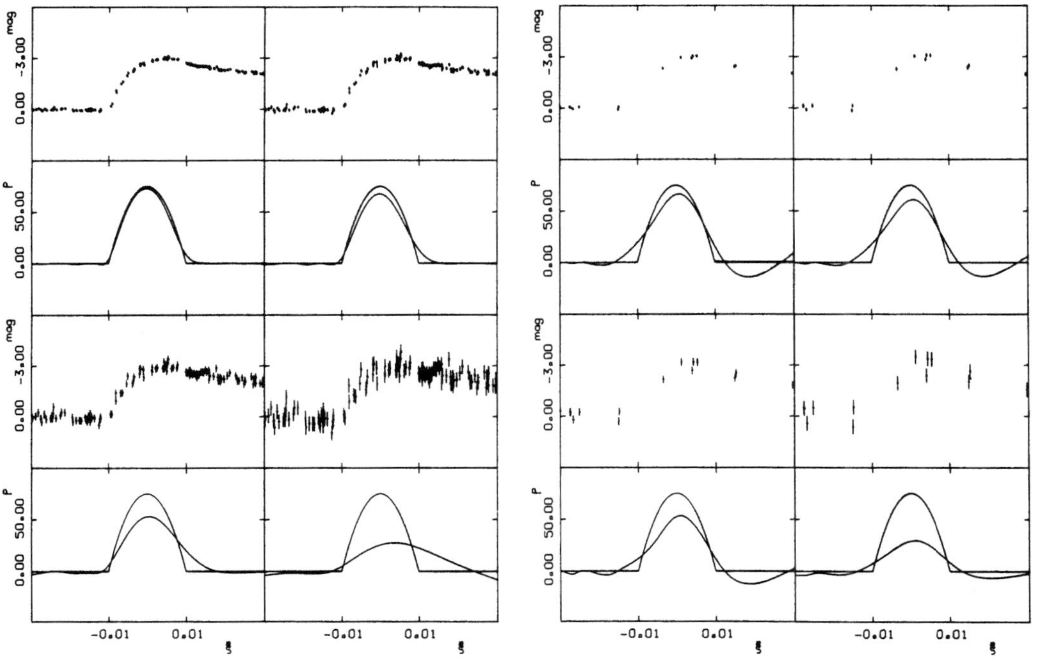

Figure 3: Some results of test calculations on simulated light curves. Below each simulated light curve the corresponding restored source profile is shown, together with the true source function. On the left the light curves consist of 40 data points per source diameter, on the right of 5 data points per source diameter.

observed brightness changes cannot be due to intrinsic variations, and therefore there can be little doubt that they are effects of a gravitational microlensing event.

If the lightcurve of this event with a timescale of about one year had been determined once every week with an accuracy of about 1/20 mag, it would be possible to obtain detailed information on the structure of the quasar with the above described method. To measure such a lightcurve of a high amplification event, frequent monitoring of 2237+0305 should be carried out. This is performed within the quasar monitoring program of the Hamburg gravitational lens group (see the contribution of U. Borgeest in this volume).

In April 1987 Xie *et al.* observed a brightness variation in the Blazar ON+231, which has a timescale of about one hour and looks very much like a HAE. A variation with such a short timescale can only be due to microlensing if the source moves with superluminal velocity. If this is the case, the source of the HAE is a knot thrown out of the Blazar. In the early stage of their life such knots are very compact and emit the most energy in the optical range. However, the deconvolution of the light curve leads to a very plausible source profile, which is shown in Fig. 4.

If a quasar is not multiple-imaged or we do not have any information about the time delay (e. g. the discussed event in ON+231), the distinction between brightness changes due to the micro gravitational lens effect and intrinsic variations could be difficult. Then

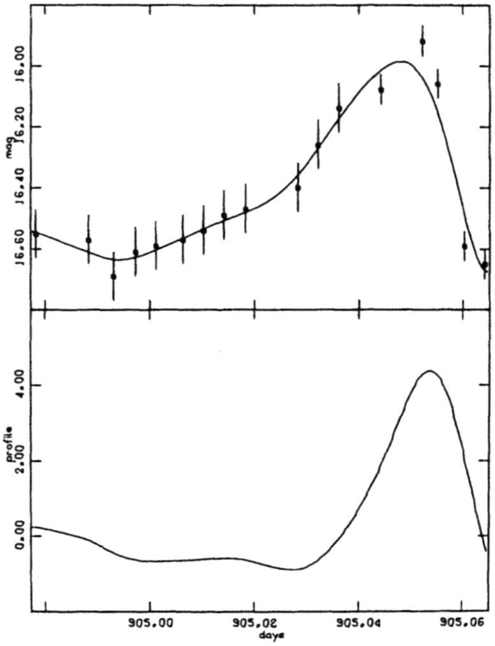

Figure 4: *Top:* The measured light curve of ON+231 together with the light curve that would be produced by the restored source profile. *Bottom:* The source profile obtained with the regularisation method (under the assumption that the brightness variation corresponds to the *disappearance* of two micro images).

it would be very usefull to have a second observer about some A. U. away, for whom amplifications by gravitational micro-lensing occur time shifted, while intrinsic variations are not affected. Besides the unambiguous detection of microlensing, this gives the possibility of the estimation of the absolut size of the source (Grieger, Kayser, and Refsdal (1986,1988)). Therefore we have made an application for quasar observations to be carried out by the interplanetary vehicle "Cassini".

References

[1] Chang, K., Refsdal, S.: 1979, *Nature* **282**, 561
[2] Chang, K., Refsdal, S.: 1984, *Astron. Astrophys.* **132**, 168
[3] Craig, I. J. D., Brown, J. C.: 1986, *Inverse Problems in Astronomy* (Bristol: Adam Hilger Ltd)
[4] Grieger, B., Kayser, R., Refsdal, S.: 1986, *Nature* **324**, 126
[5] Grieger, B., Kayser, R., Refsdal, S.: 1988, *Astron. Astrophys.* **194**, 54
[6] Kayser, R., Refsdal, S., Stabell, R.: 1986, *Astron. Astrophys.* **166**, 36
[7] Paczyński, B.: 1986, *Astrophys. J.* **301**, 503
[8] Turchin, V. F., Kozlov, V. P., Malkevich, M. S.: 1971, *Soviet Physics Uspekhi* **13**, 681
[9] Schneider, P., Weiss, A.: 1987, *Astron. Astrophys.* **171**, 49
[10] Xie, G. *et al.*: 1988, *The Astronomical Journal* **96**, 24

The Hamburg quasar monitoring program at Calar Alto: Analyses of micro-lensing variability

U. Borgeest K.-J. Schramm S. Refsdal
Hamburger Sternwarte, Gojenbergsweg 112, D–2050 Hamburg 80

In May 1989 we started an extensive monitoring program in order to find and analyse micro-lensing events in the light curves of a selected sample of quasars. We use the Max-Planck-Institute 1.23 m telescope at the German-Spanish Astronomical Centre on Calar Alto equipped with a CCD camera. For our program, 15–30% of the observation time at this telescope are available in the next five years. Fluxes are measured through different standard broad-band filters relative to comparision stars in the CCD frame. A 0.01 mag accuracy (in relative photometry) can therefore also be reached for non-photometric conditions. Standard CCD reduction, identification of the objects and photometry are carried out in about 5 min immediately after each observation by using an automatic routine (on a MicroVAX 3200 workstation) developed by us. We want to obtain up to 30 data points for each objects annual observing period with a maximum spacing of two to three weeks. It is therefore possible to recognize a micro-lensing high amplification event (HAE) at its steep wing, assuming that the transverse velocity of the foreground stars relative to the line of sight to the quasar does not exceed 1000 km/s. Since it is supposed that some 'blazars' have radiation components with relativistic transverse velocities, some members of this class will be monitored with correspondingly higher time resolution during shorter periods. When a HAE or another interesting feature will be detected, the fast reduction allows us to follow the light curve with higher accuracy and by using more different filters.

Our sample includes up to now about 50 objects with different priority:

1. *Quasars which are multiple due to macro-lensing.* These are the only objects for which micro-lensing variability can unambiguously be disjoined from the intrinsic one (*e.g.* Corrigan *et al.*, this workshop). If the macro-lens is a galaxy one may expect a relatively high optical depth for micro-lensing. Monitoring of these objects is also very important to obtain the time delays between the components (for 0957+561 cf. Vanderriest *et al.* 1989; Falco, this workshop; Schild, this workshop) from which limits on Hubble's constant (Borgeest and Refsdal 1984; Gorenstein, Falco and Shapiro 1988) and a relatively accurate estimate of the lens mass (Borgeest 1986) can be deduced.

2. *Quasars with observed foreground galaxy.* Since on very deep exposures (Tyson, this workshop) most high redshift quasars will show a foreground galaxy within 5 arcsec ($\simeq$ 20 kpc) to the line of sight, only those objects entered our sample for which one can assume a high optical depth for micro-lensing due to the surface brightness and the redshift of the foreground galaxy.

3. *Narrow absorption line quasars.* It is widely accepted that the narrow heavy element absorption systems are caused by foreground galaxies. In most cases, however, the line of sight passes through outer parts of the galaxy halo, corresponding probably to a low optical depth for micro-lensing. Only for metal-rich, low-ionization systems with lines of relative large equivalent widths one can be sure that the quasar light passes the more central part of a foreground galaxy (Blades 1988).

4. *Blazars.* It has been proposed that the violent variability of these objects is in part due to micro-lensing (Ostriker and Vietri 1985; Schneider and Weiss 1987). Blazar lightcurves seem to be composed of at least two components, a long-term variability with timescales of months to years and a short-term one with timescales of hours to days. For some superluminal sources — which are a sub-class of the blazars — it has been shown that the long-term variability is correlated with the appearance of superluminal knots (*e.g.* for 3C 345 see Babadzhanyants and Belokon 1985). When these knots originate they have their maximum intensity at optical wavelengths and can even dominate the brightness of the source. If they move behind the starfield of a foreground galaxy, the short-term variability can indeed be due to micro-lensing.

Strong micro-lensing variability will only occur if the source is smaller than the 'lensing cross-section' of the stars in the foreground galaxy, typically 10^{-2} pc. This is only two orders of magnitude more than the Schwarzschild radius of a black hole with 10^8 solar masses. Eventually, the compact continuum source in the most luminous quasars is too large for micro-lensing. Webster and Hewett (this workshop) found a number excess of foreground galaxies within 6″ – but most of it outside 3″ – around quasars in a magnitude limited sample. This effect is very probably due to the magnification bias caused by gravitational micro- or macro-lensing. In a sample of highly luminous quasars, Magain (this workshop), Swings (this workshop) *et al.* found several multiply lensed objects and cases where a foreground galaxy is placed directly in front of the quasar — both compatible with a strong magnification by macro-lensing — but no number excess of galaxies outside 3 arcsec. This can be a hint that the magnification bias due to micro-lensing is not so important at the bright end of the quasar luminosity function because of the large source sizes. Therefore the few known low-redshift (intrinsically faint) quasars with foreground galaxies have a relatively high priority in our monitoring program.

There are two ways to distinguish micro-lensing variability from intrinsic quasar variations: The first one is for the case of multiply lensed quasars (see above), the second is to search for HAEs which have a typical asymetric shape. Right- and left-hand HAEs occur with the same frequency (see Kayser *et al.* 1989 and refs. therein). The detection and identification of at least one HAE is a key to the structure of the most compact quasar radiation components. As Grieger (this workshop) pointed out, the luminosity profile of these components is reflected in the shape of a HAE. Structures with linear dimensions

$\leq 10^{-3}$ pc can be resolved by this method. How many HAEs we can hope to see during our program and whether it will be possible to make any statistics, strongly depends on the fraction of galaxy matter that is in compact star-like objects. If the dark matter in galaxies is mainly composed out of such objects (brown dwarfs, low-mass main sequence stars, *etc.*), we should be able to detect a lot of them (cf. Witt, this workshop).

References

Babadzhanyants, M.K. and Belokon, E.T. 1985, *Astrofizika* **23**, 459

Blades, J.C. 1988 in *QSO Absorption Lines* ed. by Blades, J.C., Turnshek, D. and Norman, C.A., Cambridge University Press, Cambridge

Borgeest, U. 1986, *Astrophys.J.* **309**, 467

Borgeest, U. and Refsdal, S. 1984, *Astron.Astrophys.* **141**, 318

Gorenstein, M.V., Falco, E.E. and Shapiro, I.I. 1988, *Astrophys.J.* **327**, 693

Kayser, R., Weiss, A., Refsdal, S. and Schneider, P. 1989, *Astron.Astrophys.* **214**, 4

Ostriker, J.P. and Vietri, M. 1985, *Nature* **318**, 446

Schneider, P. and Weiss, A. 1987, *Astron.Astrophys.* **171**, 49

Vanderiest, C., Schneider, J., Herpe, G., Chevreton, M., Moles, M., and Wlérick,G. 1989, *Astron.Astrophys.* **215**, 1

PHOTOMETRIC MONITORING OF 2237+0305

R. T. Corrigan, M. J. Irwin, P. C. Hewett
Institute of Astronomy, Cambridge, England

and R. L. Webster
Canadian Institute for Theoretical Astrophysics, Toronto, Canada

INTRODUCTION

The gravitational lens system 2237+0305 is a spiral galaxy at $z = 0.0394$ in which Huchra *et al.* (1985) discovered a high-redshift quasar ($z = 1.695$) coincident with the nuclear region. Yee (1988) and Schneider *et al.* (1988) resolved the central 2 arcsec of this system into four quasar components plus the galaxy nucleus. Modelling the galaxy as a smooth mass distribution reproduces the observed configuration remarkably well (Schneider *et al.* 1988, Kent and Falco 1988). The optical depth due to stars *i.e.* compact objects, through the galaxy is also expected to be significant and the 2237+0305 system provides an ideal case for the study of microlensing events due to individual objects.

In Irwin *et al.* (1989) we reported the detection of a significant brightness variation in the brightest of the four quasar components (component A - using the nomenclature of Yee (1988) here and throughout), and we interpreted that change as the first detection of a microlensing event. Here we present our analysis of images obtained by Yee before and just after those in which we detected the brightening of component A. We find our magnitude estimates of the four quasar components in excellent agreement with his, demonstrating the modelling technique used to obtain the magnitudes of the individual quasar components is robust. Confirmation of the brightening evinced by component A from analysis of all the images, employing identical procedures, confirms and strengthens the results reported in Irwin *et al.* (1989).

OBSERVATIONS

Direct images of 2237+0305 at six epochs will be described in chronological order. The first two were obtained in September 1986 and September 1987 by Schneider *et al.* (1988), and the third independently in September 1987 by Yee (1988). The published astrometric and photometric results for the four quasar images from this data are in detailed agreement, and Schneider *et al.'s* comparison of their own 1986 and 1987 data showed no evidence of variability.

On the night of August 18 1988, we obtained direct images in B and R of 2237+0305 using the new Very High Resolution camera at the prime focus of the 3.6-metre Canada-France-Hawaii Telescope (CFHT) in excellent seeing (FWHM $\sim$ 0.7 arcsec). On the night of 1988 September 16 further direct images of the system were taken in good seeing (FWHM $\sim$ 0.9 arcsec) through U B V R filters at the Cassegrain focus of the 4.2-metre William Herschel Telescope (WHT). Finally, images were taken by Yee on November 12 1988 at the prime focus of the CFHT in excellent seeing (FWHM $\sim$ 0.6 arcsec). Yee kindly made all his published and unpublished data available to us, and those frames taken in the Gunn r passband are crucial to the analysis described below.

Given such a complex system - there are four quasar components and the galaxy nucleus within 2 arcsec - some form of model fitting is necessary in order to produce reliable magnitude and position estimates (Irwin 1985). While we have argued in Irwin *et al.* that the magnitude estimates for the individual components are well determined the availability of Yee's frames allows an empirical test of the complex modelling procedure by comparing to Yee's own results, and, furthermore, the use of a consistent modelling procedure on all the frames increases confidence in the derived amplitude of the brightness change in component A. Our modelling procedure is described in detail in Irwin *et al.* (1989).

Table 1 presents our magnitude estimates for the CFHT and WHT R frames and Yee's CFHT r data. Yee's own results (1988 and private communication) are given in brackets. The R magnitudes have all been zero-pointed on to the r system to facilitate comparison. Yee uses a somewhat different iterative image decomposition technique to the one outlined in Irwin *et al.* (1989) (although both are non-parametric), thus the excellent agreement between our results and his gives considerable confidence in their validity. The visual impression (figure 1) that component A was significantly brighter (~ 0.5 mag) than component B in August 1988 is confirmed.

Table 1 R and r band Photometry of the resolved components

	r_{87}	R_{CFHT}	R_{WHT}	r_{88}
A	17.61 (17.62)	17.03	17.18	17.48 (17.47)
B	17.81 (17.77)	17.56	17.59	17.83 (17.72)
C	18.03 (18.06)	18.17	18.15	18.13 (18.10)
D	18.40 (18.40)	18.40	18.38	18.39 (18.40)

See Irwin *et al.* 1989 for discussion of errors.

INTERPRETATION

A light curve of only six points does not enable unambiguous determination of the parameters describing the lensing event, however, the potential of the 2237+0305 system for providing unique information on the mass function of compact objects in the galaxy and the size of the quasar emitting regions is illustrated by the two simple models investigated in Irwin *et al.* (1989).

The data of Yee and Schneider show that the magnitudes of the two brightest quasar components remained roughly constant over the period September 1986 to September

1987. Our CFHT data taken on 18 August 1988 show a 70 per cent increase in brightness of image A from the September 1987 data of Yee. The September 1988 and November 1988 data show an increase of 50 per cent and 16 per cent respectively. If the decreasing brightness of image A between these three observations represents the fall-time for a microlensing event, then, assuming a linear fall, t_r is approximately 100 days - a lower limit if our first observation was made during the fall time and an upper limit if the first observation was made during the intermediate amplification portion of the event. If the entire event has taken place since Yee's 1987 observations then ΔT is approximately 420 days. Once a timescale has been determined a constraint on the mass of the microlens depends on knowing how rapidly the object passes across the line of sight. Taking a range in transverse velocity based on our motion relative to the microwave background and the timescale for the lensing event of $\Delta T \sim 420$ days we can obtain an estimate for the mass of the lensing object, $0.001 < M < 0.1\,M_\odot$, implying the lensing object is possibly of sub-stellar mass.

Nadeau and Yee (this conference) present near infra-red imaging, obtained 26 days before our CFHT data, which does not show any detectable brightening of component A. If the infra-red images probe the same spatial scales as the visual data then these new observations suggest that the rise-time for the event t_r is $\sim$ 20 days which reduces the estimate of the mass of the lensing object by an order of magnitude! Alternatively if the infra-red radiation arises from a more extended region in the quasar then the microlensing effect on the infra-red light may be small and the new data does not affect the estimate of the rise-time.

We are extremely grateful to Howard Yee for making available his data and unpublished results on this object.

REFERENCES

Huchra J, Gorenstein M, Horine E, Kent S, Perley R, Shapiro I I and Smith G 1985 *Astr. J.* **90** 691

Irwin M J 1985 *Mon. Not. R. astr. Soc.* **214** 575

Irwin M J, Webster R L, Hewett P C, Corrigan R T and Jedrzejewski R I 1989 *Astr. J.* **98** in press

Kent S M and Falco E E 1988 *Astr. J.* **96** 1570

Schneider D P, Turner E L, Gunn J E, Hewitt J N, Schmidt M and Lawrence C R 1988 *Astr. J.* **95** 1619

Yee H K C 1988 *Astr. J.* **95** 1331

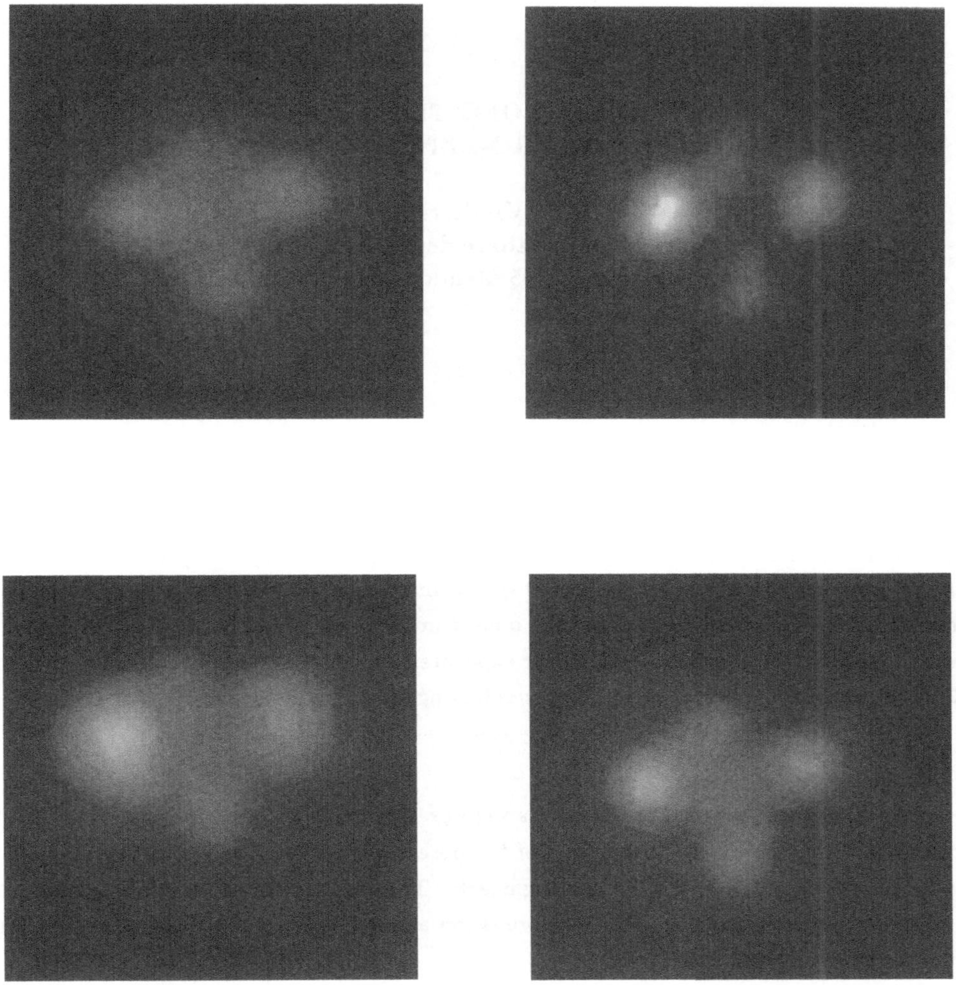

Figure 1: Red images of the quasar components and galaxy nucleus; (top left) 25 September 1987, r (Yee), (top right) 18 August 1988, R, (bottom left) 16 September 1988, R, (bottom right) 12 November 1988, r (Yee).

SPECTROPHOTOMETRY OF 0957+561 AND THE MICRO-LENSING EFFECT.

C. Vanderriest
DAEC, Observatoire de Paris- Meudon
5 place Janssen, 92195 Meudon cédex (FRANCE).

summary:

In the case of multiple gravitational imaging of a distant quasar by an interposed galaxy, "microlensing" due to individual stars could result into different observed amplification ratios for the broad emission lines and the continuum. Such an effect is observed for 0957 + 561 from data obtained at the C.F.H. telescope with the bidimensional fibre optics spectrograph SILFID. From the same set of spectra, an integrated spectrum of the main deflecting galaxy G_1 has been extracted without contamination from the quasar images. We found an improved value of the redshift $Z_{G_1} = 0.378 \pm 0.001$. With longer exposures, we could measure its velocity dispersion and, at the same time, better study the microlensing effect on the quasar.

For a long time, gravitational mirages were present only in the minds of theoricians, but since 10 years they haunt also the nights of the observers. Likewise, it seems that now be the time for being concerned with microlensing effects. They are suspected in several mirages (see for example the communications in this workshop about 2237 + 0305, H 1413 + 117, etc...) including the archetypal 0957 + 561.

-1) Evidence for microlensing on image B of 0957 + 561:

Programs devoted to the monitoring of the optical continuum of 0957 + 561 have lead to a reasonably secure value for the time delay $\Delta t(A, B)$ (Vanderriest et al., 1989; Schild, this workshop). When we form the flux ratio at the same emission time $R(t) = B(t + \Delta t)/A(t)$, we see some slight ($\leq 0.1$ mag.) but significant departure from a mean value near 1.0 (Fig. 1; see also Schild's data) that could be interpreted in term of microlensing effect. Indeed, such shallow variations are well reproduced by microlensing models of a source with very small but finite size (see, for example, Kayser et al., 1986).

Aside from this direct clue, another piece of evidence lies in the mean value itself of $R(t)$. It is well known that a sufficiently extended source would be insensitive to microlensing effects of individual stars. Nevertheless, if its size is small with respect to the scale of the *macro*-lens, it could be used to determine the magnification ratio of this macro-lens at the positions of the images. The small radio jet detected by VLBI (Gorenstein et al., 1988; Porcas et al., 1989) could do the job. In this case, as the size is larger than $c.\Delta t$, its flux could even be considered constant and the magnification ratio is given simply by: $R = B(t)/A(t)$. Gorenstein et al. found a value near 0.64 ± 0.03.

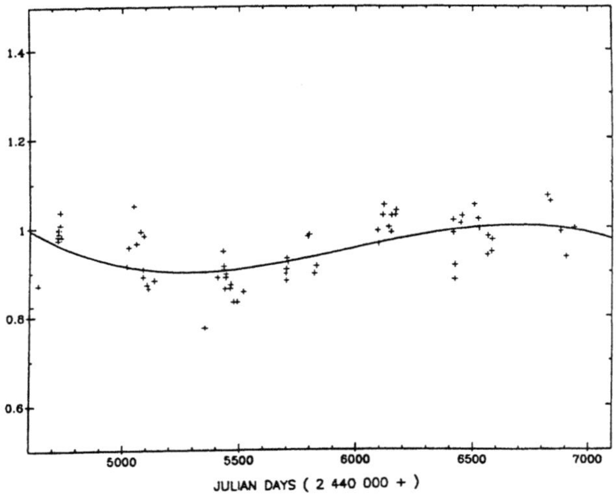

Fig. 1: Variations of the ratio $B(T + \Delta T)/A(T)$ in the blue continuum (from Vanderriest et al., 1989)

At optical wavelengths, a differential effect should be observed between the continuum and the broad emission lines, as suggested by Rees (1981). The size of the BLR is sufficient for being almost insensitive to microlensing (and the condition of being larger than $c.\Delta t$ is probably fulfilled also). Indeed, Young et al. (1981) found a ratio B/A in the emission lines 0.77 ± 0.03, and Wills and Wills (1980) about 0.7 [they found also a ratio near 0.7 for the continuum, but it was a chance combination of the intrinsic variability and the time delay].

So, the "true" amplification ratio, free of microlensing, seems to be near 0.7 and the extra 0.3 magnitude amplification seen in the continuum of image B could be attributed to microlensing. The amplitude so derived is larger than from direct inspection of $R(t)$, and suggests that several stars are involved at the same time, which is not unlikely at the position of image B, close to the center of G_1 (see Young, 1981).

In order to precise the differential effect between continuum and emission lines, we tried to obtain good spectrophotometry of the 2 images with the help of a bidimensional spectrograph.

-2) Bidimensional spectrophotometry with a fibre optics device:

The spectrograph SILFID can be used in several configurations (Vanderriest and Lemonnier, 1988). Here, it worked in its "Integral Field Spectroscopy" mode. The principle is to transform the geometry of the focal image given by the telescope, like an image-slicer would do. About 400 optical fibres are closely packed into a compact hexagonal array at one end of the device (focal plane), and they are just set in line at the other end (entrance "slit" of the spectrograph). After dispersion it gives 400 independant spectra and, by inverting the geometrical transformation, it is possible to reconstruct crude pictures of the object at any chosen wavelength or maps of the velocity field. One may also add together the spectra given by several fibres and obtain the integrated spectrum in a defined part of the field of view.

At the Cassegrain focus of the Canada-France-Hawaii telescope, each 100 μm fibre subtend 0.71 arcsec and there are 23 fibres in the largest dimension of the field. In the detector plane, the scale is 60 μm per fibre. We obtained one 30 mn exposure at a reciprocal dispersion of 235 Å/mm in December 1988 with a Thomson CCD (576 X 385 pixels of 23 μm), and two 30 mn exposures in March 1989 with a Ford CCD (512 X 512 pixels of 20 μm). So, in both cases, the CCD did not cover the whole slit length and only a part ($\simeq$ 1/3) of the initial hexagonal field was recorded (fig. 2).

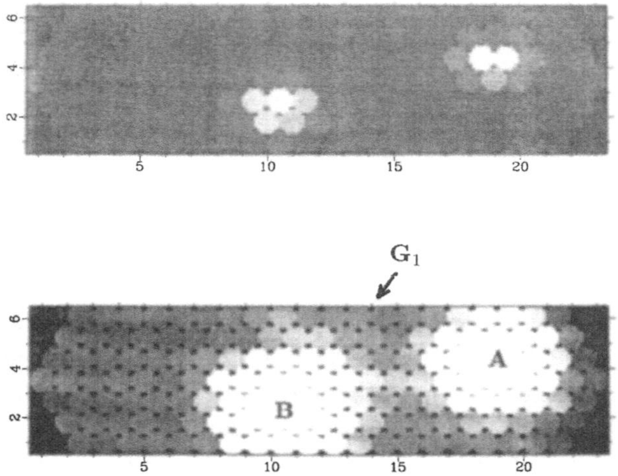

Fig. 2: Reconstruction of the field analysed by the fibre optics device. The same picture is shown with 2 different halftone scales.

The seeing was about 1.3" and, by integrating on 14 fibres for each image A and B, the light losses due to dead spaces between the fibres cannot affect the relative spectrophotometry by more than 2%. The method suppresses also the problems that could arise with classical slit spectrographs from differential atmospheric refraction and/or centering. Thus, it seems the best way for obtaining accurate spectrophotometric data.

-3) The results on images A and B:

Fig. 3 displays the processed data for December 1988. For image B, the contribution of G_1 was removed in a crude but satisfying way: A "clean" spectrum of G_1 (see next section) is first obtained by integrating on $\simeq$ 20 fibres outside image B (zone 2). Knowing the position and structure of G_1 (Stockton, 1980), we can estimate its flux in the area used for integrating the spectrum of image B (zone 1). It was found about 100% -120% of the flux in zone 2 and the correction was done accordingly. As it is only 4% of the quasar's flux at 4500Å and about 8% at 6500Å, it is not necessary to refine any further.

The apparent ratio R = B(t)/A(t) at the time of observation was 1.03 ± 0.02 in the continuum, but obviously less in the emission lines, as can be seen by plotting 1.03*A - B where they are still visible. For these emission lines, the "best" value of R is near 0.70 ± 0.08. By plotting

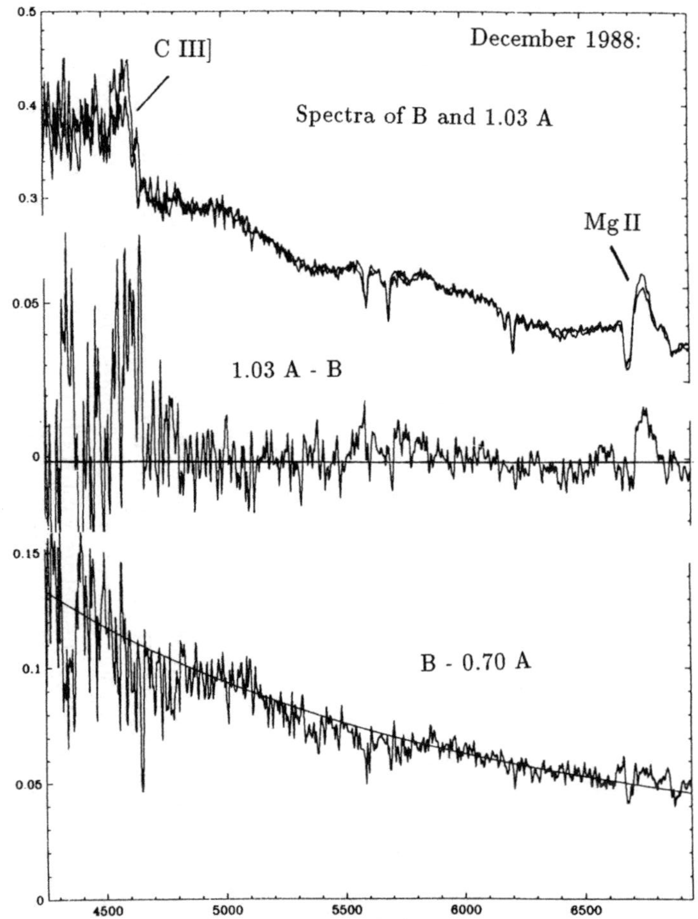

Fig. 3: Comparison of the spectra of images A and B (December 1988); best ratios for the continuum and the emission lines (see text).

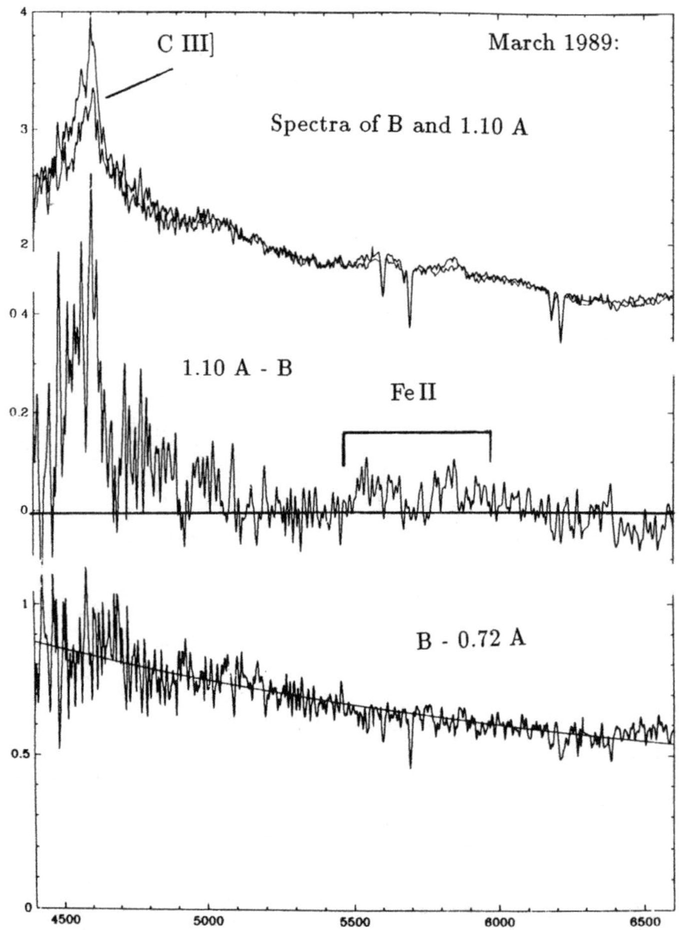

Fig. 3bis: As fig. 3 for the March 1989 data.

B - 0.70*A, we visualize the "excess" of magnification of the continuum due to microlensing. This continuum is fairly well fitted by a power law.

Fig. 3bis displays the same things from the March 1989 spectra. It is the sum of two 30 mn exposures and the S/N ratio is better; the wavelength range is smaller because of the size of the CCD. Note the very good agreement of the narrow absorption lines at $Z_{abs} = 1.391$; there is no significant difference between their equivalent widths in A and B. Here, the apparent value of R for the continuum is 1.10 ± 0.02. When plotting 1.10*A - B, the narrow absorption lines cancel out completely, as expected, and the Fe II emission lines noticed by Wills and Wills (1980) show up clearly. The best value for the flux ratio in C III] is R $\simeq$ 0.72 ± 0.02

-4) The results on the deflecting galaxy G_1:

The presence of the main deflecting galaxy G_1 is detected on reconstructed pictures (fig. 2). Then, by selecting appropriate fibres, we can obtain a representative spectrum of this galaxy without contamination from the quasar images. This integrated spectrum is used for correcting the spectrum of image B as stated before. For the March 1989 data, it is good enough to allow also a better determination of the redshift of G_1. No improvement occured since the early measurements of Wills and Wills (1980): Z $\simeq$ 0.375, Young et al. (1980): Z $\simeq$ 0.39 or Young et al. (1981): Z $\simeq$ 0.36.

Here, from the position of H, K and hydrogen lines (fig. 4) we found $Z_{G_1} = 0.378 \pm 0.001$

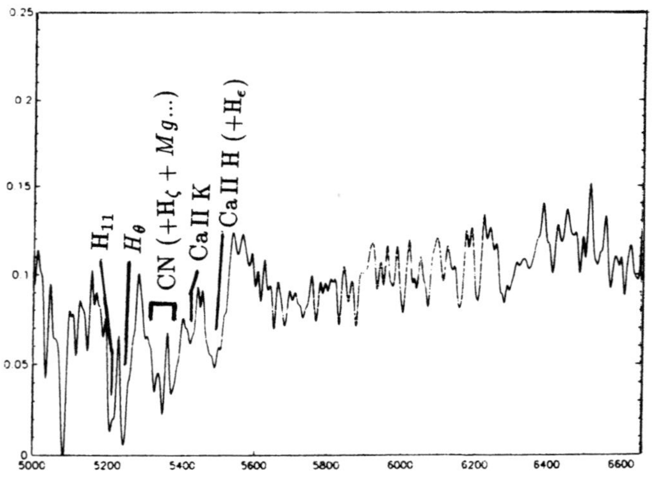

Fig. 4: Spectrum of G_1 integrated over an area without contamination from the quasar images (March 1989).

Discussing the high resolution data obtained by Crotts (1987), Caulet (1989) remarks the "anomalous" intensity of an Mg II absorption line in the spectrum of image B. She concludes that it can be explained only by "*a blend with some unidentified absorption line seen only toward image B* "; if we identify this feature with H_β in G_1, we find a redshift 0.379 in perfect agreement with the above value. H_β absorption is indeed suspected in our (poor S/N) spectrum of December 1988.

If longer exposures could be obtained by good seeing conditions, this line could be used for measuring the velocity dispersion in G_1, an important parameter of the lens models. H_β is favourable, being quite strong and isolated above a "clean" galaxy continuum (not true for H and K, for example) and at an observed wavelength where the efficiencies of CCD and spectrograph are good.

-5) Conclusions:

Our tentative observations of 0957 + 561 with a bidimensional spectrograph have provided us with a spectrum of G_1 of better quality than previously published, and with spectrophotometric data on the quasar images that show clearly an effect of microlensing. The amplification ratio for the emission lines seems to be near 0.7. As we cannot be completely sure that the intrinsic variations of these lines are negligible over a time scale ΔT, it would be more correct to form the ratio $B(T + \Delta T)/A(T)$ by observations at 2 epochs. That means necessity of *absolute* spectrophotometry with an accuracy of a few hundredths.

If the slight discrepancy between the value of R from optical lines and radio jets survives further observations, it would be an evidence for microlensing effect on the emission line region itself. The size of this region could be evaluated, and eventually found to be different for different ions (e.g. C III] and C IV). Then, with very high S/N data, one could see differences on the line profiles of A and B related to the kinematic structure of the emitting region (Nemiroff, 1988). Such observations would take several hours on a 4-meter class telescope and would be easier on Very Large Telescopes; but they are perfectly feasible from the ground.

References:

- Caulet A., 1989, Astrophys. J., **340**, 90.
- Crotts A., 1987, in: "*QSO absorption lines: Probing the Universe*" vol. **1**, Space Telescope Institute workshop, p. 119.
- Gorenstein M., Cohen N., Shapiro I., Rogers A., Bonometti R., Falco E., Bartel N., Marcaide J., , 1988, Astrophys. J., **334**, 42.
- Kayser R., Refsdal S., Stabell R., 1986, Astron. Astrophys., **166**, 36.
- Nemiroff R., 1988, Atrophys. J., **335**, 593.
- Porcas R., Garret M., Quirrenbach A., Wilkinson P., Walsh D., 1989, Conf. "*Gravitational lenses*", Lectures Notes in Physics, **330**, 82.
- Rees M., 1981, ESO Conf. "*Scientific importance of high angular resolution at Infrared and Optical wavelengths*", M.H. Ulrich, K. Kjär eds., p.423.
- Stockton A., 1980, Astrophys. J., **242**, L.141.
- Vanderriest C., Lemonnier J.P., 1988, IXth Santa Cruz workshop "*Instrumentation for ground-based Astronomy*, p. 304.
- Vanderriest C., Schneider J., Herpe G., Chevreton M., Moles M., Wlérick G., 1989, Astron. Astrophys., **215**, 1.
- Wills B., Wills D., 1980, Astrophys. J., **238**, 1.
- Young P., 1981, Astrophys. J., **244**, 756.
- Young P., Gunn J., Kristian J., Oke J., Westphal J., 1980, Astrophys. J., **241**, 507.
- Young P., Gunn J., Kristian J., Oke J., Westphal J., 1981, Astrophys. J., **244**, 736.

IR IMAGING AND THE LIGHT CURVE OF 2237+030

D. Nadeau
Dép. de Physique, Univ. de Montréal, Montréal, PQ, H3C 3J7, Canada
H. K. C. Yee
Dept. of Astronomy, Univ. of Toronto, Toronto, Ont., M5S 1A7, Canada
W.J. Forrest, J.D. Garnett, Z. Ninkov, J.L. Pipher
Dept. of Physics and Astronomy, Univ. of Rochester, Rochester, NY 14627, U. S. A.

I. Observations

Direct images of the gravitational lens system 2237+030 in the J ($1.2\mu m$), H ($1.6\mu m$), and K ($2.2\mu m$) bands were obtained at the CFHT on each of the nights of 21, 22, and 24 July 1988 (UT) using the University of Rochester 62×58 InSb infrared camera (Forrest *et al.* 1989) with a scale of $0\farcs30$ per pixel. The final images in each band for each night were formed by co-adding 5 to 12 frames selected from a large number of 42 sec exposures having seeing better than $0\farcs8$ FWHM. New results from r band images obtained at CFHT on November 12, 1988, will also be included in the discussion of the light curve.

The infrared images were preprocessed by linearization, sky subtraction, flat-field correction, and frame co-addition. The extraction of the magnitudes of the components was performed using the iterative PSF subtraction method described in Yee (1988). The criterion for proper decomposition of the quasar components is that a symmetrical and smooth residual galaxy is obtained.

II. Results

Our 1988 IR and optical photometries are similar to the 1986 and 1987 data, and are consistent with there being no significant variation. The flux ratios between A and B, the most accurately measured components, as obtained from the IR data, are identical to those obtained in the i band by Yee (1988), after dereddening. Our Nov/88 r band absolute magnitude measurements agree within the error bars with those of Sep/87, with the possible exception of a brightening of ~0.1 mag in A.

Irwin *et al.* (1989) have reported to have found evidence of a micro-lensing event by comparing the results of their observations with those of Yee (1988). They found that component A had brightened relative to B by 0.4 mag. Using the data of Yee (1988) as the starting date of the micro-lensing event, Irwin *et al.* concluded that the upper limit on the mass of the lensing object is $\sim 0.1\ M_\odot$.

The light curve of component A from IR and optical data is plotted as magnitude differences with respect to component B. Included are published r and R photometry from Schneider *et al.* (1988) for Sep/86, Yee (1988) for Sep/87, and Irwin *et al.* (1989) for Aug/88 and Sep/88. We note that the IR data, taken just 27 days before their R band data, and our r band data from Nov/88, have ratios for components A and B consistent with that observed by Yee (1988), even after correcting for the maximum amount of differential extinction possible between the bands. Hence, *if* the brightening observed by Irwin *et al.* is attributed to a micro-lensing event, then a very stringent upper limit of ~120 days can be put on the time scale of the event, giving an upper limit of 0.008 $M_\odot$ for the mass of the lensing object by assuming a conservatively high effective transverse velocity of 10,000 km s^{-1} between the lensing object and the quasar. This mass is approaching that of a large planetary body. The rise time can also be limited to ~25 days, giving a scale size of ~0.7 light-day for the quasar continuum source. Since it does appear from inspection of the images that component B in the Irwin *et al.* Aug/88 image is significantly fainter than A, it does not seem possible to explain the apparent variability by differences in the methods of extracting the magnitudes of the components. From lensing models, a maximum light-path time delay of the order of a day is expected between the different images; the probability of a rapid intrinsic variability of the quasar making component A brighter at the time of both the Aug/88 and Sep/88 observations is therefore remote. It it possible to relax the constraint on the timescale of the event, however, if the quasar continuum source size increases with wavelength, as expected from the accretion disk model, leading to smaller micro-lensing amplification in the infrared.

REFERENCES

Forrest, W. J., Pipher, J. L., Ninkov, Z., Garnett, J. D. 1989, in *Proc. of the 1989 Infrared Detector Tech. Workshop*, ed. C.R. McCreight (NASA Ames Res. Center), to be published.

Irwin, M. J., Webster, R. L., Hewett, P. C., Corrigan, R. T., Jedrzejewski, R. I. 1989, *A. J.*, **98**, in press.

Schneider, D. P., Turner, E. L., Gunn, J. E., Hewitt, J. N, Schmidt, M., and Lawrence, C. R. 1988, *A. J.*, **96**, 1619.

Yee, H. K. C. 1988, *A. J.*, **95**, 1331.

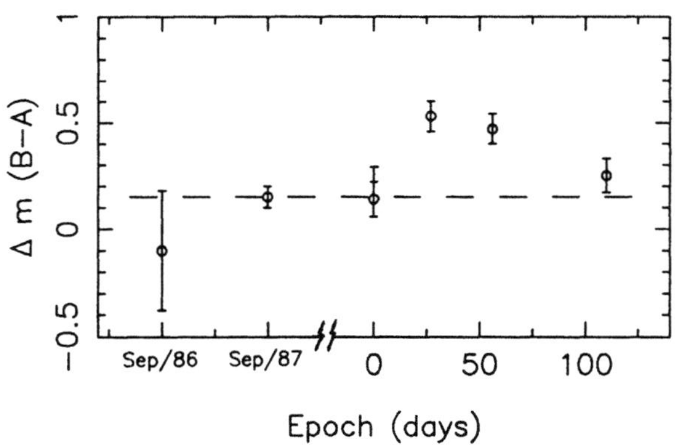

Figure 1. Light curve of component A with respect to B. The data points from left to right are from Schneider *et al.* (1988), Yee (1988), the IR data of Jul/88 (taken as epoch 0), the two points from Irwin *et al.* (1989), and the Nov/88 r band CFHT observation. The additional error bar on the IR point shows the maximum amount of correction expected from the K band to the r band due to absorption in the lensing galaxy.

IV

Rings, Arcs and Arclets

CLUSTERS OF GALAXIES : A NEW OBSERVABLE CLASS OF GRAVITATIONAL LENSES

Bernard FORT

Observatoire Midi-Pyrénées, 14, avenue Edouard Belin 31400 Toulouse, France

I. INTRODUCTION

At the time of the discovery of several giant luminous arcs in the core of several rich clusters of galaxies and of the spectroscopic proof that the one in Abell 370 was undoubtedly a gravitational image of a distant background galaxy (Soucail et al. 1988, Lynds and Petrossian 1989), Tyson et al. (1988) revealed the existence of a huge population of faint blue galaxies, whose color and surface brightness suggested a redshift larger than one. Immediately, Fort et al. (1988) showed that the cluster Abell 370 indeed contained several faint arclets, gravitationally distorted images of some objects belonging to this distant population. It became obvious that clusters of galaxies could regularly act as gravitational lenses. Moreover the high surface density of the background galaxies (sources) and the very great number of clusters (lenses) were enough to demonstrate that we potentially have a new class of gravitational lenses with an extremely great number of candidates compared to classical galaxy-QSO lenses. They were not detected earlier because the probability of detecting such gravitational images of distant galaxies falling near caustic lines of a cluster-galaxy lens (hereafter CGL) is high only if we use extremely deep photometry up to B=29 per square arcsecond, (Tyson 1989). Within the last two years this observational topic has evolved very rapidly, confirming our first guesses.

In this introductory talk to the afternoon session on "arcs in clusters", we would like to give the other observers a rapid view of the interest of observing cluster-lenses and the theoreticians a list and a short description of the candidates already observed. We know that for modeling, theoreticians are particularly eager for detailed maps of the observations with adequate photometric tables of the cluster members. Thus we urge the observers to release them as soon as their data reductions are finished.

II. THE LARGE ANGULAR SCALE OF CGL

The angular size is probably the most important characteristic of such lenses. For a singular isothermal sphere, the size of the caustics and critical curves increase as the second power of the velocity dispersion and the probability of having background galaxies produce arclets will increase as the fourth power. With such a crude approximation for a cluster center, we give in table 1 the number of arclets with a tangential stretching larger than four times the diameter of the sources. For more realistic potential, with velocity dispersion above 1500 km/s the lens still produces such arclets within a strip contour which outlines the geometry of the projected potential (fig. 1). So for such CGL we will have many pixels to describe the lens (typically equal to 1 arcsecond seeing) and we can make sense of the photometric data of the brightest galaxy members of the clusters or in the future with the mapping of the X-ray emission of the hot gas. In short the large angular scale provides both numerous gravitational images and a good possibility of describing the geometry of the lens according to the observations. Thus, we should be in a favorable position for modeling CGL candidates.

$\sigma_{cluster} \times km/s$	Einstein Radius	max. radius=4/3 E.R.	N(a/b)>4
600	6"	10"	< 1
1000 expl : A963	15"	20"	3
1380 expl : A2218	29"	40"	13
1800 expl : A370, A1689	50"	66"	28
2800 expl : 3C295(?)	130"	180"	230

Table 1 : Estimate of the number of arclets with a tangential elongation larger than 4 times the diameter of background sources for a singular isothermal cluster core. The lens is at redshift 0.374 (A370) and the sources at the mean redshift z=1.2; the number of sources is taken as 56 background galaxies per square arcminute (Tyson 1988).

III. WHAT CAN WE LEARN FROM CLUSTER LENSES?

From CGL we can derive information about the sources, the clusters and the geometry of the gravitational optics both with individual cases and from a large statistical sample of lenses.

3a) Sources

The convergence of a lens is proportional to the cosmological term $D_{ol} \times D_{ls} / D_{os}$. It increases rapidly between z=0 and z=0.15 (Blandford and Kochanek 1987) and, for any sources above z=2 passes through a maximum between z=0.2 and z=0.7 where the clusters are the most efficient lenses. However if the mean redshift of the background galaxies is not so distant, say around z=1, when we push the lenses further than z=0.5 the number of arclets will diminish drastically for the same velocity dispersion. We therefore have the possibility of estimating the lower limit for the distance of the background galaxy population which is still a matter of debate (see Grossmann or Ellis, this conference).

The arclets are faint sources and there is no possibility in the future of getting spectra and redshift for most of them. But according to the theory, with a good modeling of a cluster-lens it is in principle possible to estimate the distance of the arclets. Here it is worth noting that the best prediction of Grossmann and Narayan (1989) for the arclet A5 detected in A370 (Fort et al. 1988) corresponds to the redshift effectively measured with extremely deep spectroscopy (Soucail et al. this conference). But the ranges of variation for the free parameters used in the modeling are too large to hope getting redshift with an accuracy that we would like to have to study galaxy evolution (say dz=±0.2). In this matter, the hope could be the statistical use of the radial distribution of arclets : the images of the sources are projected in the lens plane at radii which increase slowly with z (for a given q_o). If we can observe several tens of clusters with many arclets, it could be interesting to check if there is a systematic tendency for surface brightness and color variation versus the radial distance to the cluster center.

We said the arclets are generally so faint that it is useless to hope for more than wide band photometry up to the near infrared. But such observations are particularly interesting on arcs because we can use the magnification factor of the gravitational images to significantly increase the signal to noise ratio of the photometry for a randomly selected sample of very distant galaxies. The same idea gives better results when by chance the source falls near a caustic cusp and produces a large arc because the magnification factor can be as large as 10 or more. This is the case for the arc in Cl2244 (Hammer et al. 1988). By chance its surface brightness was high enough (B=23 per square arcsecond) to get spectra with a curvilinear slit and a long exposure with the ESO Faint object spectrograph EFOSC. After summing the spectra of each point along the arc in order to increase the signal to noise ratio we got the spectrum of a very distant galaxy which turns out to have a redshift z=2.238 (Soucail et al. this conference). This method can be systematically used in the future to get spectra of distant galaxies that otherwise would have been unobservable. Such an observational program will actually use the gravitational telescope effect.

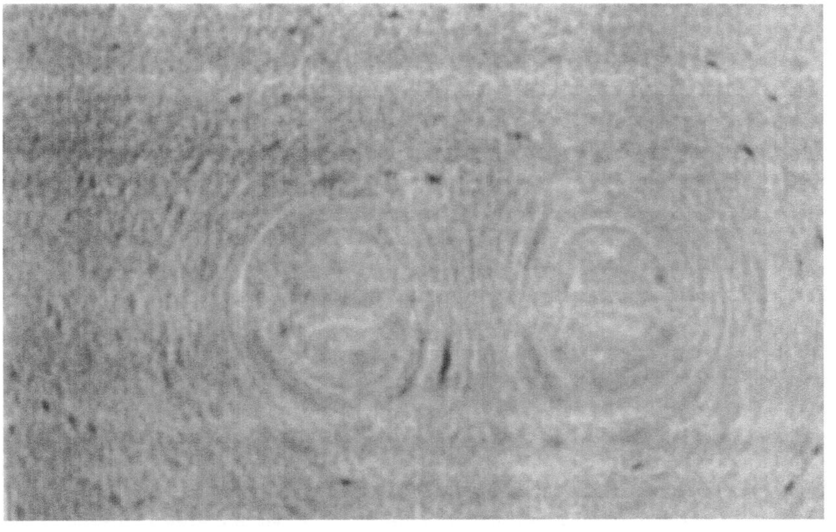

Figure 1 : Qualitative simulation with a regular grid of background galaxies showing the formation of arcs and arclets around two clumps of masses 40 arcsecond apart with σ =1400 km/s. Note that the CCD noïse is also gravitationally distorted! (Courtesy Tyson)

3b Cluster

From table 1 it is also interesting to note that there exist very rich clusters with true velocity dispersion above 2000 km/s, as was measured for 3C295 (Gunn 1989), the number of arclets should be extremely great. The observed number should give a rough but immediate estimate of the dynamical state of a distant cluster. This can be a way of knowing whether we are actually dealing with a virialized rich cluster or instead with the superimposition of several substructures having lower velocity dispersions.

In fact the CGL will probably provide an uncomparable tool for studying the distribution of the masses in the central part of a cluster of galaxies. When we observe a large arc or the mean circular locus of many arclets, it is possible to calculate the total mass inside this radius. It seems that the determination of this number is robust (not very sensitive to the potential profile). For the first candidates that we know (table 2), it seems that the mass to light ratio is about 100 for the cluster cores (luminosity expressed in the R. Johnson filter generally used for the photometry of the cluster galaxies). For Abell 370 the virial mass that we can deduce from an extensive spectroscopic survey giving the velocity dispersion within the Einstein radius pretty well matches the mass of the lens (Mellier et al. 1988).

Cluster	Z	C	σ km/s	M/R Virial	M/R Lens	Lx	Comments & references
A 370	0.374	40	1350/1700	80/100	90 ± 30	$9.7 10^{44}$	Z large arc = 0.724 (Soucail et al 88) mini arcs (Fort et al 88) Lynds & Petrosian 88
Cl 2244	0.328	-	-	-	70 ± 20	-	Lynds & Petrosian 88 Hammer et al 89 Z Large arc=2.23 Soucail et al 89
A 2218	0.171	218	1400	90	-	$6.5 10^{44}$	Mini arc Pello 87 ring around a galaxy
A 2390	0.23	72	1600	100 ± 30	-	$9.9 10^{44}$	Mellier et al 89 Z arc=0.92 Pello et al 89
A 963	0.206	134	1100	60	60	-	2 Arc around the CD Lavery & Henry 88
Cl 0500-24	0.321	200	1300	80 ± 20	100 ± 30		Giraud, Schneider et al 89
A 1689	0.181	228	1800 km/s				50/70 Arclets Tyson 89
3C 295	0.461		2800 km/s 3×1000 km/s			$9.2 10^{44}$	50/70 distorded images 10/15 Arclets Tyson 89
A 545	0.154	234					one large arc. Tyson 89
A 1525	0.181	186					one possible Arclet. Giraud 88 to be confirmed
Cl 0024+16	0.391	-	1300	-	-	-	Koo 1988

Table 2 : First cluster-galaxies lenses with gravitational arcs or arclets

With the grid of distant sources corresponding to a mean distance of about <d> = 12 arcsecond, it is possible to deduce the general shape of the potential from the mean locus of the arclets. But it is not possible to see if the total mass is clumped on smaller scales than about 30 arcseconds. However for some distant clusters with bright multipolar components like A370 it could be interesting to see if general smooth haloes always exist and if they are formed before the merging of the visible giant galaxy clumps.

For most CGL given in table 2 the locus of the arclets seems to be elliptical. Tyson (this conference) was the first to use the statistical properties and distribution of the arclets in order to constrain the potential of the clusters. He will show the comparison of various simulations with the observation (A1689, CL1409+52, Cl10024+16). He was able to derive the existence of general haloes and to deduce their core radius. These preliminary results also indicate that for these cluster cores the total mass seems to follow the visible mass distribution. Indeed as compared to the modeling of the first single luminous arcs the use of arclets extends the interest of CGL for the mapping of the dark matter (see Mellier and Tyson, this conference for a more complete discussion). The next observational step will be to obtain with the satellites AXAF and XMM a simultaneous mapping of the X-ray emission with a spatial resolution of a few arcseconds because the hot gas traces the projected potential via the measurement of the emissivity along each line of sight (see Kellog et al., this conference). At this point all the observations available will allow us to give good modeling with the relative balance between visible mass, gas mass and unseen mass for many distant clusters.

However we don't know if the dark matter in the universe is somewhat connected with X-ray gas. Searching the arclets on large field along long chains of clusters building the wall of the sponge-like universe will be a fascinating target for very wide field CCDs. If the skeleton of the large structure in the universe is made of dark matter milestone with $M/L \sim 100$, we should see the gravitational distortion they will induce on the "Tyson" population (fig. 1).

3c Cosmography

Up to now all the hopes we have placed in classical galaxy-QSO lenses for the determination of the Hubble constant have remained fruitless despite the beautiful measurement of the time delay in the double QSO. The ultimate limitation seems to be the impossibility of giving a unique modeling of the lens. The subject of CGL is so new that with this first experience in mind we cannot disregard the possibility of the same fate concerning the determination the cosmological constant from CGL. Thus we will just mention two ideas.

Kovner (1988) shows that if we follow the light curve of the multiple images of a supernova (m_R = 20 to 23) within a large fragmented luminous arc (Cl2244, A2390, Cl0024+16) we can probably determine the Hubble constant from the short time delay (days from weeks). These blue arcs are very active star forming galaxies and the probability of having a supernova can be as high as one every few years. If the potential of such a lens is reasonably known we can hope somebody will take up the challenge of such a systematic survey.

A second idea arose when we succeeded in getting the redshift of the arclet A5 (z=1.305) in Abell 370 just after the redshift of its first luminous arc A0 (z=0.724). If we are able to model the cluster potential we can write expressions like $f_{ij}(Z_{lens}, Z_{a0}, Z_{a5}, \ldots Z_{an}, \ldots, q_o) = 0$ which allows us to calculate q_o. For the time being it is fruitless to use this method because the modeling of the lens is too uncertain, but this possibility should be kept in mind for the future because with it we can measure the redshift of many arclets for a cluster properly modeled with high resolution X-ray maps. The hope of getting q_o in this way is probably still small but we should at least investigate it in the future because to date no other possible measurements can be considered seriously.

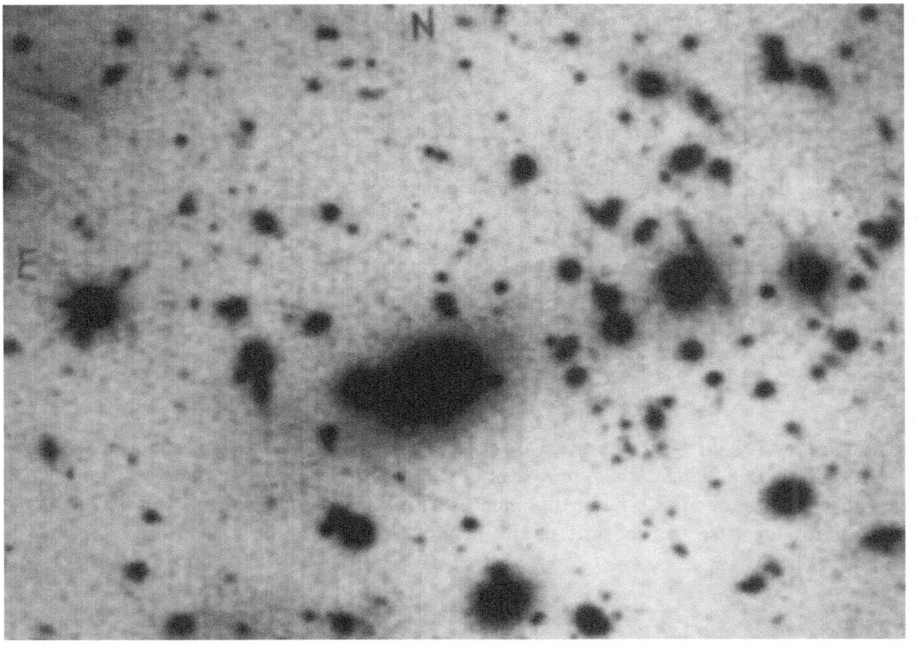

Figure 2 : the galaxy/cluster lens A2390

IV. THE OBSERVATIONAL POSITION

Table 2 presents all the CGL which have been published or presented at various conferences. At the date of this writing we already know many others and there is no doubt that the list will be extended by a large factor before many years. For each candidate we will just make a general comment about some peculiar points which could attract the attention of observers or theoreticians.

Abell 370

This is the only CGL with two arcs having a proposed redshift. Up to B=28 we detected about 27 arclets and/or gravitationally distorted images (GDI) when using the Tyson's findlens software which automatically searches for faint elongated tangential images. This very rich cluster has a strong X-ray emission and was extensively studied from a very complete spectroscopic survey (Mellier et al 1988). In this cluster the visible mass is largely bipolar within the core but it seems that the unseen mass cannot follow the visible light distribution. We have to consider a general elliptical halo with a small core radius ($r_c < 100$ kpc/h75) which could be centered on the X-ray emission (see also Grossmann and Narayan 1989). The luminous arc A_o has a thicker and tilted east end which was fairly reproduced with a source being a face-on spiral galaxy (Mellier 1989). We feel that Abell 370 still deserves extremely deep photometry "à la Tyson" in order to continue the effort of modeling.

Cl2244-02

The cluster is moderately rich with a low X-ray luminosity but with a very compact clump in the center which produces a beautiful blue arc having an angular extension of about 110 degrees (Lynds and Petrosian 1988). Hammer et al. (1989 a, b) found the possible counter image and give a model of the lens. In this conference Soucail et al. show that the arc is the image of a star-forming galaxy at a redshift of z=2.238. This is to date the spectrum of one of the most distant regular galaxies, and this result demonstrate the interest of CGL in studying very distant objects (cf 3a).

Abell 2218

Deep photometry has shown many arclets which confirms the first photometric result on this cluster (Pello et al. 1987). This candidate has also a complete optical Einstein blue ring around a galaxy near the center. This means that the dark halo of this galaxy does not have completely merge in the general cluster halo. So the presence of such a ring very close to the large giant galaxy having the arclet system is particularly interesting for the modeling of the lens and can impose some constraints on the core radius of the invisible mass associated with the giant cD galaxy. The ring should be reobserved at high spatial resolution with the HST. The reduction of all the photometric and spectroscopic data is under way in Toulouse.

Abell 963

Lavery and Henry (1988) were the first to report on the two blue arcs associated with the giant cD galaxy in Abell 963. This case is interesting for modeling the cluster center because we observe the arc and the counter-arc at large distance but still within the halo of the cD galaxy. It will be important to measure the redshift of the bright arc, the visible halo profile and the velocity dispersion of this cluster.

Abell 2390

This cluster located at z=0.231 was selected on the basis of its high X-ray luminosity. A lot of photometry and spectroscopy was obtained by Pello and co-workers (1989). It is also a fascinating

CGL for modeling (Mellier 1989) because it has a "straight" arc with several apparently parallel arclets located in the east near the second bright member of the cluster (figure 2). The large arc has a redshift of z=0.913 with an emission line (OII) visible all along the arc exhibiting a clear velocity gradient of about 370 km/s (Pello et al. 1989). The existence of such an arc geometry reflects a peculiar potential. The most obvious idea is that such straight gravitational images are produced in the saddle of a bipolar projected potential since a visible mass excess was marginally detected on the other side of the arclets. Indeed this object should be reobserved very soon with ultra deep photometry and should be also among the first target candidate for X-ray mapping.

Cl0024+1654

This rich cluster has a moderate X-ray emission and presents a very beautiful fragmented large arc and a counter-arc Koo (1988). It was observed spectroscopically by (Dressler and Gunn 1982) and reobserved with ultra deep photometry by Turner and Tyson (Tyson, this conference). It presents numerous arclets from the background population. So far we were not able to derive a secure redshift of the arc from the spectra we have, but this should be within reach of our new efficient faint object spectrographs.

Cl0500-24

This southern very rich cluster contains a bright blue arc (Giraud 1989) which was satisfactorily modelled with a general spherical cluster halo plus two important galaxies very close to the arc (Wambsganss et al. 1989). The surface brightness of the arc is bright enough to get the spectrum (during the conference P. Schneider announced a probable redshift of 0.9). No ultra-deep photometry is yet available for this CGL.

3C295

The cluster is a very rich and distant X-ray cluster. It is one of the clusters studied with ultra-deep photometry by Tyson (this conference). Long ago a very high velocity dispersion of 3000 km/s was suspected (Dressler and Gunn 1983). We notice from table 1 that for such a monster the locus of the very numerous arclets should be at large radii from the center (and possibly out of the frame of CCD prime focus field). Since then, it seems that a new spectroscopic survey has revealed the superimposition of three clumps more probably having a velocity dispersion of about 1000 km/s (Tyson 1989). This would be more in accordance with the observations of the number and locus of the arclets.

Abell 1689

This is one of the richest class 4 Abell clusters with a classical Abell number count = 228 and a strong X-ray emission. The Bahcall No count of galaxies ($m < m_3 + 2$ within 0.5 Mpc from the center) equal to 123 gives, from the classical empirical correlation law, a velocity dispersion near 2000 km/s. Gudehus (1989) calculates a radial velocity of 1800 ± 200 km/s from a spectroscopic survey of 68 galaxies. It was clear from table 1 that such a cluster was among the best candidates for finding numerous arclets. This was marvelously demonstrated by Tyson during this conference and we refer the reader to his paper (see also Tyson 1989).

Abell 545 and others...

This is one of many other CGL taken on a list established from a long term observing program (figure 5 and 6) in which we are accumulating ultra-deep photometry for the mapping of dark matter with arclets (like : A33, A483, A520, A665, A908, A1758, A2111, A2645, Cl0940+47). For the best candidates a spectroscopic follow-up of the cluster members and brightest arcs is expected. This effort will consume a lot of telescope time and is now part of a more general international US/EEC collaboration (Fort, Mellier, Leborgne, Pello, Turner, Tyson, Soucail). It will benefit from the ESO Key-program procedure in 1991.

V. THE OBSERVATIONAL STRATEGY

In order to fulfill the scientific objectives of paragraph 3 we have decided to survey about 60 clusters of galaxies of various richness, half of them being very rich X-ray emitters or having large C, No counts (these parameters are strongly correlated to high velocity dispersion). We are also planning to observe several distant CGLs at redshift larger than 0.7 to study the distance and evolution of the background population (see 3.1). We also hope to analyze this way, if at large distance, most rich clusters that did collapse in several substructures. We also consider the possibility of participating in a follow up of CGLs in the near IR to get the spectra of very distant objects (3.1) and in X-rays for the implications of better modeling of the lenses (3.2 and 3.3).

In fact any cluster of galaxies will produce distorted gravitational images of the background galaxy population. Only if we want image splitting with a lot of elongated arclets and arcs is it necessary to have very massive and compact clusters. We already have more Abell clusters to survey at $z > 0.15$ than available telescope time. Such a program will actually need very large CCDs and a dedicated large telescope because it already appears that the number of such CGLs could be potentially very great and that they will constitute a new and abundant observational class of gravitational lenses spread over very large scale in the Universe.

Acknowledgement

We are grateful to Y. Mellier, G. Soucail, T. Tyson, R. Blandford, I. Kovner, J.-F. Leborgne, R. Pello and R. Ellis for stimulating discussions. We also want to sincerely thank G. Ellis, who at an early time pointed out the importance of our pioneering observational works. It was a great encouragement for the Toulouse group.

Bibliography

FORT, B., PRIEUR, J.-L., MATHEZ, G., MELLIER, Y., SOUCAIL, G. : 1988, Astron. Astrophys. **200**, L17

DRESSLER, A. and GUNN, J.E., 1983, Astroph. Journal **270**, 7

DRESSLER, A. and GUNN, J.E., 1982, Astroph. Journal **263**, 535

GIRAUD, E. : 1988, Astroph. Journal **334**, L69

GROSSMAN, S., NARAYAN, R. : 1988a, Astroph. Journal **324**, L37

GROSMAN, S., NARAYAN, R. : 1989, Astroph. Journal **344**, 637

GUDEHUS, D. : 1989, Astrophys. J. **340, 661**

GUNN, J.E. : 1989, in "The epoch of galaxy formation", ed. C.S. Frenk (Dordrech : Kluwer)

HAMMER, F., LE FEVRE, O., JONES, J., RIGAUT, F., SOUCAIL, G. : 1989, Astron. & Astrophys. **208**, L7

KOO, D. : 1988, Private communication

LAVERY, R., HENRY, J.P. : 1988, Astroph. Journal **329**, L21

LYNDS, R., PETROSIAN, V. : 1989, Ap. J. **336**, 1

MELLIER, Y. : 1989a, Proceeding of the Space Telescope Sciences Institute Workshop, Clusters of Galaxies, preprint, Baltimore 15-17 May, USA

MELLIER, Y., SOUCAIL, G., FORT, B. and MATHEZ, G. : 1988, Astron. & Astrophys., **199**, 13

PELLO, R., SOUCAIL, G., SANAHUAJA, B., MATHEZ, G. and OJERO, E. : 1988, Astron. & Astrophys. **190**, LII

PELLO, R., LE BORGNE, J.-F., SOUCAIL, G., MELLIER, Y., and SANAHUJA, B. : 1989 Toulouse preprint, submitted to Astrophys. J.

SOUCAIL, G., FORT, B., MELLIER, Y., PICAT, J.P. : 1987a, Astron. & Astrophys. **172**, L14

SOUCAIL, G., MELLIER, Y., FORT, B., HAMMER, F., MATHEZ, G. : 1987b, Astron. & Astrophys., **184**, L7

SOUCAIL, G., MELLIER, Y., FORT, B., MATHEZ, G., CAILLOUX, M., : 1988, Astron. & Astrophys, **191**, L19

TYSON, J.A., SEITZER, P. : 1988, Astroph. Journal **335**, 552

TYSON, T., WALDES, F., WENK, R. : 1989, preprint, Astrophysical Letter, 15 january 1990, in press

WAMBSGANS, J., GIRAUD, E., SCHNEIDER, P., WEISS, A. : 1989, Astrophys. J. Letters, in press

LENSING THE BACKGROUND POPULATION OF GALAXIES

J. A. Tyson
AT&T Bell Laboratories
Murray Hill, NJ 07974

ABSTRACT

Deep CCD surveys to 29 mag from .32 to .9 micron wavelength reveal an isotropic population of very blue galaxies. There are over 200,000 of these objects per square degree per mag. The redshift of this population appears to be in the range 1 - 3. Systematic alignment of many of these faint background galaxies has been detected, centered on foreground galaxy clusters. The background galaxy population is selected using its anomalously blue B-R color. At a limiting surface brightness of 29 Bj mag arcsec^{-2} there are 30 background galaxies per square arcmin, which is sufficient to map statistically the dark matter distribution in the foreground cluster. Automated pattern recognition software generates a 2-d mass density map. There is evidence for a soft core of about 100 kpc radius for the dark matter in the clusters studied so far. Independent of other redshift constraints, the existence of such a large percentage of lens-distorted faint blue galaxies sets a lower mean redshift limit to this background galaxy population of $z > 1$.

1. INTRODUCTION

Deep CCD imaging surveys have revealed a population of faint blue resolved galaxies which cover the sky isotropically. At a limiting surface brightness of 29 B mag arcsec^{-2}, the sky is about 15% covered with faint blue galaxies. This population has an unusually blue color near B-R = 0.3. Several lines of evidence suggest that these galaxies are at redshifts between 1 and 3 (Tyson 1988a; Guhathakurta, et al. 1990). This distribution of distant galaxies is uniform over the sky to better than 10%. As reviewed by Bernard Fort in the previous paper, evidence for frequent gravitational lensing is mounting, with the recent discovery of several "arcs" in compact rich galaxy clusters (Soucail, et al. 1988). A large blue arc occurs if a distant evolving galaxy falls near a caustic. Lensed QSOs and bright arcs from caustics (as in A370) may be just the brightest examples of gravitational lensing.

Compared with galaxy-QSO lenses, the probability of galaxy-galaxy lensing is high. If all galaxies have massive dark halos sufficient to collectively close the Universe, then the systematic distortion of background galaxies within 30 arcsec of these foreground galaxies would be detectable. Automated detection of galaxy-galaxy lensing has been used to set limits to the dark halo of individual foreground galaxies (Tyson, et al. 1984; Tyson 1987). 1200 foreground galaxies with background galaxies within 30 arcsec were studied, out of a photographic sample of 50000 galaxies. No systematic background galaxy image gravitational lens distortion was found. That experiment was designed to test the hypothesis that *all* galaxies have massive dark halos, independent of absolute luminosity. Those data were sufficient to rule out that hypothesis, for isothermal halos with greater than 180 km sec^{-1} circular velocity. Background image distortions at radii larger than a few arcseconds due to a galaxy mass are small, and the measurement of isolated galaxy mass distributions using this technique must await large area deep surveys using CCD mosaics. It will be necessary to search the region within 30 arcsec of tens of thousands of foreground galaxies to faint limits, correcting in each case for the bright foreground galaxy's light profile.

The much larger mass associated with rich clusters of galaxies distorts background galaxy images over a larger area of the sky. Foreground galaxy clusters at redshifts .2-.5 with radial velocity dispersions above 1000 km s^{-1} have sufficient mass density to significantly distort background galaxies of redshift greater than 1 (see Turner, Ostriker, and Gott 1984). The 20-30 high redshift background galaxies within the central square arcminute of a foreground compact cluster of galaxies will be significantly lensed by the cluster mass distribution, creating faint arcs on several arcsecond scales, selectable by their relatively blue color. These are not the relatively rare caustic alignments: *every* background galaxy above some critical redshift will be seen as stretched along a circle centered on the lens. By going faint, the improved statistics of these distortions can help map the dark matter (DM) distribution, and thus constrain the nature of the DM (Tyson 1988b; Tyson, Valdes, and Wenk 1990). This new method is immune to the problems associated with unbound galaxy contamination and dynamical assumptions in cluster virial mass estimates. We have measured multiple background galaxy gravitational lens distortions for four clusters of known high velocity dispersion and X-ray luminosity.

Fig. 1. The left half shows a 1.7 x 2.4 arcminute part one of the deep fields produced in the CCD survey in good seeing, with the bright objects removed. The right half of this frame appears as it would with a σ_v = 1500 km s^{-1} soft core isothermal dark mass plus 33 cluster galaxies (not shown) placed at z=.5 at the right edge of the frame.

2. CCD SURVEYS OF THE BACKGROUND POPULATION

Deep CCD imaging surveys have been carried out in 13 random high galactic latitude fields to 29 B_J, 28 R, and 26 I mag arcsec^{-2} limiting surface brightness. Recently, ultra-deep CCD imaging to 30 mag arcsec^{-2} in U, B_J, and R has been obtained in three fields (Guhathakurta, Tyson, and Majewski 1990). Limits to the redshift distribution of this faint population of galaxies may be obtained with existing large telescopes, from spectrophotometry and gravitation lens tests. Observing techniques are described in Tyson and Seitzer (1988). The automated detection and photometry software FOCAS (Jarvis and Tyson, 1981; Valdes, 1982) produces a catalog of object multi-band photometric data. Figure 1 shows part of one deep survey field, with objects < 22 mag removed, and with the right half replaced by a simulation of a cluster DM lens (discussed below). Features in the surface brightness distribution of the extragalactic sky detected as discrete objects [size under 1 arcminute] constitute the majority of the extragalactic light. Deep imaging in several wavelength bands can be used to construct a rough spectral distribution for this population. The observed object size distribution is narrowly peaked in the 1-3 arcsec range. Our measured spectral distribution of these galaxies is slightly redder than flat in F_ν. This spectrum is more UV-bright than nearby galaxies and is probably due to evolution in this faint galaxy population.

After seeing deconvolution galaxies in our sample fainter than 26 B_J mag are found to have average metric exponential scale length 0.2-0.3 arcsec. Over the magnitude range of 17 to 27 B_J mag of our CCD sample there is no evidence for a significant tail past 3 arcsec radius in the galaxies, and all but two of the faint galaxies with central surface brightnesses fainter than 24 B_J mag arcsec^{-2} have scale lengths less than 1 arcsec. The blue galaxies in our deep sample have considerably higher ellipticities than those in clusters. An upper redshift limit for this population of faint blue galaxies is immediately apparent from its spectral energy distribution. Virtually all which are detected in B_J are detected in U, with no significant drop in flux. If a significant fraction of these galaxies were at redshift greater than 2.8, the Lyman break would be shifted through the U passband, causing these galaxies to either drop out in the U band or to have a drop in flux between B_J and U. Lyman breaks from most stars in these primeval galaxies are probably at least a factor of two. In addition, these galaxies would have even more hydrogen than present galaxies, which would absorb all the Lyman continuum photons, causing these galaxies to be black in the U-band for redshifts larger then 3 (Guhathakurta, Tyson, and Majewski 1990).

3. CLUSTER LENS DISTORTIONS

Preprocessing and FOCAS automated detection, splitting and photometry reduced the multi-color images to a position-matched catalog (Tyson 1988a). Lens-distorted background blue galaxies become mini-arcs along circles centered on the cluster lens. A lower redshift limit for this faint population can be derived from its response to a gravitational lens placed in front of it. A good test is provided by the dark matter lens in the clusters 1409+52 (z=.46) and 0024+16 (z=.39). Roughly the same number of faint blue galaxies per square arcminute are seen behind these red clusters, but most of their images are stretched by the gravitational lens and are aligned orthogonal to the vector to the cluster center. The distorted images of background galaxies appear as mini-arcs, small parts of circles centered on the foreground cluster. A significant excess of faint blue background galaxies are aligned tangent to circles centered on the foreground cluster lens, by the distortion of the gravitational lens. The excess in the tangent alignment bin can only occur if most of these galaxies have redshifts greater than 0.9. Figure 2 shows the critical surface mass density for lensing vs the redshift of the background population of galaxies, for cosmological density parameter .2 and 1, for the cluster 1409.

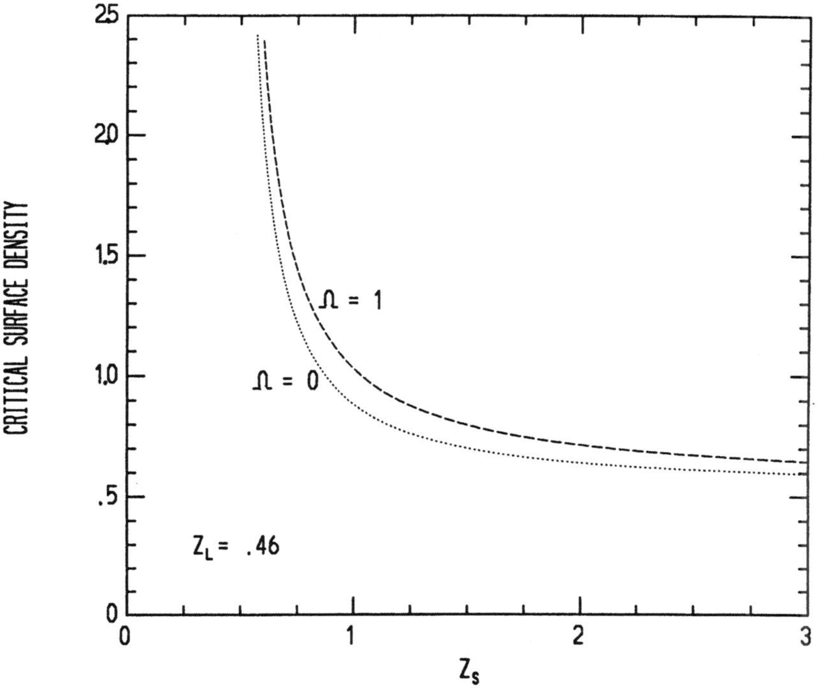

Fig. 2. The critical surface mass density for a gravitational lens at redshift 0.46 to focus a background galaxy as a function of redshift is shown. The asymptotic surface density at high redshift, only slightly dependent on Ω, is about 0.8 g cm^{-2}. For background redshifts much lower than 1, unphysically large mass densities would be required in the cluster lens for this case.

3.1 Image Distortion Statistic

Several faint arcs are seen in our blue images. However, the coherent alignment of the more numerous faint blue (background) galaxy ellipticities gives a stronger gravitational signal. The effect of a cluster gravitational lens is to stretch the apparent size of the image tangent to a circle centered on the lens. This stretch is measured using two scale lengths: the intensity weighted second moment of the galaxy image (part of the FOCAS catalog; Tyson et al. 1984, 1990) orthogonal and along the radius relative to the lens center. To detect the coherent stretching from the lens and cancel the intrinsic ellipticities of the galaxies, the scale lengths are averaged over the background population. From the two mean scale lengths we form a dimensionless quantity similar to an ellipticity, the difference of the lengths divided by the sum of the lengths, which we call the net tangential alignment. A random population of galaxies will give a net alignment of zero while a population of lensed galaxies will give a positive value.

To detect if a gravitational lens is present, the alignment statistic is computed using a grid of positions as candidate lens centers and the faint blue background galaxies with ellipticities > 0.2 within 300 kpc of that point (the radius is arbitrary and not critcal). A real lens produces an alignment value which is a maximum at the true center. The alignment strenths at the lens centers are highly significant (> 99% likelihood) as determined by comparison to a large number of random fields created by randomizing the sample galaxy positions without changing their measured shapes.

We study the alignment strength as a function of radius and compare it to the light distribution. The alignment statistic is computed using galaxies at various radii from the cluster center. Figure 3 shows the cluster red luminosity profile compared with the alignment strength of faint blue galaxies in CL1409. Log red intensity vs log radius is plotted along with the radial dependence of the lens image distortion.

The gravitational image distortions appear to extend radially at least as far as the cluster red light. The lens distortions are roughly correlated with cluster red light both in central position and radial extent. This implies that the dark matter distribution follows the *smoothed* cluster luminosity, with most of the mass interior to 100 h_{75}^{-1} kpc. A null result s found in eleven control fields, each with a surveyed area similar to that of our cluster data. Statistical significance of the lens alignments (7-10σ per cluster) and particularly the integral distortion statistic is high. This systematic lens distortion is not possible for background galaxies below a critical redshift, depending on the lens mass and redshift (Turner, et al. 1984): an unreasonably large mass for the lens would be required. Outside the cluster core, the area density of aligned blue galaxies found in the z = .46 cluster 1409+52 field is nearly equal to the number of blue galaxies found in our deep sparse field surveys.

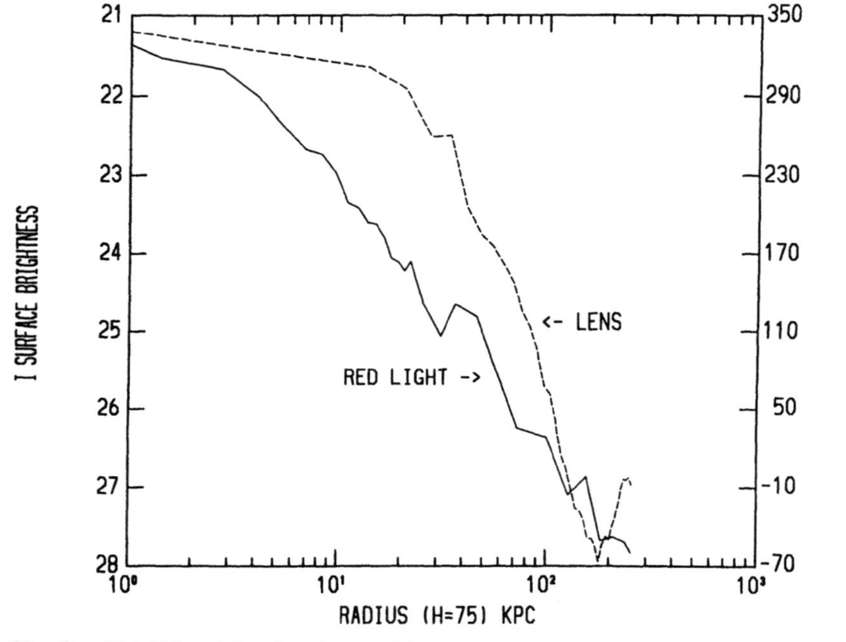

Fig. 3. CL1409 red luminosity profiles compared with strength of tangential alignment of faint blue background galaxies. Observed log red intensity vs log radius is plotted along with the radial dependence of the net tangential alignment.

4. Cluster Lens Simulations

We have simulated cluster lens distortions of background galaxies, by ray-tracing real images, as a function of cluster and DM parameters and compared them with the observations (Tyson, et al. 1990). The background galaxy fields are taken from actual deep sparse field imaging on the same observing run. These background galaxies are assigned redshifts in three model schemes: all at z=1, all at z=2, and equally at z=1 and 2. The cluster galaxies (found by their color, brightness, or redshifts) are assigned truncated isothermal mass distributions obtained from individual galaxy surface photometry to 28 B_J mag, K-corrections derived from the observed colors, and the Faber-Jackson relation σ_V = 220 dex [(20.2 - B_J)/10]. These cluster galaxies make a small contribution to the background galaxy image distortions beyond several arcseconds from the cluster galaxies. A soft core isothermal mass distribution is superposed on the cluster galaxies. The velocity dispersion (500-2500 km sec^{-1}) and core half-surface-density radius (25 to 100 h_{75}^{-1} kpc) are varied. To compare the observations to the simulations the net tangential alignment of background galaxies over the extent of the cluster is computed for both the observations and the simulations in exactly the same way (including FOCAS processing). This is paticularly necessary because this distortion statistic is wealy dependent on the parent background galaxy ellipticity distribution.

The alignment strength is found to be proportional to the radial velocity dispersion and weakly dependent on DM core size ($R_{core} < 100\ h_{75}^{-1}$) and the background redshift (z_{bg}>1). The best agreement with the observed lens statistics for CL1409+52, CL0024+16, and A1689 is obtained for $25 < R_{core} < 50\ h_{75}^{-1}$ kpc, $\sigma_{V,DM}$ = 1100-1500 km sec^{-1}, for z_{bg}=2. Note these "DM" parameters refer to all mass, dark + luminous, *other* than that associated with the modeled 20-50 brighter cluster galaxies. We are experimenting with other distortion statistics and their relation to the surface mass density. This statistic is insensitive to a gentle outer mass cutoff. Based on these initial clusters, we find that the DM distribution is similar to the smoothed red cluster light. This is consistent with baryonic DM or baryonic - non-baryonic dissipative coupling. A study of a large sample of galaxy cluster fields can constrain the nature of the dark matter.

We are grateful to A. Cronin, P. Guhathakurta, C. Christian, P. Waddell and P. Boeshaar for help with some of the observing and data preprocessing. We had helpful discussions with E. Turner, J. Gunn, J. Huchra, B. Fort, Y. Mellier, and C. Jones. This research is now part of a collaboration with the Toulouse group to study galaxy clusters and arcs with spectroscopy and imaging.

REFERENCES

Guhathakurta, P., Tyson, J. A., and Majewski, S. 1990, *Astrophys. J.*, in press.

Jarvis, J. F., and Tyson, J. A. 1981, *Astron. J.*, **86**, 476.

Soucail, G., Mellier, Y., Fort, B., Mathez, G., Cailloux, M. 1988, *Astron. Astrophys.*, **191**, L19.

Turner, E. L., Ostriker, J. P., and Gott, J. R. III 1984, *Astrophys. J.* **284**, 1.

Tyson, J. A., Valdes, F., Jarvis, J. F., and Mills, ,A. P., 1984, *Astrophys. J.*, **281**, L59.

Tyson, J. A., 1987, in *Theory and Observational Limits in Cosmology*, W. R. Stoeger, ed. (Specola Vaticana), p. 441.

Tyson, J. A., 1988a, *Astron. J.*, **96**, 1.

Tyson, J. A., 1988b, *Bull. Amer. Astron. Soc.*, **20**, 980.

Tyson, J. A. and Seitzer, P. 1988, *Astrophys. J.* , **335**, 552.

Tyson, J. A., Valdes, F., and Wenk, R. A. 1990, *Astrophys. J.*, in press.

Valdes, F. 1982, Proc. *SPIE*, **331**, 465.

THE REDSHIFT DISTRIBUTION OF FAINT GALAXIES – IMPLICATIONS FOR GRAVITATIONAL LENSING

Richard S. Ellis

Physics Department, Durham University, U.K.

SUMMARY. *The redshift distribution derived from complete samples suggests that the majority of blue galaxies dominating the faint field counts are moderately low redshift $z<1$. Additionally, new results in distant clusters suggest the presence of numerous old red galaxies at $z \simeq 1$. I question whether these observations are consistent with the first results on the predominantly blue population of very faint galaxies background to $z \simeq 0.4$ clusters as deduced from lensing studies.*

1. Introduction

The study of faint "field" galaxies has undergone great strides in the last decade. Measuring machine number-magnitude counts of deep prime focus photographs revealed an excess of faint galaxies when compared to expectations based on the properties of present day galaxies (Kron 1978, Peterson *et al* 1979, Tyson & Jarvis 1979), and subsequent studies in several passbands indicated a general bluing of the galaxy population with increasing apparent magnitude (Koo 1981). Most authors have claimed these results represent some form of galaxy evolution, a common interpretation being that some subset of the population was more luminous and blue in the recent past because of enhanced star formation activity (e.g. reviews by Koo 1985, Ellis 1987). In general, these conclusions have been reinforced by deeper CCD imaging surveys of Tyson (1988), Metcalfe *et al* (1988) and Cowie & Lilly (1989) where a substantial surface density of blue "flat-spectrum" objects have been detected. With these objects it has been proposed that we are witnessing directly the process of metal formation in normal galaxies (Cowie *et al* 1989).

Recently, through the advent of multiple object spectrographs it has at last become practical with long exposures to gather *redshifts* for galaxies at limits where the excess population is seen. Redshift distributions for complete magnitude-limited samples directly address the epochs over which the bulk of the normal galaxy population is evolving, as well as providing valuable data for large scale structure questions.

In this paper I will review the results of the redshift surveys and discuss how studies of gravitational lensing of the faint galaxy population might offer complementary information. Already some important results in this area have been presented by the Toulouse group and Tyson and co-workers (see articles in this volume) using rich clusters of known redshift as lenses. It is interesting, therefore, to see to what extent the results of the two different approaches agree, and to consider the cosmological implications.

2. Spectroscopic Surveys of Faint Galaxies

The major breakthrough in spectroscopic surveys of faint galaxies was undoubtedly the implementation of multi-object spectrographs on our largest telescopes (see the review by Ellis & Parry 1988 for technicalities). Several surveys have been completed or are underway; these are listed in Table 1. I will concentrate here on the surveys conducted at the AAT with my collaborators at Durham, although the conclusions are broadly similar to those found by Koo and co-workers.

Table 1

Deep Spectroscopic Surveys

Instrument	N_{sp}	m_{lim}	R Å	Reference
AAT+FOCAP	187g	$20< b_J <21.5$	4	Broadhurst *et al* 1988
AAT+LDSS	87g+33s	$21< b_J <22.5$	13	Colless *et al* 1989,1990
AAT+LDSS	>150	$b_J <24$	45	Colless *et al**
KPNO+Cryocam	>400g	R<22	15	Koo & Kron (1988)*
CFHT+FOS	>6g+2s	$23<b_J <24$	50	Cowie & Lilly (1989)
ESO 3.6m+EFOSC	–	–	–	de Lapparent *et al**

* – in progress

The Broadhurst *et al* (BES) survey was motivated by Kron's suggestion that the faint end of the galaxy luminosity function had been seriously underestimated in the local surveys and that the faint count "excess" might simply be due to a plethora of low luminosity blue dwarf galaxies. This seemed possible given the abundance of dwarfs in nearby clusters (Binggeli *et al* 1985) although one might have expected many to be classed as stars at faint magnitudes.

The BES survey, conducted with the AAT fibre positioner *FOCAP* (Gray 1985), reached b_J=21.5 where the counts are $\sim$30% above the no evolution prediction. However, the survey showed *no* evidence for a departure from no evolution (NE), the mean survey redshift $\overline{z}=0.216\pm0.016$ (1σ scatter from 5 independent fields) agreeing well with that (0.22) for the NE model. Models with a steep extension of the galaxy luminosity function or an additional dense population of low-luminosity dwarfs such as would be required to steepen the counts, were ruled out since they would lower $\overline{z}$ well below 0.2, in conflict with the survey. Too few low luminosity galaxies were found (none with z<0.03) for these to be a major contribution to the faint counts.

The absence of luminous high z galaxies to b_J=21.5 might not be surprising given the small count excess at that magnitude. However, a fainter survey completed with our new wide-field multislit spectrograph, the Low Dispersion Survey Spectrograph (LDSS, Wynne & Worswick 1989) reveals a similar result to b_J=22.5 (Colless *et al* 1989,1990). Spectroscopic targets were chosen with no regard to image classification – both galaxies *and* stars were surveyed to ensure compact objects were not misclassed. Here $\overline{z}=0.312\pm0.022$ (1σ from 3 independent fields), again close to the no evolution prediction of 0.29, yet the surface density of galaxies at the LDSS limit is almost *twice* the NE prediction. The combined dataset BES+LDSS represents ~300 galaxies with 20< b_J <22.5 and the mean redshift ($\bar{z}$) - apparent magnitude relation is in good agreement with the trend predicted by the NE model (Figure 1). Few galaxies to b_J=22.5 have z>0.5 and the mean redshift is only slowly increasing with apparent magnitude.

Incompleteness cannot seriously bias the observed N(z). Only ≃15% of the AAT sample remains unidentified and thus *regardless of their redshift the bulk of the "excess" population seen in the counts must be at low z.* [O II] 3727 Å emission is present in 85% of the identified galaxy spectra and since the observed N(z) declines well before [O II] shifts out of the observing spectral window (z=0.9 for the LDSS data), it is unlikely that the unidentified objects have 0.6<z<0.9. Thus, either the redshift distribution is bimodal (with the unidentified 15% having z>0.9) or the observed N(z) is representative of that for the entire population. Colless *et al* (1990) were able to use this point quantitatively by considering their survey to be limited in terms the s/n *in the emission line*, rather than in apparent magnitude.

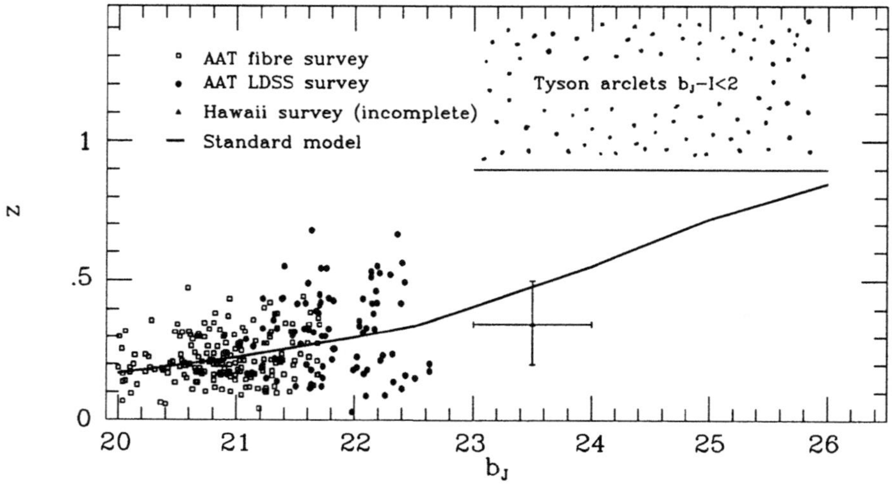

Fig. 1: Redshift versus apparent magnitude for the two AAT field redshift surveys (BES, Colless et al), with preliminary data from Cowie & Lilly (1989). The solid line shows the predicted mean relation for magnitude limited samples assuming no evolution in galaxy properties. The latter curve is extrapolated for illustration purposes to limits equivalent to Tyson's imaging surveys, and suggests that the majority of b_J ≃25 galaxies would not lie substantially behind a z=0.4 cluster.

The conclusion that most of the excess galaxies cannot be at high z provides a strong constraint on many forms of luminosity evolution originally postulated to explain the steep count slope. Strong evolution of the entire galaxy luminosity function, for example, is ruled out. The apparent paradox of a non-evolving N(z) that fails to predict the absolute number of galaxies to a given magnitude was resolved by BES who proposed a luminosity-dependent evolution where *only the low luminosity galaxies get brighter with look-back time*, at least for z<1. This has the effect of increasing the number at a given magnitude limit at the expense of little change in $\bar{z}$.

An important feature of BES' hypothesis was their explanation of the increasing proportion of faint spectra showing unusually strong [O II] emission. Unlike the spectra of nearby spirals which are accurately modelled by near-constant star-formation rates, the faint galaxies are too blue for their Balmer absorption line strengths and can only be understood in models incorporating short intense bursts of star formation. These galaxies are, in effect, luminous HII regions, examples of which can be found at low redshift but with an abundance far lower than in the faint surveys. The optical and rest-frame uv luminosities of such a galaxy increase markedly during an active burst of star formation. BES suggested the 30% excess population to b_J=21.5 would be too faint to be detected in their quiescent state and that the steep count slope originates from a gradual increase with look-back time in the occurrence of these short bursts of star formation.

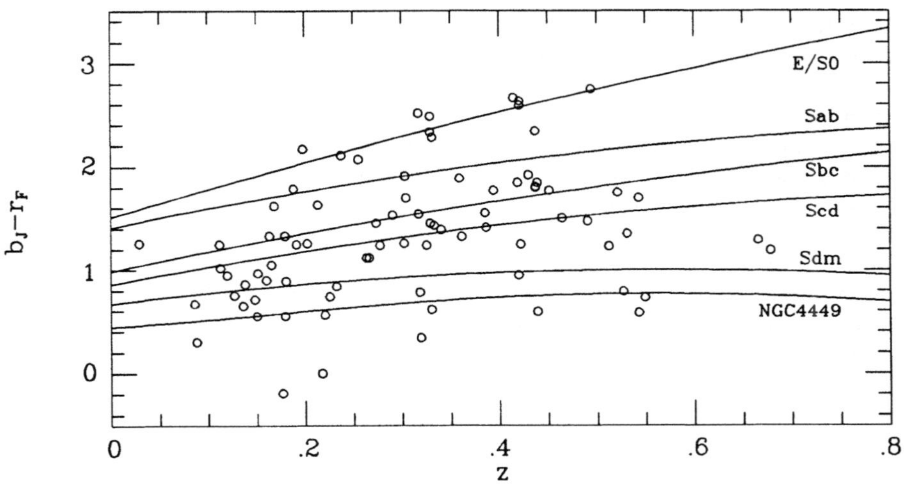

Fig. 2: The distribution of b_J-r_F colour with redshift for 81 galaxies from Colless et al (1990). The curves represent unevolved colours as a function of redshift for various types including NGC4449, a Sm/Irr exemplary of intense short-term star formation.

Colours from the deeper LDSS survey supports the burst hypothesis further (Figure 2). No blueing at the red end of the colour distribution is observed; to at least z∼0.5 galaxies are found with colours similar to present day old E/S0s. Thus general evolution in the entire population is minimal. Stricter constraints are now available from similar arguments in high redshift clusters (Hamilton 1985, Persson 1988, Aragón Salamanca *et al* 1990) indicating the presence, even at z≃1, of truly old (>2-3Gyr) galaxies. Additionally, the bluest galaxies are not exclusively at high redshift; the most extreme examples, representative in colour of a significant fraction of the *flat-spectrum* population in the CCD counts (Cowie & Lilly 1989) are *at low z.*

In conclusion therefore, spectroscopic evidence to $b_J \sim 23$ suggests that the majority of the excess galaxies seen in the number magnitude counts lie below $z \simeq 0.5$. The proportion of high redshift galaxies, say with $z>1$, is very small ($<15\%$) and consistent with zero. Additionally, there is growing evidence that the bluest objects may be the *least* likely to be at high redshift, and that many red galaxies may be truly old even at $z \sim 1$. The identification faint blue≡luminous high z may not hold true.

3. Flat Spectrum Galaxies in Deep CCD Surveys

The above spectroscopic studies lie firmly in the photographic domain. Whilst imaging with CCDs has pushed considerably fainter, grafting the results of the new CCD data to the well-established photographic data has not been easy. In the overlap region at $b_J \sim 24$, the properties of ~20,000 photographic galaxies on a single plate have to be compared with <~100 on one of Tyson's CCD frames (and barely 10 on one of Cowie & Lilly's frames). Both Tyson (1988) and Cowie *et al* (1989) claim to have identified the onset of a *new* population of flat-spectrum objects, but the fact that this emerges at an apparent magnitude close to the intersection of the two techniques, and that the two teams disagree on the rapidity of onset, leads one to suspect these claims. Cowie & Lilly (1989) report the flat-spectrum onset is steep ($\propto 10^{1.5m}$ – 3 times that of the normal population). Yet Tyson (1988) refers to the population's "sub-Euclidean counts (which) suggest they are neither a new class of local object ($z<0.3$) nor dwarf galaxies".

Is this flat spectrum population a continuation of the gradual bluing seen at brighter magnitudes – in which case one might imagine most to be modest z bursting galaxies, or does it represent a genuine population of new distant primaeval galaxies? Lensing studies offer an unique opportunity to resolve this important question.

Spectroscopic redshifts for genuinely flat-spectrum objects have been reported for only a handful of objects by Cowie & Lilly (1989) within the range $23<b_J<24$ (all have $z<0.5$) and 7 are seen in the larger LDSS sample to $b_J=22.5$ (of which 4 are identified, all with $z<0.35$). Although larger and more complete samples are desirable, it appears so far that the bulk of these blue galaxies are also at low z.

4. Inferences on High Redshift Galaxies from Lensing Studies

At this workshop we have witnessed the exciting new results from deep imaging surveys (c.f. articles by Tyson, Grossman and the indefatigable Toulouse group). The spectroscopic confirmation that the giant arcs seen in modest z clusters are indeed gravitationally lensed background objects (Soucail *et al* 1988) opens up a new tool in observational cosmology. Traditionally, the only way of studying objects with $z>1$ has been via absorption line spectroscopy of QSOs or use of rare (and probably untypical) classes of radio galaxies. Our discussion above (c.f. Figure 1) shows that, so far at least, it is proving extremely difficult to systematically identify high z galaxies efficiently in the field. Using rich clusters as *gravitational telescopes* we may be able to determine the properties of high redshift galaxies unobtainable any other way. Furthermore, since objects are lensed only because they happen to lie serendipitously behind a totally unrelated cluster, selection effects are minimal; this is a most important consideration.

Two basic results have struck me at this meeting. Firstly, most of the arcs identified spectroscopically with z>1 galaxies are surprisingly blue. New redshifts have been presented for 3 such arcs at this meeting; the A5 arc in Abell 370 has z_{arc}=1.3 (Soucail), that in Cl2244 has z_{arc}=2.238 (Mellier *et al*) and the "straight" arc (terminology here needs attention!) in A2390 has z_{arc}=0.92 (Pelló *et al*). Although the photometric data on many of these arcs is not yet adequate, the indications are that each of these high redshift objects is extraordinarily blue. Aragón Salamanca and Ellis (this meeting) show from precision infrared photometry in A370 that the colours of the A5 arc are *at least* as blue as the most active star forming galaxies today. Similar results may hold for Cl2244 which has yet to be detected at 2μm.

The second result concerns the smaller distortions (arclets) found in deep images of moderate z clusters. In the case of those detected visually from colour subtraction techniques (e.g. Fort *et al* 1988), these are typically as blue as the interesting object A5 mentioned above. In the case of statistical elongation searches (Tyson *et al* 1989), a significant excess of elongations orthogonal to the cluster radius vector is *only* found for the bluest galaxies. The surface density of such galaxies in and out of the cluster frames indicates that at least 70% of the galaxies with b_J-I<2 (a criteria somewhat less strict than used for defining the flat-spectrum component in §3) at b_J ~24-25 must lie substantially behind the clusters, i.e. z>1.

Thus, in contrast to the spectroscopic data of §2-3, the lensing studies suggest the majority of galaxies with z>1 serendipitously-found by arc-hunters are very blue - as blue, if not more so, than any present day galaxy. The inferred star formation rate for the only one studied in detail (A5 in Abell 370) is at least as high as that seen in the bluest Magellanic irregulars today. Additionally, the surface density of high z objects inferred from the lensing technique suggests *most* of the faint flat-spectrum sources must be genuinely high z objects of considerable luminosity.

5. Discussion

Although giant arcs can be recognised fairly straightforwardly in deep images, they are relatively rare. For example, a preliminary analysis of sets of coadded frames for ≃20 rich clusters with <z>≃0.3 in the Danish supernova search programme (see Nørgaard-Nielsen *et al* 1989) revealed no arcs longer than θ=15 arcsec apart from that in Abell 370 (which happened to lie in the sample) (Smail *et al* in preparation). For 5< θ <15 arcsec, distinguishing arcs from edge-on galaxies becomes very subjective in crowded fields. Colour subtraction e.g. B-R, has been employed by Fort *et al* and Tyson and co-workers, but this must produce a bias in favour of detecting *blue* features against the predominantly red cluster galaxies. The colour-redshift relation for early-type objects flattens at high z and a background elliptical would be far harder to distinguish from a cluster member from colour alone. Whilst isolated features such as A5 (A370) and the curvaceous arc in Cl2244 are genuinely blue, a proper statistical sample *with accurate optical/infrared photometry* is urgently needed before one can conclude from such data that the vast majority of serendipitous z>1 galaxies are blue. As Aragón Salamanca and Ellis (this volume) show, infrared data is vitally important to understand properly the significance of the optical colours; what appears to be "blue" in B-R at z>1 may merely reflect the presence of a small number of hot stars in an otherwise red object.

Given the predominance of *red* objects in clusters to z~1, it is surprising that not a single red arc has been seen. Grossman *et al* (1989) in an analysis similar to that of Tyson *et al* claim to find distortions in the red not blue populations but their data was considerably shallower (R~24) and has poorer spatial resolution (their seeing was rarely better than 1.5 arcsec).

In Figure 1 I summarise, rather schematically, the possible conflict between the complete redshift samples and Tyson *et al*'s conclusion regarding the background nature of blue objects at fainter limits. Although, the orientation test data reaches considerably fainter, it seems one cannot reconcile the two datasets unless there is a transition in the redshift distribution (and hence nature) of blue objects beyond b_J ~23. If correct, this is a most important result.

Attempts to 'fill in the gap' with more spectroscopic redshifts are ongoing in Hawaii and at the AAT (see Table 1). As discussed in §2 preliminary data from Cowie & Lilly (1989) aggrevates the discrepancy. Meanwhile, we require accurate optical/infrared photometry for a good sample of arcs likely to have z>1 and a thorough examination of the orientation statistics in rich clusters with accurate velocity dispersions. Systematic surveys in large numbers of clusters are required. The prospects for determining the most active era for the formation of stars in normal galaxies have never been better, and the exciting results of this meeting have shown clearly that we have more than one way ahead.

References

Aragón-Salamanca, A, Ellis, R S, Couch, W J & Carter, D 1990, in preparation.

Binggeli, B, Sandage, A & Tammann, G 1985 *Astron. J.*, **90**, 1681.

Broadhurst, T J, Ellis, R S & Shanks, T 1988 *M.N.R.A.S*, **235**, 827.

Colless, M M, Ellis, R S & Taylor, K 1989 in *The Epoch of Galaxy Formation*, eds. Frenk *et al*, p 359, Kluwer.

Colless, M M, Ellis, R S, Taylor, K & Hook, R N 1990 *M.N.R.A.S*, in press.

Cowie, L L & Lilly, S J 1989, in *The Evolution of the Universe of Galaxies* ed. Kron, R, ASP Conference Series in press.

Cowie, L L, Lilly, S J, Gardner, J & McLean, I S 1989 *Ap. J.*, **332**, L29.

Ellis, R S 1987 in *Observational Cosmology*, IAU Symposium 124, eds Hewitt, D *et al*, p367, D Reidel.

Ellis, R S 1989 in *The Evolution of the Universe of Galaxies* ed. Kron, R, ASP Conference Series in press.

Ellis, R S & Parry, I R 1988 in *Instrumentation for Ground-Based Astronomy*, ed. Robinson, L, p192, Springer-Verlag.

Fort, B, Prieur, J-L, Mathez, G, Mellier, Y & Soucail, G 1988 *Astr. Astrophys.*, **200**, L17.

Hamilton, D 1985 *Ap. J.*, **297**, 31.

Gray, P M 1986 *Proc. S.P.I.E*, **627**, 96.

Koo, D C 1981 Ph D thesis, University of California, Berkeley.

Koo, D C 1985 in *Astronomy from Measuring Machines*, eds. Reid, N & Hewett, P C, RGO Publications.

Koo, D C 1989, in *The Evolution of the Universe of Galaxies* ed. Kron, R, ASP Conference Series in press.

Koo, D C & Kron, R 1988 in *Towards Understanding Galaxies at High Redshift*, eds. Kron, R and Renzini, A, p209, Kluwer.

Metcalfe, N, Shanks, T & Fong, R 1987 in *High Redshift & Primaeval Galaxies*, eds. Bergeron, J *et al*, p37, Editions Frontieres.

Nørgaard-Nielsen, H U, Hansen, L, Jørgensen, H E, Aragón Salamanca, A, Ellis, R S & Couch, W J 1989 Nature **339**, 523.

Persson, S E 1988 in *Towards Understanding Galaxies at High Redshift*, eds. Kron, R & Renzini, A, p251, Kluwer.

Peterson, B A, Ellis, R S, Kibblewhite, E J, Bridgeland, M T, Hooley, T & Horne D 1979 *Ap. J.*, **233**, L109.

Soucail, G, Mellier, Y, Fort, G, Mathez, G, Cailloux, M 1988 *Astr. Astrophys.*, **191**, L19.

Tyson, J A 1988 *Astron. J.*, **96**, 1.

Tyson, J A 1989 in *The Evolution of the Universe of Galaxies* ed. Kron, R, ASP Conference Series in press.

Tyson, J A & Jarvis, J F 1979 *Ap. J.*, **230**, L153.

Wynne, C G & Worswick, S 1989 *Observatory*, **108**, 161.

The Theory and Practice of Radio Ring Lenses

Christopher S. Kochanek
Department of Astronomy
University of California, Berkeley
Berkeley, CA 94720

INTRODUCTION: The formation of closed ring images by gravitational lensing was predicted by Einstein in a Science article published in 1936. Einstein considered only lensing by stars in the galaxy, and it was not until the following year that Zwicky first made the suggestion of gravitational lensing by galaxies. The possibility of ring images around galaxies was certainly considered as a theoretical possibility, but it was not necessarily expected in observed samples of lenses – primarily because little thought has been devoted to multiple imaging of extended sources. In 1986, when Hewitt *et al.* discovered the radio ring lens MG1131+0456, the lensing community's reaction was less one of astonishment than one of delight in a new and beautiful phenomenon. A second radio ring MG1634+1346 was found in 1989 by Langston *et al.* as part of the ongoing MIT-Green Bank (MG) survey which found MG1131. The fact that the rings are found with radio observations of extended sources which have both better resolution, and better contrast between lens and source than those observed at optical wavelengths allows inversion procedures of unprecedented accuracy to determine the parameters of the lensing potential and the form of the source (Kochanek *et al.*, 1989). In this brief review, I will first cover the observational details of the two rings, followed by a description of the inversion technique developed by Kochanek *et al.* with its application to MG1131 and MG1654. I will also discuss the weaknesses of the technique and several ideas for improving its speed and accuracy. Finally, I will discuss the probability of ring images relative to that for lensed point sources.

MG1131+0456 was selected as a lens candidate on the basis of its radio morphology in a 5 GHz "snap-shot" image from the VLA. The radio map shows a nearly complete ring of radio emission approximately one arc-second in radius connected to two compact cores, one inside, and one outside the ring. It has been observed at 5 GHz with a spatial resolution of 0.″3 and a signal-to-noise ratio of $S/N \simeq 100$, and at 15 GHz with a spatial resolution of 0.″1 and a signal-to-noise ratio $S/N \simeq 10$. Polarization maps were also made at 5 GHz with $S/N \simeq 10$. Contour maps of the object at both frequencies and the polarization map are shown in Figures 1, 2, and 3. An optical counterpart to the radio source is found with $m_R \simeq 22$, but all attempts to determine a redshift have failed because the object is so faint. The major axis of the optical source appears to be aligned with the major axis of the radio source and Tyson reports (in a comment at the conference) that the blue optical source shares the orientation of the red source, and appears to be a ring. This suggests that the optical counterpart is dominated by light from the source rather than from the lens.

MG1634+1346 has been observed at 5 Ghz (resolution 0.″30, $S/N \simeq 50$), 8 GHz (resolution 0.″15, $S/N \simeq 20$), and 15 Ghz (resolution 0.″10, $S/N \simeq 5$), with polarization measurements at 8 GHz (Langston *et al.*, 1989ab). The underlying radio source is a triple – one radio lobe and the core are unlensed, and the second radio lobe is converted

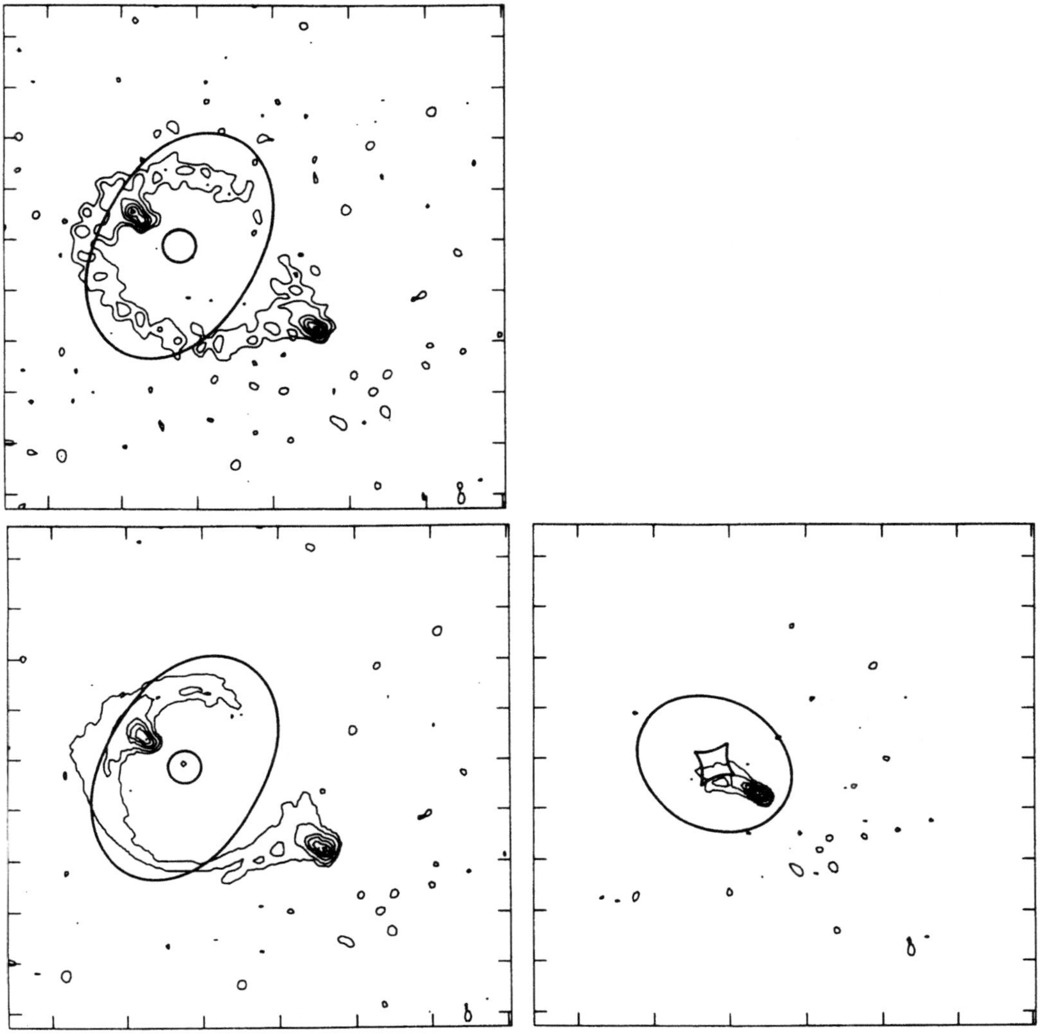

Figure 1. The 15 GHz map of MG1131+0456 (top) shown with the inversion for the source (lower right) and the inverted image (lower left). The inversions have been smoothed over a kernel three pixels wide (0."1) to eliminate numerical pixel-to-pixel noise. Contour levels are 10, 20, 40, 60, 80, and 95% of the peak flux density. The noise level is at approximately 10% of the peak flux density. All three panels show the same region of the sky, with North at top and East to the left. The sky coordinates of the map centers are $\alpha = 11^h31^m56^s.40$ and $\delta = 04°55'49''.40$ (J2000.0). Tick marks are at 0."5 intervals on the declination axis, and at 0^s.05 intervals on the right ascension axis. The critical curves are superimposed on the image plane (top and lower left) and the caustics are superimposed on the source plane (lower right).

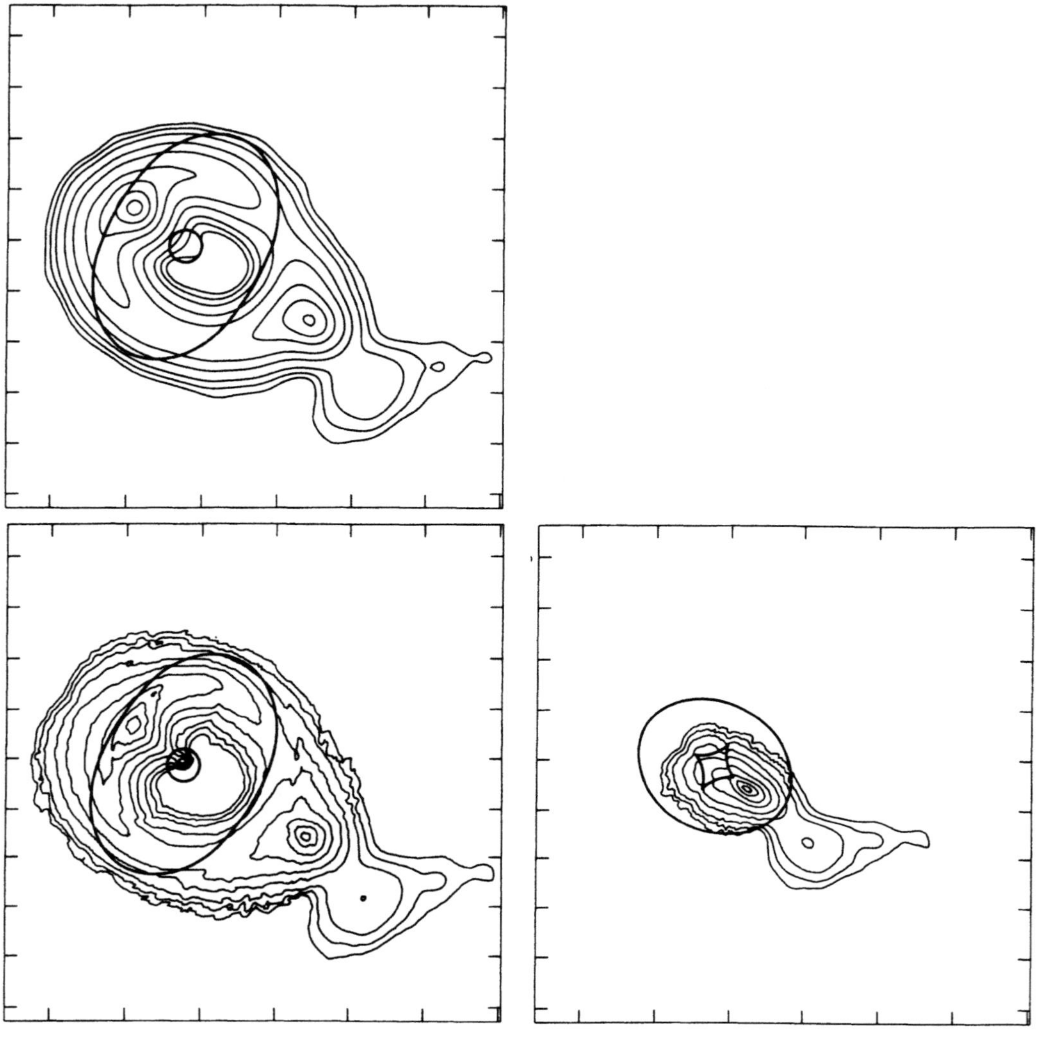

Figure 2. The 5 GHz map of MG1131+0456 (top) shown with the inversion for the source (lower right) and the inverted image (lower left). The inversion is smoothed over a kernel three pixels wide (0.″1). Contour levels are 1, 2, 4, 10, 20, 40, 60, 80, and 95% of the peak flux density. The noise level is at approximately 1% of the peak flux density. Note that these inverted maps were generated using the inversion computed for the 15 GHz map. The region of sky shown is identical to that in **Figure 1.**

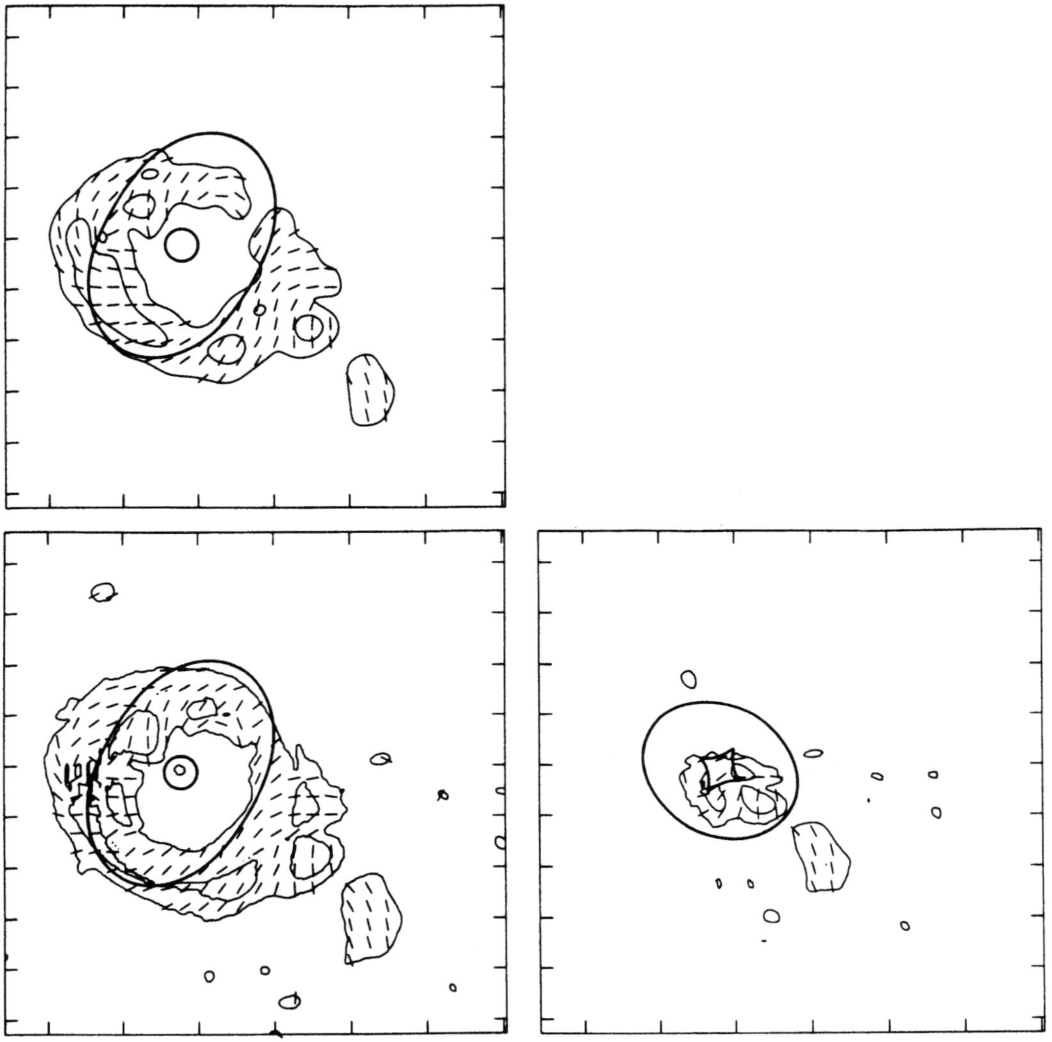

Figure 3. Polarized intensity $(Q^2+U^2)^{1/2}$ map of MG1131+0456 at 5 GHz (top), shown with the inversion for the source (lower right) and the inverted image (lower left). The inversion is smoothed over a kernel three pixels wide (0."1). Contour levels are 15 and 50% of the peak polarized flux density. Superimposed line segments of constant length show the local direction of the electric field. The noise level is at approximately 10% of the peak polarized flux density. The Q and U maps were inverted separately, using the inversion computed for the 15 GHz map, and then summed and smoothed to form the polarized intensity map. In inverted image (lower left), the inner contours in the southwest quadrant are at the 15% level, while the corresponding contours in observed image (top) are at the 50% level. This conversion of a ridge in the observed image to a trough in the inverted image is at least partly due to the low resolution of the polarized map. The region of sky shown is identical to that in **Figure 1.**

into a ring approximately one arc-second in radius. The lensed lobe has two principal components: A which lies towards the core, and B which lies away from the core. (See Figure 4) If we assume that the total flux in the lensed lobe is comparable to that of the unlensed lobe, we can estimate the magnifications of the A and B components to be $M_A \simeq 17$, $M_B \simeq 12$ and $M_A/M_B \simeq 1.4$. The radio core is coincident with an optical QSO of $m_r \simeq 21$ and redshift $z_q = 1.74$. A galaxy is seen centered on the lenses radio lobe, with $m_v = 18.7$ and redshift $z_g \simeq 0.25$. The principal axis of the galaxy appears to lie at 95 degrees (N through E) and the major to minor axis ratio is 0.7 although considerable uncertainties remain in the ellipticity and orientation.

The Inversion Procedure: The key to the Kochanek *et al.* inversion procedure is the fact that gravitational lenses conserve surface brightness. If we have a resolved image of a lensed source, than points of the image which originate from the same point on the source, must have the same surface brightness. The lensing equations represent a one-to-one map of the image plane onto the source plane independent of the choice of the lensing potential. This means that if we assume a lensing potential, the surface brightness distribution of the image can be mapped onto the source plane. If portions of the source are multiply imaged, the true lens potential only connects image points with equal surface brightness to a given source point, and *the extent to which the assumed potential does not map points of equal surface brightness gives a measure of the error in the lens potential.* The inversion strategy is as follows: (1) assume a parametrized form for the lensing potential, (2) measure the error associated as a function of the values of the parameters, where the error is defined to be the average r.m.s. dispersion in the surface brightness of multiply imaged source pixels, (3) the best model for a given parametrization are those values of the parameters which minimize the error function. In practice there are some additional complications which are discussed in Kochanek *et al.*, but this is the essential feature of the technique – what remains are tricks for finding the global minimum rather than a local minimum. The particular beauty of the method is that no source shape is ever assumed – it is a natural consequence of the method. The technique can be generalized to operate on several images at once (different frequencies, polarizations) but in practice it is best to simply use the highest resolution map available.

Results for MG1131+0456: The potential model we use is a pseudo-isothermal sphere,

$$\phi = b\left(s^2 + (1-\epsilon)x'^2 + (1+\epsilon)y'^2\right)^{1/2}$$

where $b = 4\pi(\sigma/c)^2 D_{LS}/D_{OS}$ is the radius of the critical line if the potential was circular and singular, s is the core scale length, ϵ is the ellipticity, and x' and y' are related to the coordinate axis x and y by a translation and a rotation. ($x' = (x - x_0)\cos\theta + (y - y_0)\sin\theta$). The quality of the current data does not justify the use of more complicated "realistic" potentials, and this simple model contains all of the qualitative features of real galaxy potentials. We define the uncertainty in a parameter to be the change in the parameter under which the error doubles, if all other parameters are varied so as to minimize the increase in the error. There appeared to be a problem with this solution because the lensing galaxy's major axis is perpendicular to the ring, while the optical image is parallel to the ring. The optics, however, require this alignment to generate the

TABLE 1. Fit Parameters and Errors for MG1131

Parameter	$x_0('')^2$	$y_0('')^2$	ϵ	θ^3	$b('')$	$s('')$
Value	−0.70	+0.11	0.14	150	0.97	0.07
Variation[1]	0.15	0.16	0.04	12	0.11	0.20

[1] The variation is an estimate of the change in the parameter which will lead to a 50% increase in the error measure, when all other parameters are varied so as to minimize the error due to the change.
[2] The position is measured as an offset from the center of the map.
[3] Position angle (north through east) of the major axis (x axis) of the potential.

observed gap in the ring. The alternate orientation can generate the gap only if there is a gap in the radio emission between the cores and the ring. The observation of a hole in the blue optical image suggests that this interpretation is correct. If we assume a lens redshift of 0.5 and a source redshift of 2, these results imply a galaxy velocity dispersion of 250 ± 20 km s^{-1} and a time delay of $(0.19 \pm 0.05)h^{-1}$ years. Further improvements of the model can be made if the object can be remapped with better signal to noise at 15 GHz. If the signal to noise in the ring were to be doubled from the current ratio of 2:1, the shape of the potential could probably be constrained much more tightly. The inverted sources and images are shown in Figures 1, 2, and 3.

Results for MG1634+1346: The inversion of MG1634 is in progress, and we have only preliminary results. The best current results have $b = 0\farcs98 \pm 0\farcs10$ with an ellipticity of approximately 0.12 aligned in the direction of the optical image. The lens center lies within the centroid of the optical image. If we convert b into a velocity dispersion, we get $\sigma = 227 \pm 11$ km s^{-1} compared to 216 km s^{-1} derived from the Faber-Jackson relation. The current best-fit results are shown in Figure 4.

Problems and Potential Improvements: The problems with the inversion method fall into the two categories of resolution and convergence problems. It is clear that neither ring is resolved in the 5 GHz maps, but what about the 8 and 15 GHz maps? A gravitational lens does not introduce new extrema into the surface brightness distribution, it can only create multiple images of existing extrema. This means that the two brightest points on opposite sides of the ring are images of a single extrema, and the extent to which the measured surface brightness at these points differ is a measure of whether or not the lens is resolved. In particular, the two brightest peaks in the A and B images of MG1654 have flux ratio of 1.06, a difference which is ten times larger than the noise level of the maps. Hence, the images are smeared out by the beam, and the assumption that the radio map is an intrinsic surface brightness map is only approximately correct. The map resolution is also the limiting factor in the accuracy with which the parameters of the inversion can be measured. The general rule seems to be that changes in the lens position or critical radius which are comparable to the beam size will cause a doubling of the error measure. Within the current implementation of the technique, this problem cannot be solved except by higher resolution observations.

The second problem is that multi-dimensional, non-linear minimization techniques are time consuming and erratic. The convergence technique tends to find local extrema,

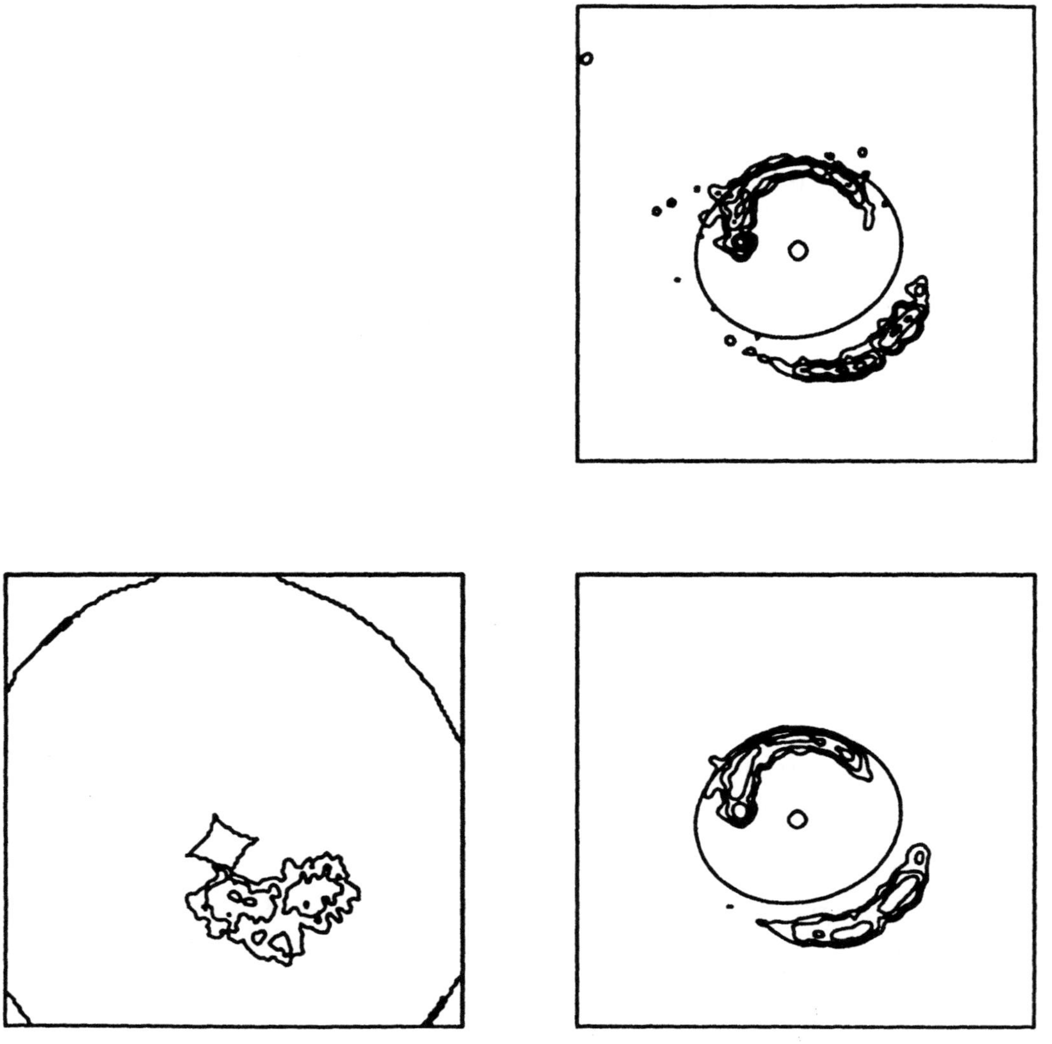

Figure 4. The 8 GHz map of MG1654+1346 (top) shown with the inversion for the source (lower left) and the inverted image (lower right). The location of the optical core coincides with the point of radio emission near the top left corner of the 8 GHz map. The critical curves are superimposed on the image plane (top and lower right) and the caustics are superimposed on the source plane (lower left). The source plane is magnified relative to the image plane by a factor of 3.3.

most of which correspond to highly fragmented sources, which led us to introduce various weighting functions into the error measure to bias the method against finding fragmented sources. It typically takes 500 iterations per convergence run, which requires approximately 30 minutes on a Convex mini-supercomputer, and then many convergence runs are required to be sure a true global minimum has been found. This makes the method computationally intensive even though we are using a minimalist potential model. The best hope for reducing the level of difficulty is to reduce the effective number of parameters through the introduction of external constraints. We know that the two images of the core in MG1131 must come from the same source, so that the gradient of the time delay evaluated at the positions of the two cores must be equal. If we could measure the positions of the cores perfectly, this would give two constraint equations, reducing the number of parameters in our simple potential from six to four, leading to an enormous reduction in computation. In practice, of course, we cannot determine the positions exactly, so that we can only reduce the volume of parameter space from a full, six dimensional volume to a thin layer about a four dimensional sheet in the six dimensional space. The thickness of the sheet depends on the error in eliminating the two parameters because of the uncertainties in the positions of the image points.

How Many Rings Are There? The MIT-Green Bank survey has now found four lenses: two point source lenses (2016+112, by Lawrence *et al.*, 1984, and 0414+0534 by Hewitt *et al.*, 1989) and two extended source lenses (1131+0456 and 1654+1346). What balance do we expect between these two general morphology classes? A preliminary examination of this problem (Kochanek and Lawrence, *in preparation*) indicates that ring-like images should overwhelmingly dominate the total number of lenses. This occurs because the typical radio source found in the MG survey is not small compared to the cross section for multiple imaging, so that the effects of finite source size are critical to the statistics of the MG survey. In a very crude statistical model, the relative probability for lensed points sources to lensed extended sources is 1:6. Most of the extended source lenses will be rings found inside or on the edges of extended radio lobes, with typical optical depths for a large, distant radio lobe approaching 0.006, compared to 0.001 for typical point sources. The size of the rings in extended sources will typically be quite small – the average image separation for point sources is r_0, where r_0 is the radius of the critical line for a source at infinite redshift, while for the extended sources the average separation is $r_0/2$. For lensing potentials modeled as a singular isothermal sphere, $r_0 = 1\overset{''}{.}4\sigma_{220}^2$ where σ_{220} is the velocity dispersion of the lensing potential in units of 220 km s^{-1}. If the typical ring has a diameter of only $r_0/2 = 0\overset{''}{.}7$, while the resolution of the survey maps can only resolve rings larger than approximately $2\overset{''}{.}0$ (VLA at 5 GHz in B array) or $0\overset{''}{.}7$ (VLA at 5 GHz in A array) we will miss most of the lenses simply because they cannot be resolved. Our calculations indicate that in a sample of 642 objects (the original MG sample) the expected number of lenses is approximately 4.3, while with the VLA you would find 2.5 at 5 Ghz in the A array, 0.2 at 5 Ghz in the B array, and 3.5 in 8 Ghz in the A array. The original survey contained a mixture of A and B array observations, and would be expected to find 0.4 lenses.

These numbers are suspect because they contain some unjustifiable assumptions concerning the redshift distribution of the sources, but while the absolute value of each estimate is probably uncertain to a factor of two, the ratios of the estimates are much less uncertain. Hence we can summarize the rules of how and where to look for lenses in the radio:

(1) Always look at the most distant objects. The optical depth for a source at redshift $y = 1 + z_S$ scales as $(1 - 1/\sqrt{y})^3$ (Turner, Ostriker, and Gott, 1984).
(2) Always use the highest resolution you can. If the ratio of beam FWHP to asymptotic ring radius (r_0) is f, the value of y in the optical depth scaling of the first law is replaced by $y[1 + f(\sqrt{y} - 1)]^{-2}$. For radio surveys using the VLA, this means avoiding the B array, and low frequencies. Ideally, you want resolutions which can easily resolve a ring with diameter $r_0/2 = 0\farcs7\sigma_{220}^2$.
(3) If the survey contains extended objects, you will find that they are lensed much more frequently than point sources at the same distance. If the object has major and minor axis ℓ_1 and ℓ_2 the optical depth scales as $1 + 5(\ell_1 + \ell_2)/2r_0 + 10\ell_1\ell_2/r_0^2$. Note that for the value of r_0 given above, the optical depth doubles if a circular source is only $0\farcs2$ in radius, and that if the source radius equals r_0 the optical depth is 16 times higher!

Conclusions: The radio surveys are probably the most best understood of all the lens surveys from the point of view of selection criteria and completeness. Even so, there are serious questions about whether or not the MG survey is finding all of the lenses present in their sample. It is quite possible that the survey has missed three out of every four lenses (or worse) because of the combined problems of resolution, and the difficulty of recognizing lensed features as part of a larger, extended radio structure.

Even so, the extended radio lenses are the best candidates for sophisticated inversion techniques leading to tight constraints on the structure of the lens and the source. Current optical lenses (with the possible exception of the cluster arcs) have separations which are too close to the resolution of the observations. Moreover, all examples of isolated, optical lenses consist of multiply imaged point sources for which there are only a limited number of constraints. A system of five point images contains only 15 constraints even if the image magnifications are uncontaminated by the effects of microlensing. The typical number of multiply imaged pixels in the radio rings is two thousand! Moreover, the rings sample the shape of the potential all around the lens rather than only at a few isolated points.

This makes the rings the best candidates for measuring the Hubble constant. They are very highly constrained, and "clean" optical benches compared to objects like 0957. The time delay for the typical ring is conveniently short – approximately 90 days – compared to 400 days for 0957. Unfortunately neither of the current rings will serve for this purpose: MG1131 has two lensed, compact cores which are likely to vary, but we do not know the lens and source redshifts, while MG1634 has known redshifts, but radio lobes do not vary. We would, ideally, like to find a lens system which combines the two objects, an MG1131 with redshifts. It is well worth the effort, however, to monitor MG1131 for variations in the cores. If a convincing time delay can be measured, the difficulty in finding spectra can be overcome. An enormous amount of telescope time is dedicated to determining the Hubble constant by techniques with far greater systematic problems than those in a lensing determination. Even a small fraction of this time, if devoted to obtaining the spectrum of one object, would be enough.

Acknowledgements: I would like to thank Jackie Hewitt and Glen Langston for giving us the original VLA maps of MG1131 and MG1654. This research has been supported by a fellowship from the AT&T Foundation, NSF grants AST86-15325 and AST86-11121, and the Theoretical Astrophysics Center at U.C. Berkeley.

References

Einstein, A., 1936, *Science,* **84**, 506.

Hewitt, J.N., Turner, E.L., Schneider, D.P., Burke, B.F., Langston, G.I., and Lawrence, C.R., 1988, *Nature,* **333**, 537.

Hewitt, J.N., Burke, B.F., Turner, E.L., Schneider, D.P., Lawrence, C.R., Langston, G.I., and Brody, J.P., 1989, in *Gravitational Lenses*, J.M. Moran, J.N. Hewitt, and K.-Y. Lo, eds. (Springer-Verlag).

Kochanek, C.S., Blandford, R.D., Lawrence, C.R., and Narayan, R., 1989, *Mon. Not. R. astr. Soc.,* **238**, 43.

Langston, G.I., Schneider, D.P., Conner, S., Carilli, C.L., Lehar, J. Burke, B.F., Turner, E.L., Gunn, J.E., and Hewitt, J.N., 1989, *Astr. J.,* **97**, 1283.

Langston, G.I, 1989, *private communication.*

Lawrence, C.R., Schneider, D.P., Schmidt, M., Bennett, C.L., Hewitt, J.N., Burke, B.F., Turner, E.L., and Gunn, J.E., 1984, *science,* **223**, 46.

Turner, E.L, Ostriker, J.P. and Gott, J.R., 1984, *Astrophys. J.,* **284**, 1.

Zwicky, F., 1937, *Phys. Rev. Lett.,* **51**, 679.

LENS AND SOURCE MODELS FOR ARCS IN CLUSTER ABELL 370 AND CL 2244

by
Vahé Petrosian and Anton G. Bergmann
Center for Space Science and Astrophysiocs
Stanford University, Stanford, California 94305
and
Roger Lynds
Kitt Peak National Observatory, NOAO
Tucson, Arizona 85726

I. INTRODUCTION

We discuss two main aspects of the modeling of the arcs in clusters of galaxies as gravitational lens images. The first aspect deals with the lens and the determination of the amount of dark matter (or the mass to light ratio, M/L), its distribution and its relation to the distribution of the visible matter. This is done by construction of images of a source (assumed to be a simple circular or elliptical disk of constant surface brightness) that fit the observed shape and size of the well defined arcs in clusters Abell 370 and Cluster 2244-01. The second aspect has to do with the reconstruction of the source using the best fit lens model and the details of the surface brightness distribution of the observed images.

Before this, however, we present in Figure 1 the old (circa 1975) video camera images of the above-mentioned cluster arcs and the pieces of tangentially elongated images in Abell 2218 and compare these with the more recent (1985) CCD data, all obtained using the NOAO 4-meter telescope at Kitt-Peak. The CCD pictures of Abell 370 and Cl2244 were published in Lynds and Petrosian, 1989, hereafter LP89. The earlier data, in addition to being limited to a smaller area around the cluster centers, are of lower quality than the more recent data. In particular, for Abell 2218 the general impression of circular symmetry due to the presence of many tangentially elongated images in the CCD picture (even more obvious in the larger CCD frame not shown here) is not evident in the video camera data. For Abell 370 the pronounced deviation of the eastern knot from the circular shape of the rest of the arc (see LP89 for more details) is not clearly evident in the old data. Finally, in the early data the arc in Cl2244 is barely detected and does not show the pronounced surface fluctuations seen in the CCD data. The degree of the circularity of this arc could not have been as clearly demonstrated with the old data as done in LP89 using the CCD data. All these features revealed by the new data have been very important in establishing the gravitational lens model as the most viable model and in the determination of the characteristics of the lens (cluster of galaxies) and the source (a distant galaxy), which we describe in the next two sections, respectively.

II. IMAGE CONSTRUCTION

The details of our work on the description of the lens characteristics derived from accurate matching of model images to the data can be found in Bergmann, Petrosian and Lynds, 1990, some of which were presented at the Cambridge Workshop on Gravitational Lenses (Petrosian 1989). We summarize these results here. In general, we find that with plausible values for the parameters describing the lens and the source we can fit the observations satisfactorily. More specifically we find that:

1) The arcs provide clear and conclusive evidence for the existence of dark matter in clusters with a mass to blue light ratio within the range

$$200 < M/L_B < 1000(M_\odot/L_\odot)(H_0/(50\mathrm{kms}^{-1}\mathrm{Mpc}^{-1}).$$

2) In the case of Cl2244 the dark matter can be associated either with galaxies or distributed smoothly throughout the cluster, and a source consisting of a simple circular disk is sufficient to reproduce the general features of the arc.

3) Abell 370 arc, on the other hand, cannot be produced by a dark matter distribution that closely follows the distribution of the visible matter. In this case we can constrain the shape of the distribution of the dark matter assumed to be centered half way between the two giant ellipticals with central surface density, Σ_0, and size, r_0, (or velocity dispersion, σ_0). We find that a singular isothermal gas distribution truncated at $\geq$ 100 kpc is acceptable. Non-singular and non-truncated distributions can fit the observation only if most of the dark mass is concentrated within the arc ($\leq$ 100 kpc). More spread distributions require an improbable equality between the central surface density of the dark matter Σ_0 and the critical density Σ_c (a number which depends on the redshifts and the cosmological models) and a very large mass to light ratio $M/L_B \geq 10^3 M_\odot/L_\odot$.

4) In general, the required velocity dispersion of our models is compatible with that calculated based on X-ray observation of other similar clusters or the measured velocity dispersion of the galaxies in clusters.

5) The deviation from circularity at the eastern end of the arc in Abell 370 can be accounted for by an elliptical source of major to minor axis ratio of about 5.

6) The surface fluctuations of the arc in Cl2244 may be produced either by minilensing due to small ($< 10^{10} M_\odot$) and therefore invisible ($L_B \leq 10^8 L_\odot$) galaxies, or by fluctuations in the dark matter distribution on the scale of $10^{11} M_\odot$. A second possible source for these fluctuations is surface brightness variation in the source, which we will discuss elsewhere.

III. SOURCE RECONSTRUCTION

In the image construction just described, we assume a simple source (an elliptical or circular disk of constant surface brightness) and adjust the mass distribution in the cluster to obtain images which match the shape and size of the arc but not the details of its surface brightness distribution. This procedure gives a good approximation to the structure of the lensing clusters. We can now use acceptable models of the lens and the observed surface brightness distribution of the arc to construct the details of the source. This is a more straightforward analysis because the mapping from the image or lens plane to the source plane does not require the iterative solution of the lens equation that was necessary for image construction described above.

This procedure requires a good knowledge of the structure of the arc all along its length. Our data on Abell 370, obtained with better seeing than that for CL2244, is better suited for this task. Consequently, here we describe some reconstruction for Abell 370 and leave the same for Cl2244 to a future publication. Unfortunately, the image of the arc in Abell 370 suffers from considerable contamination by images of cluster galaxies. Therefore, not all image points on the arcs can be used for this mapping. The contour map of the arc and its surrounding area is shown in figure 2. The contamination mentioned above is more clearly evident here. It is clear that without an accurate subtraction of galaxy images most of the data cannot be used for this reconstruction. Such a subtraction on a low signal-to-noise ratio data is risky because of our lack of knowledge of the surface brightness distribution of these galaxies.

We, therefore, limit our analysis only to portions of the arc which are not contaminated by the images of the galaxies. The eastern knot is the best candidate for this. Source images obtained from mapping of this portion of the arc using one of our best models for the lens are shown in figure 3. As can be seen, a highly elliptical source is the result, but because of our truncation of the arc as we approach the first encroaching galaxy, the source appears truncated in the north. A complete image of the arc is required to complete this section of the source.

We have also mapped the few uncontaminated points on the arc between the second and third (from the west) encroaching galaxies. The result of these mappings is shown by the filled circles, which is in agreement with the contours but does not help in extending the image of the source to the north. It should be noted that this last mapping is sensitive to the mass of the nearby encroaching galaxies. For example, increasing the mass of the encroaching galaxies by a factor of 5/3 moves the image points to that shown by the open circles.

This demonstrates the usefulness of the procedure and that through source reconstruction one can constrain the masses and therefore the M/L ratios of the galaxies with greater certainty than through image construction described in section II. Further details on this and image reconstruction of Cl2244 will be described elsewhere.

Figure 1. Comparison of old (left) video camera and new (right) CCD data for Abell 2218, Abell 370 and Cluster 2244, respectively, from top to bottom. North is at the top and east to the left. Scale: $\approx$ 11 arc sec/cm.

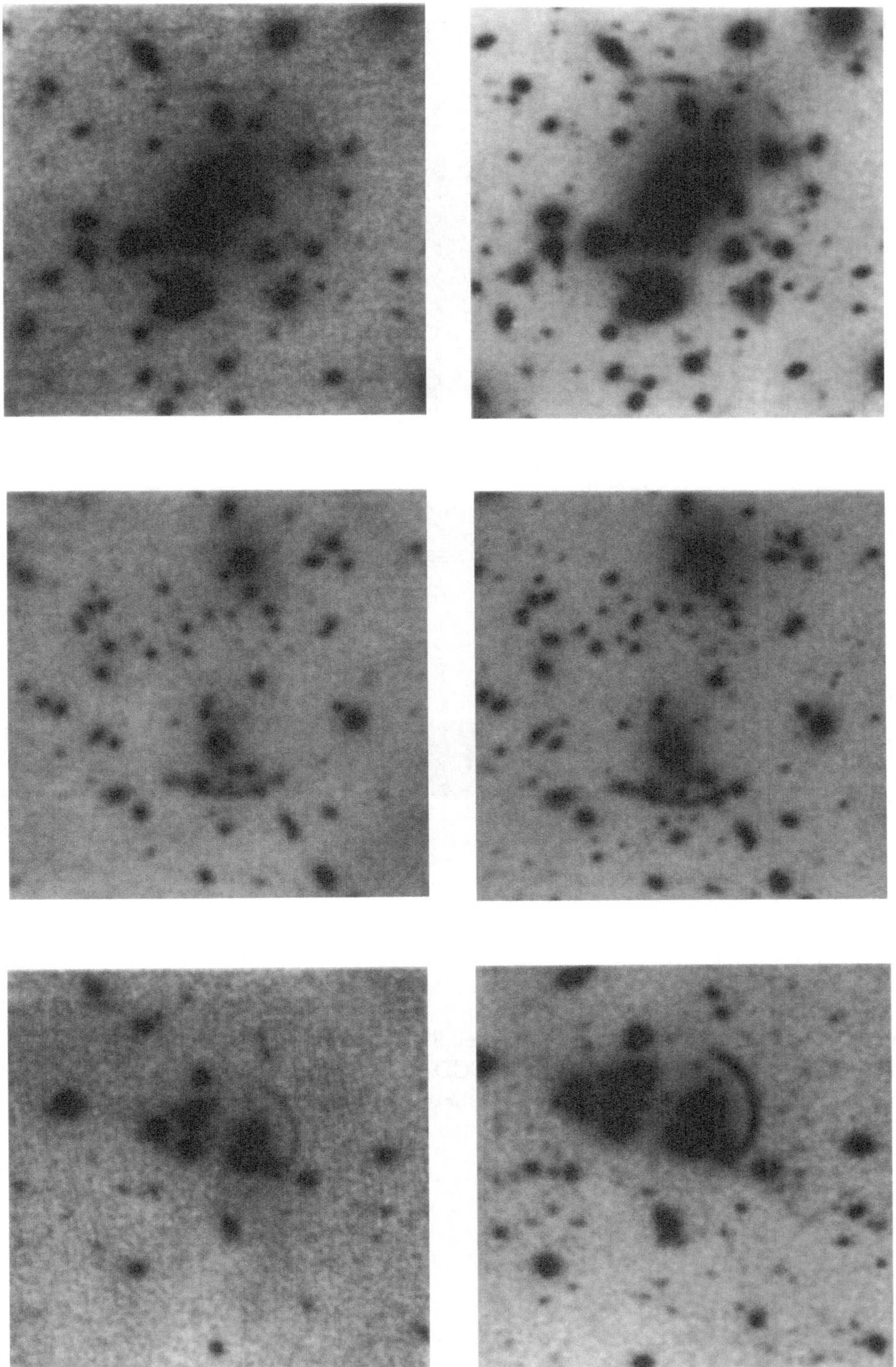

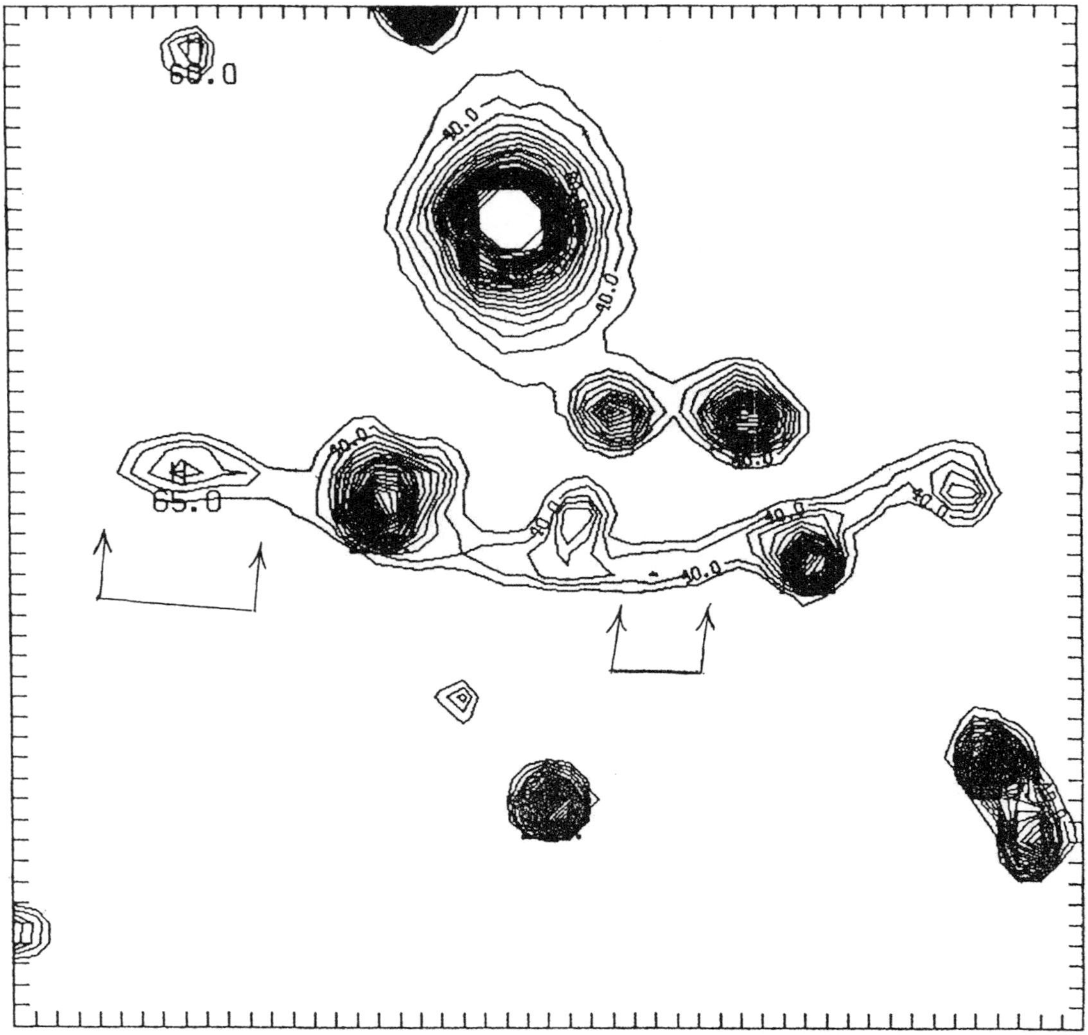

Figure 2. Contour map of the arc in Abell 370 and its surrounding area form an average V, B. R and I images of data from RCA CCD at the 4-meter at Kitt Peak. The portions of the arc used for mapping are the region between the arrows.

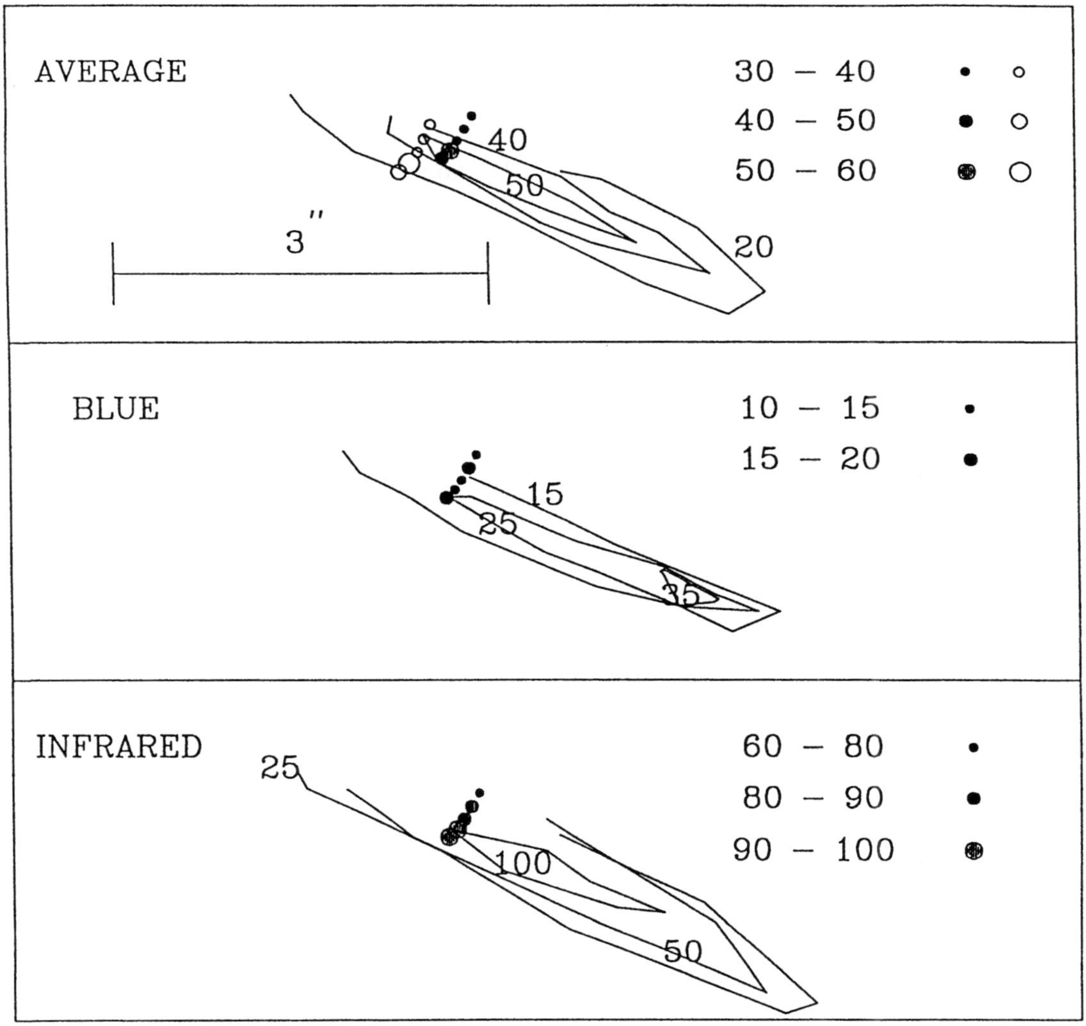

Figure 3. Source image from the reconstruction of the portions of the arc. The three contours with indicated relative surface brightnesses (in linear not magnitude scale) are obtained from the mapping of the eastern knot. The filled circles are obtained from the mapping of the other portion of the arc shown in Figure 2. The open circles are the same obtained when the masses of the nearby galaxies are increased by a factor of 5/3.

REFERENCES

Lynds, R. and Petrosian, V. 1989, *Ap.J.*, **336**, 1.

Bergmann, A., Lynds, R. and Petrosian, V. 1990, (*Ap.J.*. **350**.

Petrosian, V. 1989, **Gravitational Lenses, Lecture Notes in Physics, Vol. 330,** eds, J.M. Moran, J.N. Hewitt, K.Y. Lo (Berlin: Springer-Verlag), 238pp.

Modeling the Giant Arcs in A370 and A2390

Y. Mellier, G. Soucail, B. Fort, J.F. Le Borgne

Observatoire Midi–Pyrénées
14, avenue E. Belin
31400 Toulouse - FRANCE

and

R. Pello

Departimento de Fisica de la Atmosfera, Astronomia
and Grup d'Astrofisica del IEC
Universidad de Barcelona. Diagonal, 647
08028 Barcelona - SPAIN

1. Introduction

The discovery of arcs in distant clusters of galaxies (Soucail *et al.*, 1987; Lynds and Petrosian, 1986; Lavery and Henry, 1988) and the spectroscopic confirmation that they are lensed background galaxies by the cluster itself (Soucail *et al.* 1988, 1989; Pello *et al.*, 1989, 1990) opened a new field of investigation to probe distant clusters of galaxies and the deep universe. By now, a tenth of cluster–galaxies lenses are known (see Fort, 1989 for a review). They can be used as a tool either to diagnostic the distribution of the very distant blue galaxies, the distribution of dark matter within clusters of galaxies, or to infer cosmological parameters (though this point remains uncertain). However, in addition to observational data, previous investigations have shown that good modeling of the lenses is absolutely necessary.

The aim of this presentation is to discuss the modeling of the two clusters of galaxies A370 and A2390 which are the most extensively observed both in photometry and spectroscopy.

Though clusters lenses may be of a fundamental interest for galaxy evolution, our primary intent was to study cluster dynamics using multicolor photometry and a large spectroscopic survey of the clusters members (Mellier *et al*, 1988, 1990; Le Borgne *et al.*, 1990).

In principle, it is possible to infer from lens modeling the total mass, the projected mass density profile, and therefore the amount and location of dark matter within the deflector. This approach is independant of standard techniques used to study clusters of galaxies which are based either on photometric and spectroscopic data of the cluster members or on X-ray analysis. Basically, the standard analyses start from questionable hypotheses (for instance sphericity of the cluster, isotropy of the velocity tensor, isothermal X-ray gas, etc...) to infer dynamical or structural parameters. However, The and White (1986) pointed out how uncertain the models are, even in the case of the well known Coma cluster. The contradictory results of Cowie *et al.*(1987) and The and White (1988) clearly show that an independant analysis is crucial (see also Fitchett, 1989 for a review). Therefore, we may have an opportunity to check a posteriori the legacy of some assumptions made in standard dynamical analysis of clusters by comparison with the gravitational lensing approach.

In practice, modeling such clusters with giant arcs is a difficult challenge because of the large number of free parameters needed to define the source and the lens. It follows that even for the largest single arcs, a realistic diagnostic about dark matter is extremely difficult even when a maximum of observational constraints are available. This will be shown through a discussion of different solutions for the giant arc in A370 and the straight arc in A2390.

Hereafter, we will use the following parameters: $H_0 = 50km/s/Mpc$ and $q_0 = 0$.

2. Basic assumptions

The models presented here were obtained from an upgraded version of the code first developped by I. Kovner (Kovner, 1987). Basically the fitting technique to the observational data is a Monte Carlo sorting of parameters within a given realistic range. The main point is the shape of the two dimensional potential that we restrain to the class of distorted pseudo-isothermal elliptical models:

$$\Phi(x,y) = f\sqrt{b^2 + x^2(1-\varepsilon) + y^2(1+\varepsilon)}$$

where:

$$f = \kappa(0)\ b = 4\pi(\frac{\sigma_{los}}{c})^2\frac{D_{ls}}{D_{os}}$$

$\kappa(0)$ is the convergence ($= \Sigma(0)/\Sigma_{crit}$). It is greater than 1 in case of multiple imaging. b is the core radius, D_{ij} is the angular distance from i to j and ε is the ellipticity.

We do not use the velocity dispersion as an observational constraint in order to keep the mass we will compute independent of the virial mass. In addition, A370 has a strong velocity gradient (Mellier *et al.*, 1990; in preparation) which lead to an apparent increase of the measured velocity dispersion.

3a. A370:

The giant arc in A370 as well as the cluster were intensively described in previous papers (Soucail *et al.*, 1987; Mc Laren *et al.*, 1988; Mellier *et al.*, 1988) so we skip the description of the cluster itself.

Since Hammer (1987) presented the first point mass model to diagnostic A370, several models were published in which the departure from circularity of the potential well or the microstructures within the arc itself were included (Kovner, 1988; Hammer and Rigaud,

1989; Grossman and Narayan, 1989). Though those models include spectroscopic and photometric observations of galaxies, none of them include global structural parameters infered by numerical density or luminosity isocontours neither the information given by the spectral energy distribution of the source galaxy. However, in view of these data, A370 is clearly a north-south elliptically elongated cluster. The resolution given by optical or X-ray isocontours is rather poor (Mc Millan *et al.*, 1989), so it is quite difficult to say wether subclustering exists on a scale less than 1 arcmin ($\sim 400kpc$). However, the sharply elongated core is largely dominated by two bright cluster members. This encouraged us to check wether A370 could be modeled as a bimodal cluster with density peaks located near the position of the two first rank galaxies. This hypothesis is compatible with the elliptical shape of the low spatial resolution of the X-ray and the isopleth maps.

The spectrum of the arc obtained by Soucail *et al.* (1988) shows an unresolved emission line – which corresponds to the forbidden $[OII]\lambda 3727$ in the rest frame – and some typical Balmer absoption lines. This suggest that the source is a Sc or Sd spiral galaxy. Despite we cannot disgard the possibility that the source is a star-forming early-type galaxy (E-shape galaxy), the assumption that we are observing a spiral galaxy is probable and realistic since spirals represent more than 50% of field galaxies. Consequently we decided to test this alternative *bimodal cluster + spiral source* model and to compare it we previous works where a single potential and an elliptical source have been tested.

Figure 1 shows the best model we have obtained for the giant arc in A370. It matches very well the detailled geometry of the arc even in details: variation of thickness, curvature and the tilt of the eastern extremity. The model has the following caracteristics: 1) The two potential wells have velocity dispersion of 900 km/sec, which lead to a mass of $2 \times 10^{14} M_{\odot}$ within a circle of 26" centered on the cluster center. This is comparable to the total mass given by Hammer and Rigaud (1989) and corresponds to a mass-to-light ratio of $M/L \sim 100$ in agreement with the value given by Mellier *et al.* (1988). 2) There is no counter arc since the 2 elliptical potentials give an anisotropic mass distribution. 3) The eastern part of the arc which escapes from the circularity is an image of the eastern arm of the source, whereas the western arm gives a thinner shape structure on the other side. 4) In this model, no break is clearly visible near the galaxy ♯37. However, the image of the source become thinner on the left side of ♯62 before to increase and finally to give a broad image near ♯37. The thinning of the arc corresponds to the discontinuity between the image of the western arm and the western end of the bulge. Conversely the broad image in the western side of the bulge is an effect of magnification by the galaxy ♯37. This broad image is not so obvious on the CCD frames. But substracting B-R CCD image clearly show an enhancement (see fig. 2 in Fort *et al.*, 1988) exactly at this position. If we consider the broadness of the arc near ♯37 as an observational constraint, we then have to assume a velocity dispersion of about 270 km/s for this galaxy.

The bimodal+spiral model fits convincingly the observed data but it has a strange property: the only possibility to reproduce the arc is to have a first potential peak located on the giant galaxy ♯20 but the second potential well 10" away form the giant galaxy ♯35. Such a disymmetric distribution of the dark matter between the two brightest cluster members is not satisfactory unless one assumes that the galaxy ♯35 is excentered from the potential well generated by a more diffuse component of dark matter and/or X-ray emitting gas. In summary the comparison with other models seems in favour of a general single elliptical potential with possibly two small clumps associated with the two giant galaxies and a spiral source. The bimodal model for A370 does not support the hypothesis that dark matter in clusters is only associated to galaxies.

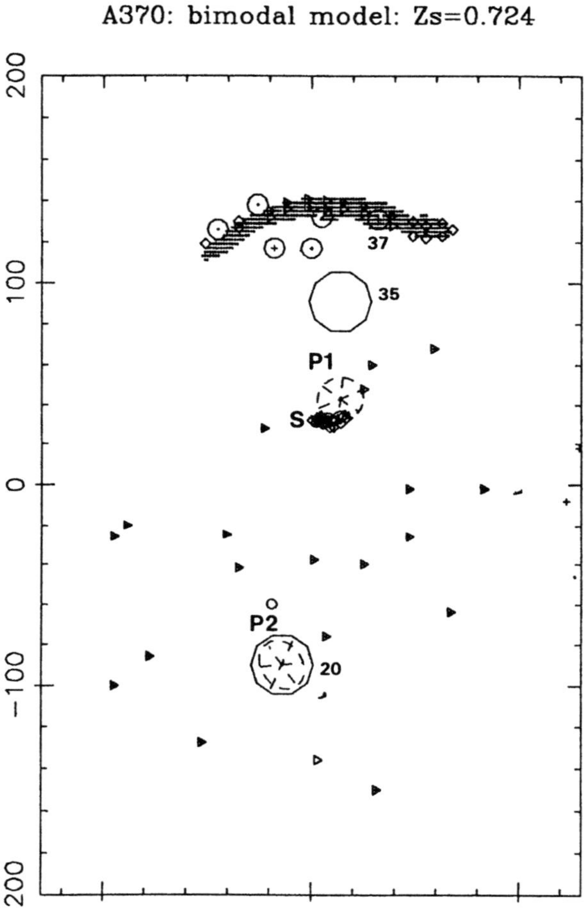

Fig. 1 : bimodal model + spiral source for the giant arc in A370. P1 and P2 are the position of the potential wells. S is the location of the source in the source plane. The polygons indicate the positions of the galaxies. The disturbing galaxies ♯20, ♯35 and ♯37 are referenced (see text). Note the difference between the position of the galaxy ♯35 and P1.

3b. A2390:

A2390 is a rich X–ray cluster of galaxies (McMillan *et al.*, 1989) at a redshift of 0.23 and dominated by a bright radiogalaxy. From optical observations (Le Borgne *et al.*, 1990) it appears as slighlty elongated along the main axis defined by the first rank radiogalaxy.

Pello *et al.* (1989) first observed this arc and measured its redshift. The spectroscopic data were obtained at the William Hershell 4.20m Telescope and the 3.60m ESO Telescope. A strong emission line identified as $[OII]\lambda3727$ in the local frame leads to a redshift of 0.913 for the lensed background galaxy. The arc extend over 15" and has a width of 1.3" . An estimate of the curvature radius gives a lower limit of 4.6'. Therefore it can really be considered as a straight arc. In addition, it has two breaks symmetric to the central position of the nearest galaxy which separate the arc in three components a,b,c (see fig. 2). So far this straight arc is probably the most intriging case of lensing we know: nowhere in the literature "straight arcs" have been predicted or even investigated as special configurations.

As for A370, the constraints will be based on the optical observations since they have a better resolution than X-ray data. In particular, we assume that the cluster is constituted by a main potential centered on the brightest galaxy. Additional minor potentials can be included if necessary. The isopleth map shows that the clusters galaxies are distributed along an axis inclined at 133 degres along the east-west direction (see Pello *et al.*, 1990).

The spectroscopic data concern the galaxies around the arc: for the purpose of this model, we only included the brighest neighbouring galaxies and in particular, the nearest one to the arc which is a cluster member at a redshift of 0.231 (Le Borgne *et al.*, 1990).

The redshift difference observed in the spectrum itself corresponds to a velocity difference of about 300 km/sec from west to east of the arc (Pello *et al.*, 1990). It can reflect the rotation of the source and therefore is a strong argument in favour of an egde–on spiral. However, it is possible that we are observing a pair of interacting background galaxies.

The preliminary results we obained are summarized in fig. 2 and 3. In the first class of models, the source is a spiral galaxy and the potential is a unique dark halo associated to the central radiogalaxy. The resulting image is a straight structure composed by 3 sub-arcs which roughly looks like the arc in A2390. The model seems too thin and too long but it is possible to fit those data using a thicker bulge with shorter spiral arms. However, we cannot fit the data so well. In particular, though we observe 2 discontinuities at the junction of the arms with the bulge, we are unable to reproduce 2 breaks. In addition, we have no good model with a potential well matching the orientation of the cluster. Therefore we need to arbitrary add a secondary small potential located on the arc itself which does not correspond to the location of any cluster galaxies. Despite the large number of free parameters we have, we are unable to obtain a nice straight arc with two breaks.

In a second class of model, we imposed the orientation of the potential in accordance with the isopleth map and tried to obtain a straight arc with an elliptical–shape source.

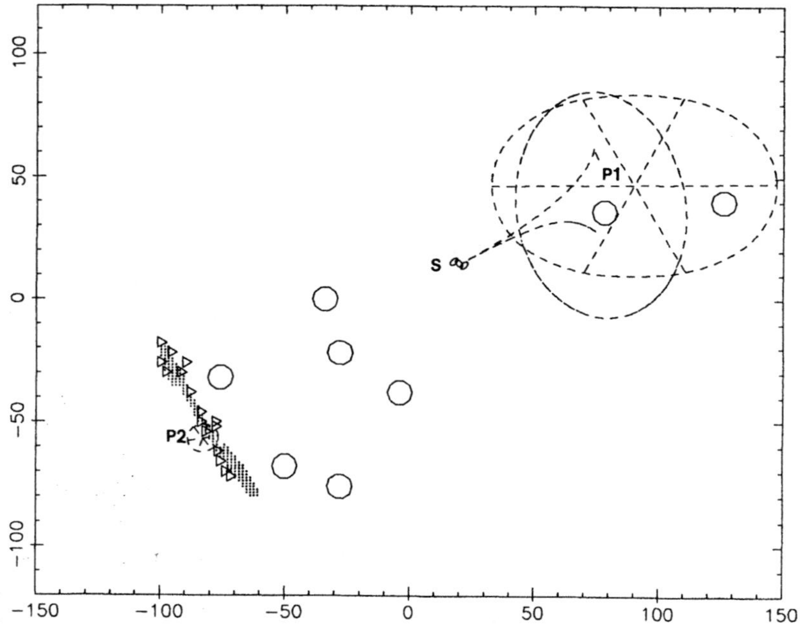

Fig. 2 : A2390, first model with a large potential centered on the radiogalaxy and a spiral source. Though 2 discontinuities are visible the arc is not broken.

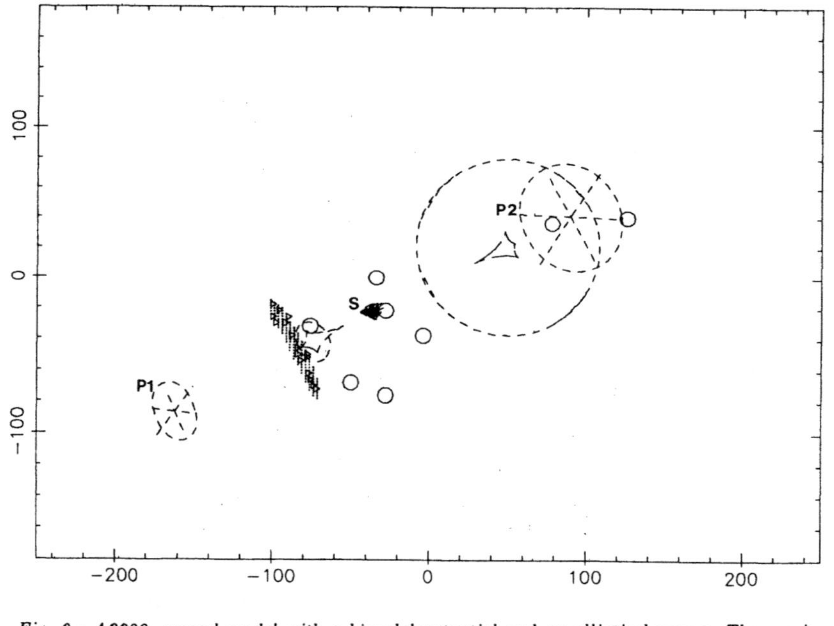

Fig. 3 : A2390, second model with a bimodal potential and an elliptical source. The arc is rather straight but no break is visible. The observational constraints are so few and a rather large number of bimodal models are possible. This one is only an example.

In principle, this should be possible if the source itself is exactly located on the cusp caustics. In such a casse three images of the source are merging. As it is shown in fig. 9 of Blandford and Narayan (1986), one can expect to observe three sub-arcs corresponding to the three merging images. This could be ideally the configuration we are observing in A2390. In such a case, a nearly straight arc can be obtained if the source has an elliptical shape but only if a secondary potential well is added on the other side of the arc. This model give a saddle configuration which is is appealing since it can easily explain why we observe several parallel arclets only on this side of the CCD frame. But, on the other hand, we do not see any excess of visible mass on the location of the secondary potential well. A possible interpretation is that this mass is mostly out of the small field of view of the CCD or that it exists a group of faint background galaxies inobservable without using deep CCD imaging. In addition, an elliptical source is not compatible with the velocity variation observed from the spectrum, unless we assume that we are observing an edge–on spiral galaxy, and the straight arc we obtain has not any discontinuity at all.

Concluding remarks:

We have presented several models for the arcs in A370 and A2390. Our main concern was to use a maximum of observational constraints (except the velocity dispersion): exact geometrical shape of the arc, isopleth map of the cluster and the spectroscopy of arcs to infer the morphology of the source from its internal kinematic and/or its stellar content. This leads to try a bimodal model plus a spiral source for A370 which reproduces fairly well the arc but only if the dark matter does not follow the visible light distribution.

For A2390, despite the number of free parameters we cannot find any satisfactory simple model. The best one would correspond to an edge–on spiral galaxy, but only if we include a secondary potential on the opposite side of the arc. Even in that case, the preliminary simulations do not reproduce the two breaks observed on the arc. In conclusion, to reproduce the delailled geometrical and spectroscopical properties of the arc in A2390 seems much more difficult than for A370.

In A2390 the conclusions that we can derive are rather puzzling and the best model reclaims a secondary potential well which could be associated to a faint group of more distant galaxies or possibly to a clump of dark matter. Our unseccessfull attempt to model this arc is in fact very exciting. Certainly our models are too simple. In particular, the remarkable symmetrical position of the two breaks in comparison with the nearest galaxy could be an indication that this galaxy locally disrupts the arc. However, much more observable are requiered, mainly in deep photometry. For instance, figure 4 show a simulation of what we could observed through a rich cluster of galaxies as A2390 if we go deeper. We assumed that all the sources are on a regular matrix at redshift of 1.2 The surface number density correspond to what it should be observed at a limiting magnitude of $B_j = 28$ in a square of 2'x2' and if we follow the Tyson's galaxy counts (1988). Many details of the projected mass density are then revealed in this picture: the density of arc and arclets depends on the density of the cluster core and a bimodal structure would be clearly shown by arclets. On the other hand, we should be also able to infer the nature of the source using bidimensional spectroscopy (Monnet *et al.*, 1990, in preparation). These are rather time consuming but this is a posible way to provide a large amount of observational constraints and derive the unique solution for the potential of the cluster.

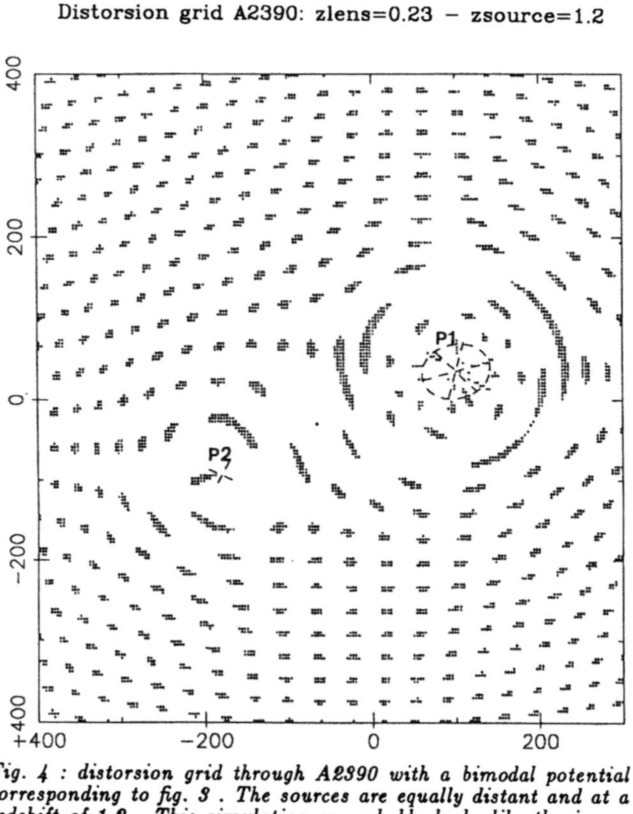

Fig. 4 : distorsion grid through A2390 with a bimodal potential corresponding to fig. 3 . The sources are equally distant and at a redshift of 1.2 . This simulation remarkably looks like the image of the cluster given in Pello et al., this proceeding.

Ackowledgments

We are very gratefull to I. Kovner (Weizmann Institute) for provinding a version of his code to model the arcs. Thanks to R. Blandford, C. Kochanek, I. Kovner and G. Mathez for the many discusions we had on this subject, mainly on the fascinating straight arc in A2390.

References

- Aragon, A., Ellis, R.S., 1989: in " The Toulouse Workshop on Gravitational Lensing". Toulouse September 13-15 1989. Y. Mellier, G. Soucail, B. Fort Eds. Springer Verlag 1990.

- Blandford, R.D., Narayan, R., 1986: Astrophys. J. **310**, 568

- Broadhurst, T.J., Ellis, R.S., Shanks, T., 1988: M.N.R.A.S. **235**, 827

- Cowie, L.L., Henriksen, M., Mushotzky, R., 1987: Astrophys. J. **317**, 593

- Fitchett, M., A., 1989: Proceedings of the Space Telescope Science Institute on " Clusters of Galaxies". Baltimore, USA, may 15-17 1989. M. Fitchett editor.

- Fort, B., Prieur, J.L., Mathez G., Mellier, Y., Soucail, G., 1988: Astron. Astrophys. **200**, L5

- Fort, B., 1989: proceedings of the " 12th Conference on General Relativity and gravitation". Boulder, Colorado, USA, july 2-8, 1989.

- Grossman, S.A., Narayan, R., 1989: Astrophys. J. **344**, 637

- Hammer, F., 1987: in "High Redshift and Primeval Galaxies". Proceedings of the Third IAP Workshop. Institut d'Astrophysique, Paris june 29 - july 3, 1987. J. Bergeron, D. Kunth, B. Rocca–Volmerange and J. Tran Thanh Van Eds.

- Hammer, F., Rigaud, F., 1989: preprint

- Kovner, I, 1988: Proceedings of the NATO Advanced Study Institute on " The Post-Recombination Universe". Cambridge, U.K.; july 27 – August 7, 1987. N. Kaiser and A.N. Lasenby Eds.

- Lavery, R.J., Henry, J.P., 1988: Astrophys. J. **329**, L21

- Le Borgne, J.F., Mathez, G., Mellier, Y., Pello, R., Sanahuja, B., Soucail, G., 1990: in preparation.

- Lynds, R., Petrossian, V., 1986: Bull. AAS, **18**, 1014

- MacLaren, I., Ellis, R.S., Couch, W.J., 1987: M.N.R.A.S. **230**, 249

- MacMillan, S.L.W., Kowalski, M.P., Ulmer, M.P., 1989: Astrophys. J. Supp. Series, **70**, 723

- Mellier, Y., Soucail, G., Fort, B., Mathez, G., 1988: Astron. Astrophys. **199**, 13

– Pello, R., Le Borgne, J.F., Mathez, G., Mellier, Y., Sanahuja, B., Soucail, G., 1989: in " The Toulouse Workshop on Gravitational Lensing", Toulouse, September 13-15 1989. Y. Mellier, G. Soucail, B. Fort Eds. Springer Verlag, 1990.

– Pello, R., Le Borgne, J.F., Soucail, G., Mellier, Y., Sanahuja, B., 1990: submitted.

– Soucail, G., Fort, B., Mellier, Y., Picat, J.P., 1987: Astron. Astrpohys. **172**, L14

– Soucail, G., Mellier, Y., Fort, B., mathez, G., Cailloux, M., 1988: Astron. Astrophys. **191**, L19

– Soucail, G., Mellier, Y., Fort, B., Mathez, G., Cailloux, M., 1989: in " The Toulouse Worshop on Gravitational Lensing". Toulouse, September 13-15 1989. Y. Mellier, G. Soucail, B. Fort Eds.. Springer Verlag 1989.

– The, L.S., White, S.D.M., 1986: Astron. J. **92**, 1248

– The, L.S., White, S.D.M., 1988: Astron. J. **95**, 15

– Tyson, J.A., 1988: Astron. J. **96**, 1

– Tyson, J.A., 1989: in "The Toulouse Workshop on Gravitational Lensing"; Toulouse, September 13-15, 1989. Y. Mellier, G. Soucail, B. Fort Eds. Springer Verlag, 1990.

NEW RESULTS FROM THE GRAVITATIONAL TELESCOPE

F. HAMMER

D.A.E.C. Observatoire de Meudon, 92195 Meudon Principal Cedex
France

Abstract. New improvements in extragalactic research have recently increased the number of sources at high redshift, QSOs and also galaxies. Gravitational magnification or lensing by foreground matter are able to significantly alter such observations. Giant luminous arcs are probably high gravitational magnification events on relatively common sources, i.e. distant "normal" galaxies.
Recent observations and realistic mass distribution models of the arc-like structures (in A370 and Cl2244-02) allow us to describe well the observations and to set new and independent conditions on the material content of the deflecting clusters. For A370 cluster, we find that more than 75% of the deflecting matter is associated with the large structures ($r_{eff}\approx$1Mpc), the rest may lie into galaxy haloes. They also allow detailed studies of sources which are at the top of the gravitational telescopes. The source related to the Cl 2244-02 arc is predicted to be a nearly face-on spiral galaxy, for which the bulge and the arms are spatially resolved in one direction, and seen as patches along the arc.

1. Introduction.

Since the discovery of the QSOs and for more than 15 years after, the studies on the very distant sources were limited on such peculiar objects. Since the discovery of the first plausible case, QSO 0957+56 by Carswell et al (1979), gravitational lensing had been reputed to be the field of the study of a very small number of very peculiar configurations, i.e. gravitationally splitted QSOs. In the beginning of the 80s, technical and observational improvements had allowed the emergence of a new class of sources at very large distance, since redhifts more than one had been measured for several radiogalaxies. Nowadays, a sample of ≈40 radiogalaxies at high redshift is available, with an higher limit (z=3.8, Chambers et al, 1989) rather close to the highest redshift of QSO known today (z=4.7, Gunn, 1989). Gravitational lensing on galaxies had been suggested for the first time on a possible companion of the gravitationally splitted QSO 2016+112 by Schneider et al (1986) and on the giant radiogalaxy 3C324 by Hammer et al (1986) and Le Fèvre et al (1987). This was following by the discovery of the giant luminous arcs by Soucail et al (1987) and Lynds and Pétrosian (1986) which was the starting point of a new field in the frame of gravitational lensing. Indeed they are probably images of very distant sources highly gravitationally magnified by deflecting clusters of galaxies. Since galaxies are very more common than QSOs (about 100 more), gravitational lensing and high gravitational magnification events on galaxies should be more numerous than on QSOs. Recent counts of galaxies at very faint levels (B_J=27 and complete for B_J=25.5 and R=25, see Tyson, 1988) have revealed a very large surface density of galaxies on the sky. They probably correspond to distant galaxies, although they may be less distant than previously predicted by Tyson (1988) since almost complete redshift counts up to B=22.5 show no galaxy with redshift higher than 0.7 (Colless et al, 1989). Gravitational lensing/magnification could affect the current counts of galaxies, by the combination of two factors linked to the magnification factor M (ratio of the magnified surface of the source to the virtual surface of the source):

-(i) a magnification of the source luminosity by $\delta m=-2.5 \log M$

-(ii) a decrease of the surface behind the lens due to the convergence of the light beam $S/S_o = <1/Amp>$.

Assuming a density-magnitude relation with a (logarithmic) slope k, the resulting source surface density (n) is given by the following relation (see e.g. Hammer and Nottale, 1986):

$\log n/n_o = (2.5k-1) \log M$, where n_o is the unaffected surface density of sources. (1)

Tyson (1988) found k≈0.45 for B_J blue counts and k≈0.39 for the R red counts, which implies that there are more high magnification events (M>>1) acting on blue faint galaxies than on red ones. This simply explains why most of the arcs have very blue color indices. For samples of sources selected above a given flux limit, sources could enter the sample only because their (radio or optical) fluxes have been gravitationally magnified. This could occur especially if the flux limit is very high, as for the high luminous QSO sample (HLQ, see e.g. Surdej et al, 1988) or for the high-z 3CR galaxies or QSOs (see Hammer et al, 1986). Another condition requested from equation (1) is a steep slope k of the density-magnitude relation.

Giant luminous arcs are potentially among the best gravitational lensing configurations which will be studied to test the universe. Firstly, they partially draw the infinite magnification curves, that are the signatures of the lens mass distribution. Hammer and Rigaut (1989) have shown in the frame of a multi-point mass model that the determination of the lens mass distribution could be derived from the solution of a system of n quadratic equations, where n could reach the number of pixels along the arc. Secondly, since high magnification events are statistically uncommon and sources related to giant luminous arcs are probably normal galaxies at large distance. Real source size is directly proportional to the arc thickness by a factor equal to 2(1-K) (Hammer and Rigaut, 1989), where K is the well-known matter term (see Hammer, 1985). It leads to a magnification factor which does not depend strongly on the model of lens mass distribution. Real source luminosity will be rather well known as well as its detailed shape if one will reconstruct the source from eventual microstructures detected along the arc. Conbined with spectroscopic measurements, it would provide an unique way to investigate properties of common distant galaxies, which may be a crucial test to the galaxy evolution models and even to cosmology.

Here, I will briefly show which conditions can be set on the deflector mass distribution, i.e. on the optical properties of the gravitational telescope, and I will also present an example of how the gravitational telescope would bring to us crucial and detailed informations on the background sources that lie just on its top.

2. Model of the A370 arc and the deflecting mass distribution.

The giant arc had been found by Soucail et al (1987). Model based on gravitational lensing of a background galaxy by several deflecting point-masses in the cluster core had been developed by Hammer (1987). This simple model was able to reproduce rather well the observations and to describe most of the properties of high magnification events, such as the attenuation of the secondary image due to the effect of a secondary lens (a cluster galaxy), the break of the arc due to the presence of another secondary lens, and the differential thickness of the arc due to the elliptical shape of the source. Refinements of this model towards more realistic mass distribution inside the deflecting cluster core have been recently made by Grossman and Narayan (1989) and by Hammer and Rigaut (1989). Grossman and Narayan (1989) models the A370 cluster mass distribution by an elliptical generalization of non-singular isothermal sphere and considers the galaxies as perturbations while the matter is hierarchically distributed in cluster, groups and galaxies which are assumed to follow an $r^{1/4}$ law in the Hammer and Rigaut (1989) 's paper. Both studies reproduce pixel by pixel the observations and predict that the arc is a single image of the source, which is probably a nearly edge-on spiral galaxy at z=0.72 (while the cluster has z=0.377). These results are emphasized by the fact that the number of the input parameters is almost balanced by the number of observational constraints.

Moreover, we (Hammer and Rigaut, 1989) have been able to constraint the mass distribution inside A370. Firstly, we assumed that the cluster mass is distributed inside a single structure having a large effective radius (the cluster with $r_{eff} \approx$ 1Mpc) and several structures having smaller effective radii (galaxies with r_{eff} from 10 to 100 kpc). Secondly, we set conditions on the masses of both giant galaxies which lie in the A370 core. One of them (#35 from the notations of Soucail et al, 1988) lies at less than 10 arcsec from the arc, which implies a mass lower than 4.2 10^{13} M_O inside a radius of 10 arcsec, which has to be decrease to $M/L_R < 60$ $H_o/50$ after accounting for the arc curvature. There is also a background galaxy (#21 with z=0.466 from Soucail et al, 1988) which is projected between the two giant galaxies, at only 6.7 arcsec off the giant cluster galaxy #20. #21 galaxy is expected to be gravitationally distorted by the A370 cluster and its galaxies, especially the #20 one. Since #21 is an elliptical from its spectrum and luminosity, we found from an analytical relation that for #20, $M/L_R < 90$ $H_o/50$. Finally, it results that more than 75% of the deflecting mass has to be associated with the cluster ($r_{eff} \sim$ 1Mpc), rather than to the galaxies ($r_{eff} <$ 100 kpc).

3. Model of the Cl 2244-02 arc and the nature of its source.

The arc had been discovered by Lynds and Pétrosian (1986). Discovery of two probable additional images (called T and S) related to the arc had been made by Hammer et al (1989) who had also made accurate B, V and R photometry of the cluster core. Model of the arc was firstly presented by Hammer et al (1989) and described in detail by Hammer and Rigaut (1989). It assumes the same kind of mass distribution than for the A370 core, i.e. $r^{1/4}$ law for cluster, groups and galaxies. The model reproduces pixel by pixel the arc and the two additional images T and S: it is interesting to note that if T and S are really related to the arc, the input parameters in the model are less numerous than the observational constraints. The Cl2244-02 core is strongly dominated by two clumps of galaxies, while the cluster itself seems rather poor from galaxy counts and the moderate X-ray luminosity. This seems consistent with the presence of an image (T, see Figure 1) close to the arc curvature center and just between the two clumps, which is also predicted in case of lensing due to two main deflectors. A configuration for which the source is splitted into five images is rather probable. Hammer et al (1989) have shown that the arc thickness is not spatially resolved down to 0.5 arcsec, which implies that the source radius has to be less than ≈0.3 arcsec. They also detect the presence of six patches, or peaks of surface brightness (see Figure 1), distributed all along the arc, for which the location seem independent of the filter frame (B, V, R). Their V and especially R imageries show at the lowest isophotal levels, that the arc is shared into two twin parts with three patches on each one. This leads to interpret the arc as made by two images of the source, that is consistent with the distribution of the luminous matter. The other images may be S and T, the fifth one being attenuated or blended in a clump galaxy. An alternative may be the arc made by three images, one of which being enough attenuated to be nearly undetectable : S would be unrelated galaxy and this solution would corresponds to a dark matter distribution very discrepant from the luminous matter distribution.

Interpretation of the patchy structure is also given by Hammer and Rigaut (1989): if the source of the arc is a nearly face-on spiral galaxy, with a minor axis ≈0.3 arcsec, the thickness of each images of the arc will be roughly 0.3 arcsec and then spatially unresolved, which means that the surface brightness will be locally proportional to its integrated surface brightness along its real width. Then the two central patches of the two twin images will correspond to the bulge of the source, while the outer ones will correspond to its arms. This analysis is strengthened by two facts: (i) all the patches have similar intensity peaks in the blue image, while the outer patches are less prominent in V, and nearly disappear in R, which is understandable since star formation occurs mainly in the arms; (ii) an U imaging and photometry by Wlérick et al (1989) reveal that T, S and the arc have similar color indices from U to R, and the detection of the six patches at the same location that in B, V, R. High resolution imaging by Space Telescope will verified such a prediction, for which the arc sub-structures will allow the detailed analysis of the arms and bulge of an extremely distant spiral galaxy.

References

Chambers, K. C., Miley, G. K., van Breugel, W., 1989, in preparation.
Colless, M., Ellis, R. E., Taylor, K., Hook, R. N., 1989, M. N. R. A. S., in press.
Grossman, S. A., Narayan, R., 1989, Astrophys. J., in press.
Hammer, F., 1985, Astron. Astrophys., **152**, 262.
Hammer, F., 1987, in Third IAP Astrophys. Meeting, "High redshift objects ", eds Frontières, p. 467.
Hammer, F., Le Fèvre, O., Jones, J., Rigaut, F., Soucail, G., 1989, Astron. Astrophys., **208**, L7.
Hammer, F., Nottale, L., 1986, Astron. Astrophys., **155**, 420.
Hammer, F., Nottale, L., Le Fèvre, O. 1986, Astron. Astrophys., **169**, L1.
Hammer, F., Rigaut, F. 1989, Astron. Astrophys., in press.
Le Fèvre, O., Hammer, F., Nottale, L., Mathez, G. 1987, Nature, **326**, 268.
Lynds, R., Petrosian, V., 1986, Bull. Amer. Astron. Soc. **18**, 1014.
Soucail, G., Mellier, Y., Fort, B., Picat, J.P., 1987a, Astron. Astrophys., **172**, L14.
Soucail, G., Mellier, Y., Fort, B., Cailloux, M., 1988, preprint.
Surdej, J., et al, 1988, P. A. S. P., in press.
Tyson, J.A., 1988, Astron. J., **96**, 1.
Walsh, D., Carswell, R. F., Weymann, R. J., 1979, Nature, **279**, 381.
Wlérick et al, 1989, in Toulouse Workshop on Gravitational Lensing, held in September 1989 in Toulouse, in press.

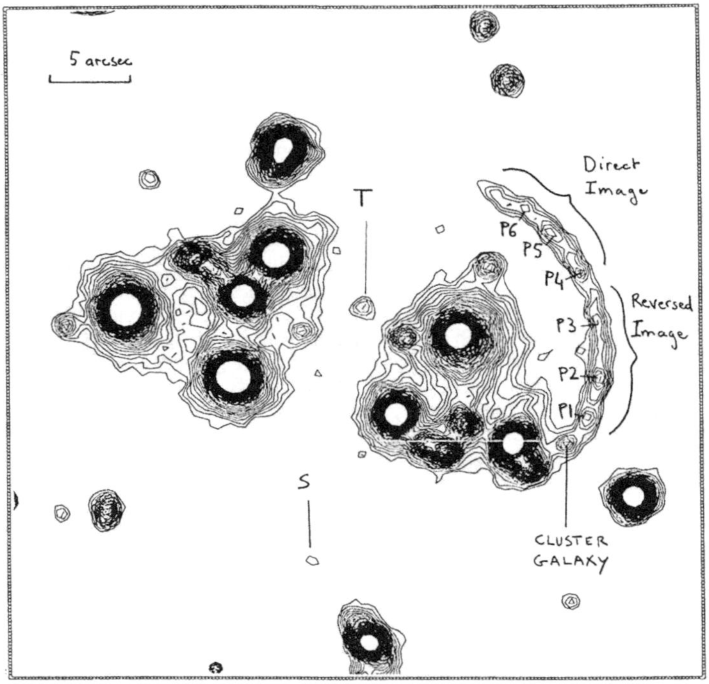

Figure 1 : The core region of Cl 2244-02 showing the two clumps of galaxies and the arc in V filter. This 1800 s exposure time image was obtained at CFHT, by Hammer et al (1989) under a seeing of 1.1 arcsec. The lowest isophotal level is 0.8% of the sky, and the two faint objects quoted as S and T are significantly above the noise. They both present strikingly similar photometric properties with respect to the arc in U, B, V and R filters and they are likely additional images related to the arc. Six patches on the arc are also detected along the arc in the four filters, and are quoted from P1 to P6. The break between P3 and P4 is more striking in the R imagery. Note how the two images which compose the arc are similar.

PROBING RICH GALAXY CLUSTERS WITH MINI-ARCS

Scott A. Grossman
Steward Observatory, University of Arizona
Tucson, AZ 85721 USA

Introduction. Luminous arcs have been observed in ten galaxy clusters (see Fort, these proceedings for a summary of observations). These arcs are particularly interesting and useful to the study of the mass distibution of galaxy clusters because results derived from gravitational lensing calculations are independent of most of the assumptions involved in virial calculations. The long arcs in A370, A963, Cl2244-02, and Cl0500-24 (Soucail *et al.* 1987; Lynds and Petrosian 1989; Lavery and Henry 1988; Giraud 1988) have been modeled in detail (Bergmann *et al.* 1989; Grossman and Narayan 1989; Hammer and Rigaut 1989; Kovner 1989; Wambsganss *et al.* 1989; see also Petrosian, Mellier, and Hammer, these proceedings). These studies have demonstrated that detailed reconstruction of these arcs provides useful constraints on the lensing galaxy clusters, with much stronger constraints possible if the arc redshifts are measured. However, such models only constrain the mass inside the arcs and only have the potential to probe small scale (galaxy-sized) substructure very near the arcs (see Narasimha and Nair, these proceedings).

Tyson (1988) has observed a faint population of galaxies densely covering the sky, indicating that there should be many sources which can be lensed behind each rich galaxy cluster core (Fort 1988). In addition, Grossman and Narayan (1988) have predicted that there should be several smaller arcs (a few arcseconds long) for each long arc observed. Indeed, such small arcs have been observed in A370 (Fort *et al.* 1988) and A2218 (Pello-Descayre *et al.* 1988). Very deep imaging of rich galaxy clusters should yield an abundance of mini-arcs, images of high redshift galaxies which have only been mildly distorted and are not necessarily morphologically distinguishable from cluster members. Such mini-arcs may be found well beyond the lenses' tangential critical lines, and their greater abundance provides the possibility of studying small scale structure throughout the clusters.

Figure 1 shows what a very deep image of A370 might look like if 106 source galaxies per square arcminute could be seen. The tangential stretching of background galaxies which lie beyond the tangential critical line and the circular distribution of images is a characteristic of gravitational lens optics. Identification of such a pattern of images in a rich cluster is evidence that the cluster is acting as a gravitational lens and that the distorted images are at high redshift. Applying this idea, I present evidence for the existance of mini-arcs in rich galaxy clusters and discuss some of the implications for the evolution of high redshift galaxies.

Evidence of Mini-Arcs. Grossman *et al.* (1989) imaged ten rich galaxy clusters ($z \approx 0.2-0.3$), tabulated by Butcher *et al.* (1983), with central giant elliptical or cD

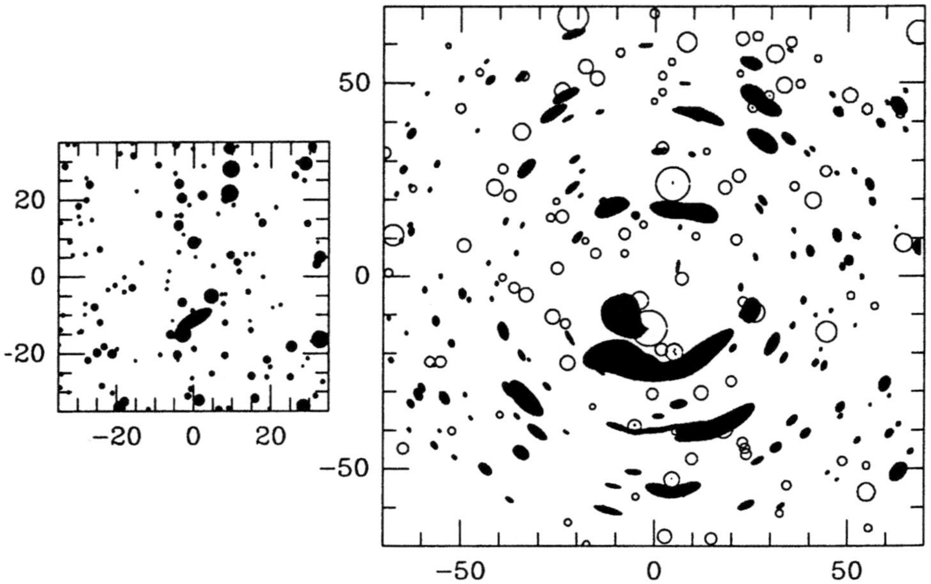

Fig. 1. A simulation of lensed images in A370, with as many sources as should be seen to B=27. Simulations of the observed arcs are also present and are shown the way they would appear if fainter isophotes could be seen. (Taken from Grossman and Narayan 1989.)

galaxies using R and I filters with a 2.3m telecscope. Images were detected to $R \approx 24$ and $I \approx 23$ with good completeness.

To select background galaxies from cluster members, it is useful to construct color-magnitude diagrams for each cluster. Figure 2 shows $R-I$ plotted against R for the objects in A1961. A ridge of galaxies is evident in the diagram, most of which are probably elliptical or SO galaxies in the cluster. The galaxies in the diagram are divided into the cluster sample, which is defined by the ridge, and the blue and red samples, which are bluer or redder than the ridge, respectively. We would like to know whether the blue or red samples are dominated by high redshift galaxies.

One can measure the orientation angles of the images with respect to the cluster centers, defined to be the central galaxies in the clusters. The orientation angle is defined so that a 90° angle means the major-axis of the image is tangential with respect to the cluster center, and a 0° angle implies a radial orientation. Grossman *et al.* combined the data of the blue, cluster, and red samples of all ten clusters they observed in order to obtain statistically meaningful results, and they examined the cumulative distribution of orientation angles for each sample (see Fig. 3). The distributions are based upon 339 cluster members, 111 blue sample members, and 73 red sample members.

In Figure 3, a straight line is a perfectly random distribution, and the blue and cluster sample galaxies are consistent with this. However, a Kolmogorov-Smirnov test shows that there is only a 0.06% chance that the observed distribution of red galaxies was selected from a random distribution. The fact that the cluster sample is randomly distributed suggests that no systematic errors are biasing the red samlple. We conclude

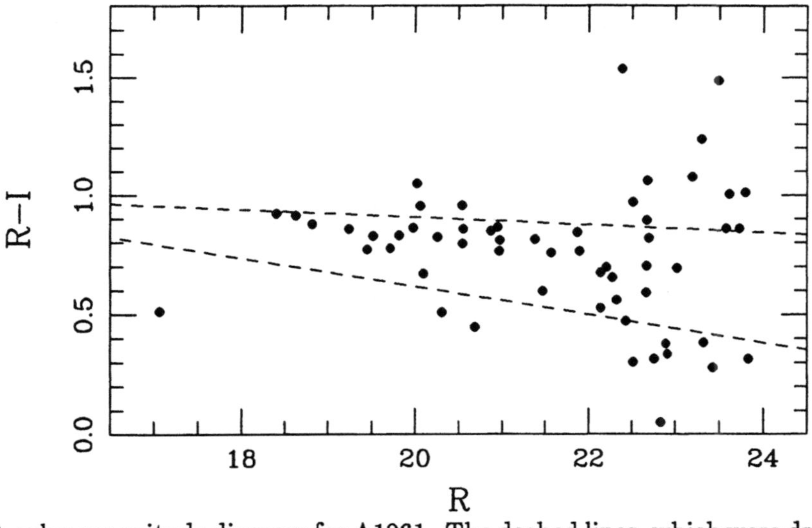

Fig. 2. A color-magnitude diagram for A1961. The dashed lines, which were drawn by visual inspection, separate the red, cluster, and blue samples of galaxies.

that there are few galaxies brighter than $R \approx 24$ which are blue in $R-I$ and at high redshift. However, most galaxies in the red sample are probably at high redshift.

Tyson *et al.* (1989) (see also Tyson, these proceedings) have also identified a sample of mini-arcs. They find about 60 mini-arcs in each of two clusters to a limiting magnitude of $B \approx 27$. Their sample of background images is blue in $B-R$, and the orientations of those images are tangentially biased with respect to the cluster centers.

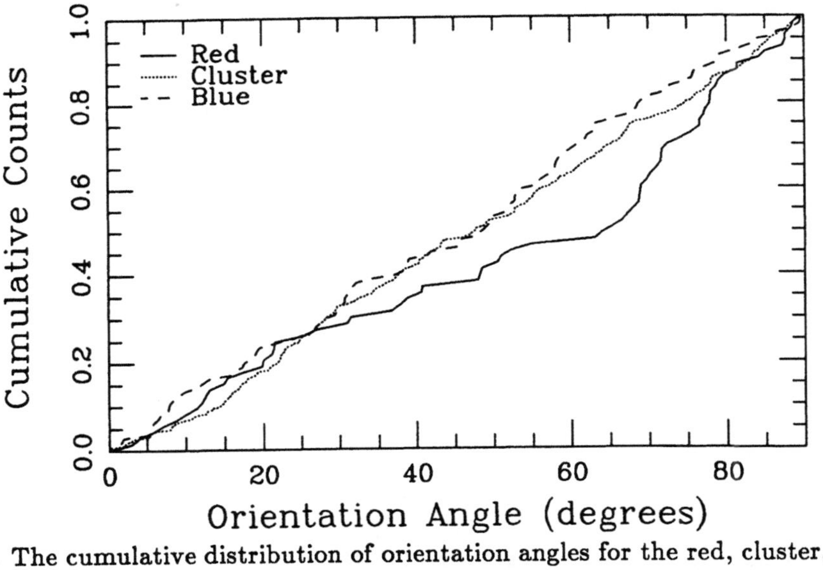

Fig. 3. The cumulative distribution of orientation angles for the red, cluster, and blue samples. The cluster and blue distributions are consistent with a random distribution, described by a straight line, but the red sample deviates from a random distribution with 99.9% significance. (Taken from Grossman *et al.* 1989.)

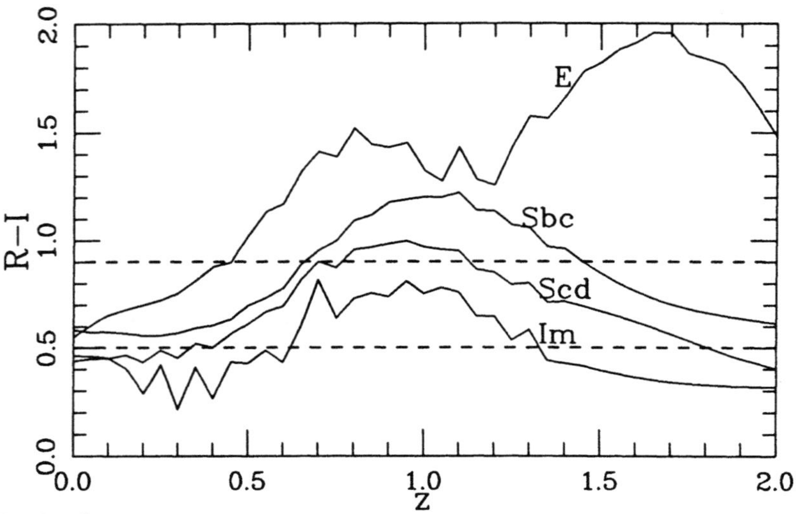

Fig. 4. $R-I$ colors versus redshift for non-evolving spectral energy distributions. The color becomes redder as the 4000Å break moves through the R band. E galaxies, which have the strongest 4000Å break, and therefore the oldest dominant stellar population, become the reddest. The dashed lines roughly separate the blue, cluster, and red, background sample members. (Taken from Grossman *et al.* 1989.)

Colors of High Redshift Galaxies. The red sample galaxies must be behind the lensing clusters, with $z \gtrsim 0.5$ being a reasonable lower limit. Furthermore, because of the $(1+z)^4$ decrease in flux from high redshift galaxies, $z \lesssim 1.5$ is a reasonable upper limit for the red, background sample. Therefore, the color $R-I$ essentially measures rest frame near UV/optical colors, which are insensitive to small, young stellar populations which may dominate the UV flux. The $R-I$ color is sensitive to the strength of the 4000Å break for this redshift range, which is an indicator of the age of the dominant stellar population.

Figure 4 shows the K-corrected $R-I$ colors as a function of redshift for non-evolving spectral energy distributions. The cluster sample galaxies fall between the dashed lines, while the red, background galaxies have $R-I > 0.9$. The background galaxy colors are consistent with the colors of ellipticals and red spirals, indicating that the optical spectral energy distribution is dominated by an old stellar populution. The red colors of the background sample are also consistent with the colors the arcs in A2218 (Grossman *et al.* 1989) and the primary arc in A370 (Lynds and Petrosian 1989). The faint field galaxies observed by Tyson (1988) also have red colors, with $R-I \approx 1$.

The red $R-I$ colors of the background sample can be reconciled with the blue $B-R$ colors of faint field galaxies observed by Tyson (1988) by recognizing that $B-R$ colors are sensitive to the UV flux of galaxies in the redshift range $0.5 \lesssim z \lesssim 1.5$. Thus, a small population of hot, young stars may dominate the UV flux of high redshift galaxies, which would account for their blue $B-R$ colors and their unevolved optical spectral energy distributions.

Conclusions. Mini-arcs are images of high redshift galaxies which have been tangentially elongated by gravitational lensing. The simulation of a very deep image of A370 shows that there should be an abundance of mini-arcs projected on rich galaxy clusters. We present observational evidence for the existence of mini-arcs, namely the tangential bias of images which are red in $R-I$. These mini-arcs must be at high redshift, and their red colors, which are consistent with non-evolving spectral energy distributions of late type galaxies, imply there are large populations of old stars in high redshift galaxies.

The problem of inverting the mini-arc images to obtain the lensing mass distribution will require the development of new algorithms. However, as observations of clusters with a hundred or more mini-arcs become available, it should be possible to construct models of clusters with enough parameters even to address the issue of substructure in the lenses. Furthermore, if many clusters are observed so deeply, then it may be possible to put constraints on the redshift distribution of high redshift galaxies. The identification and study of mini-arcs bears on several important astronomical issues, and the initial results of Grossman *et al.* (1989) and Tyson (1989) provide encouragement that this study will yield answers to the questions discussed above.

Acknowledgments. I would like to thank my collaborators, Ramesh Narayan, Richard Elston, and Dennis Zaritsky, for their contributions to the ideas and results presented here. This work was supported in part by N.S.F. grant AST 88-14725.

References.

Bergmann, A. G., Petrosian, V., and Lynds, R. 1989, *Ap. J.*, in press.

Butcher, H., Oemler, A., and Wells, S. C. 1983, *Ap. J. Suppl.*, **52**, 183.

Fort, B. 1988, in *Astronomy, Cosmology, and Fundamental Physics–Third ESO/CERN Symposium, Bologna, May 16–20, 1988* (D. Reidel Publishing Company, Holland).

Fort, B., Prieur, J. L., Mathez, G., Mellier, Y., and Soucail, G. 1988, *Astr. Ap.*, **200**, L17.

Giraud, E. 1988, *Ap. J. (Letters)*, **334**, L69.

Grossman, S. A. and Narayan, R. 1988, *Ap. J. (Letters)*, **324**, L37.

Grossman, S. A. and Narayan, R. 1989, *Ap. J.*, **344**, 637.

Grossman, S. A., Zaritsky, D., and Elston, R. 1989, submitted.

Hammer, F. and Rigaut, F. 1989, *Astr. Ap.*, in press.

Kovner, I. 1989, *Ap. J.*, **337**, 621.

Lavery, R. J. and Henry, J. P. 1988, *Ap. J. (Letters)*, **329**, L21.

Lynds, R. and Petrosian, V. 1989, *Ap. J.*, **336**, 1.

Pello-Descayre, R., Soucail G., Sanahuja, B., Mathez, G., and Ojero, E. 1988, *Astr. Ap.*, **190**, L11.

Soucail, G., Fort, B., Mellier, Y. and Picat, J. 1987, *Astr. Ap.*, **172**, L14.

Tyson, J. A. 1988, *A. J.*, **96**, 1.

Tyson, J. A., Valdes, F., and Wenk, R. A. 1989, *Ap. J. (Letters)*, in press.

Wambsganss, J., Giraud, E., Schneider, P., Weiss, A. 1989, *Ap. J.*, **337**, L73.

GRAVITATIONALLY LENSED ARCS AS GALAXY REDSHIFT INDICATORS

Robert J. Nemiroff

Racah Institute of Physics, Hebrew University of Jerusalem,
Jerusalem, Givat Ram, ISRAEL

and

E. O. Hulburt Center for Space Research, Code 4121.6,
Naval Research Laboratory, Washington DC, USA

ABSTRACT: For the observed giant arcs to be artifacts of the gravitational lens effect of clusters of galaxies on normal background galaxies, the galaxies must be behind the clusters. Since the redshifts of many of these clusters are known, they may be taken as the smallest possible redshifts for the background galaxies. These limits may give important redshift information for dim galaxies near the brightness detection limit of modern techniques. These limits may be in conflict with simple extrapolations of modern galaxy redshift surveys and some cold dark matter scenarios where galaxies form very late.

I. INTRODUCTION

This paper is a summary of a short comment I presented to the Durham Workshop on the Epoch of Galaxy Formation held in Durham, England in the summer of 1988 (Frenk et al. 1989). It seems that my comment, not written up in the proceedings of this workshop, may be particularly relevant to the quest for an accurate magnitude-redshift relation for dim galaxies and to placing limits on the formation of galaxies. As there was some discussion concerning this matter at the Toulouse Workshop on Gravitational Lensing (see papers in these proceedings by Ellis 1990, and Tyson 1990), I therefore thought it would be appropriate to present a written version of this original comment. Where appropriate, additional references, background material, and more recent determinations of the observational parameters have been included, as well as a discussion of possible errors. Much of the work I presented at this Toulouse conference can be found in the following references (Nemiroff 1989; Nemiroff and Dekel 1989, hereafter ND; and Nemiroff and Bistolas 1990).

Gravitational arcs were first reported by Lynds and Petrosian (1986). Since then, many more arc systems have been discovered (as discussed by other authors in this volume). Indeed, almost every rich cluster is expected to show an arc if observed down to a dim enough magnitude (ND, Tyson 1990). Much more common than arcs are the expected number of less dramatically lensed galaxies – which would appear as unusually elongated galaxies (ND). This discussion will focus on the best studied of the currently known arcs: the largest arc in Abell 370.

Much of the theoretical work on the gravitational arcs has been an effort to model the effect of the lens to create the observed image distribution (see for example: Grossman and Narayan 1988, Deguchi 1990, Mellier 1990, Soucail 1990). The present work is not so much concerned with the detailed properties of the lens, but, assuming some simple lens properties, what can be said about the redshift distribution of the source galaxies. The lensed galaxies are assumed to be normal field galaxies that are by chance placed so that they lie behind the cluster as seen from earth. This seems a more likely hypothesis than that these galaxies are peculiar, as it is unlikely that only peculiar galaxies would be positioned behind clusters. The first estimate of the probability of arc lensing was given by Kovner (1987).

The most important assumption necessary for every lens model is that the background lensed galaxy lie behind the observed cluster. Paczynski (1987) predicted gravitational lensing as a possible explanation of the arcs, and suggested looking to redshift as confirmation. Soucail et al. (1988) have measured the redshift of the arc in A370, and indeed found it behind the cluster, with the cluster at a redshift of about $z = 0.37$ and the background galaxy at $z = 0.72$. The accuracy of the galaxy redshift is contingent, however, on the identification of a single spectral line.

In §II will be given a brief analysis of the probability of arc lensing and how this can relate to the redshift distribution of galaxies. In §III this analysis will be applied to A370 and a discussion will be given on the information this may give. In §IV concluding comments are made.

II. ANALYSIS

Assume an arc has been measured behind a cluster of redshift z_c and singular isothermal velocity dispersion σ_c. Following ND, it is possible to theorize about a hypothetical cone existing behind the cluster where a background galaxy must fall to be lensed into an arc of axial ratio $A = a/b$ or more by the cluster. I will therefore assume that this cone defines a solid angle, and that galaxies falling into this solid angle and behind the cluster are lensed so as to be detectable to the observer. The angular area of this solid angle is (ND, equation 8)

$$\Omega = 2 \times 10^{-4} \deg^2 (\sigma_c / 1000 \, \mathrm{km \, sec^{-1}})^4 (A-1)^{-2}. \tag{1}$$

Assume that the observer can see galaxies down to a given limiting apparent magnitude m. Galaxy counts (see, for example Tyson and Jarvis 1979) give the number density $n(m)$ of galaxies brighter than m per solid angle. If it is assumed that a fraction of these, $\beta(m, z)$, is behind the cluster and thus eligible to be lensed, then the probability that a given cluster will show a gravitational arc is

$$P = 2 \times 10^{-4} \deg^2 (\sigma_c / 1000 \, \mathrm{km \, sec^{-1}})^4 (A-1)^{-2} \, n(m) \beta(m, z_c). \tag{2}$$

The only unknown in Eq. 2 is $\beta(m, z_c)$, the fraction of galaxies brighter than a given magnitude m that is at a redshift greater than z_c. Since the redshift that comes into the problem is z_c, it is therefore possible to get information on the redshifts of dim galaxies from the (more easily attainable) redshifts of clusters and statistical arguments. Put more simply – if you want to know the redshift of dim galaxies of a particular magnitude, look for them as arcs near clusters. If they make arcs, they must be further away than the clusters. If they do not make arcs, then they must be closer than the clusters. More distant clusters are more interesting because they can constrain the redshifts of galaxies the most.

III. APPLICATION TO ABELL 370

It is possible to make a first estimate of β at the distance of A370. First the cluster velocity dispersion is taken to be on the order of 1500 km/sec. The magnitude limit of Lynds and Petrosian (1989) was roughly $m = 24$ (in the blue, see ND), and the axial ratio of the arc was on the order of 15. Next, Lynds and Petrosian (1989) inspected 29 clusters to find 2 arcs, so I will take P (at this m) to be on the order of 0.07. The sky density of galaxies of magnitude $m = 24$ is roughly 10^4 galaxies per degree^2 (Tyson and Jarvis 1979). If one plugs these numbers into equation 2, one sees that $\beta(24, 0.37)$ is on the order of unity. This implies that most 24th magnitude galaxies have redshifts greater than 0.37.

This estimate may be in conflict with linear extrapolations of the results of Broadhurst, Ellis, and Shanks (1988, hereafter BES). Although their redshift results have only been obtained for galaxies brighter than $m = 21.5$, a linear extrapolation would predict a mean redshift of about 0.35 for magnitudes between 23.5 and 24. This redshift would place the lensed background galaxy just *in front* of the A370 cluster – a place where it is impossible to get the kind of gravitational lensing observed.

The situation diverges further if one considers two other important points. The first is that the redshift of this arc/background galaxy has actually been measured and found to be at a significantly greater redshift than both the cluster and linear extrapolations of the BES relation. A recent determination is $z_{arc} = 0.725$ (Lynds and Petrosian 1989). If this was a randomly selected galaxy of (unlensed) magnitude 24, then its redshift is well away from a naively expected value.

Secondly, as shown in the analysis of ND, the galaxy must actually be far behind the cluster in order to be distorted so much. The previous discussion assuming that the galaxy could be as close as the lens was overly optimistic. In fact, there is a limiting distance behind the cluster that a galaxy of finite radius must be in order to be distorted more than a given amount. This implies that the β measured by the lens effect is not $\beta(m, z_c)$, but actually $\beta(m, z_x)$ where z_x is determined by lens criteria and can be a significantly greater redshift than z_c. Figure 3 of ND shows that if the source galaxy was an L_* galaxy, then z_x is approximately 0.7, which is quite close to the actual redshift of the arc.

If there is significant spread in redshift about the mean, or a linear extrapolation of the BES relation is not applicable, then the above two estimations of the redshift of $m = 24$ galaxies can be brought into agreement. If better lens statistics are used to determine $\beta(m, z)$, then this function could be used to constrain the abundance of nearby galaxies, and hence put a limit on cold dark matter theories where galaxy formation at low z is expected.

IV. DISCUSSION AND CONCLUSIONS

This calculation is overly simplified and as a consequence somewhat inaccurate. This does not dissuade the author, however, from believing the results generated here are approximately correct and appropriately interesting.

One inaccuracy involves equation 1, whose derivation assumed that all galaxies behind the cluster are in the hypothetical cone of limiting A. As shown in ND, however, the cone would normally have its vertex at the cluster and not at the observer – and so it does not simply correspond to a solid angle on our sky. Also, as mentioned above, the cone is usually truncated at a redshift greater than the cluster because galaxies close to the cluster lens cannot be distorted to the necessary A value. Therefore it is possible for a galaxy to be behind the cluster, inside the solid angle relevant to the above analysis, and yet not be inside the relevant cone of constant A. The result is that galaxies close behind the cluster are distorted less than in the approximation in equation 1. A more comprehensive calculation would therefore find fewer high A events than postulated here.

Another inaccuracy is that the model of a cluster as a single isothermal sphere is clearly too simple. This probably works, however, to increase the probability of high A events. To include even slight lens asphericity, however, would significantly complicate the analysis, and I speculate that it would not significantly alter the character of the disagreement. At small A, however, the approximation of the sphericity of the lens is reasonable.

Next, the magnitude limit of arc survey techniques may be somewhat uncertain, leading one to use an inaccurate value of n. Lastly, the later analysis included only one data point – the largest arc in A370. Clearly better statistics are needed to get a more accurate answer.

In sum, arc lensing can be used to constrain dim galaxy redshifts. Preliminary estimates lead to an apparent disagreement between between linear extrapolations of known galaxy $m - z$ relations valid at m brighter than 21.5 and those deduced from the arcs at m brighter than 24.

I thank A. Dekel and M. Bithel for useful discussions. I am grateful to the organizers of the Durham Workshop on The Epoch of Galaxy Formation for allowing me to present a comment to the workshop. I am happy to acknowledge the support of a Golda Meir Postdoctoral Fellowship while at the Hebrew University of Jerusalem and an National Research Council Research Associateship while at the Naval Research Laboratory.

REFERENCES

Broadhurst, T. J., Ellis, R. S., and Shanks, T. 1988, *M.N.R.A.S.*, **235**, 827 (BES).

Deguchi, S. 1990, *Toulouse Workshop on Gravitational Lensing*, elsewhere in this volume.

Ellis, R. S., 1990, *Toulouse Workshop on Gravitational Lensing*, elsewhere in this volume.

Frenk, C. S., Ellis, R. S., Shanks, T., Heavens, A. F., and Peacock, J. A. 1989, *The Epoch of Galaxy Formation*, Kluwer, Dordrecht.

Grossman, S. A., and Narayan, R. 1988, *Ap. J. (Letters)*, **324**, L37.

Kovner, I. 1987, *Nature*, **327**, 193.

Lynds, R., and Petrosian, V. 1987, *Bull. A. A. S.*, **18**, 1014.

Lynds, R., and Petrosian, V. 1989, *Ap. J.*, **336**, 1.

Mellier, Y. 1990, *Toulouse Workshop on Gravitational Lensing*, elsewhere in this volume.

Nemiroff, R. J. 1989, *Ap. J.*, **341**, 579.

Nemiroff, R. J., and Bistolas, V. G. 1990, *Ap. J.*, submitted.

Nemiroff, R. J., and Dekel, A., 1989, *Ap. J.*, **344**, 51 (ND).

Paczynski, B. P. 1987 *Nature*, **325**, 572.

Soucail, G. 1990, *Toulouse Workshop on Gravitational Lensing*, elsewhere in this volume.

Soucail, G., Mellier, Y., Fort, B., Mathez, G., and Cailloux, M. 1988, *Astr. Ap.*, **191**, L19.

Tyson, A. J., 1990, *Toulouse Workshop on Gravitational Lensing*, elsewhere in this volume.

Tyson. A. J. and Jarvis, J. F. 1979, *Ap. J. (Letters)*, **230**, L153.

SPATIAL CORRELATION IN GRAVITATIONALLY LENSED IMAGES

Shuji Deguchi

Nobeyama Radio Observatory
National Astronomical Observatory
Minamimaki, Minamisaku
Nagano 384–13, JAPAN

1. Introduction:

Luminous blue arcs in clusters of galaxies (Soucail *et al.*, 1987) are produced by gravitational lensing of light due to a cluster of galaxies with a mass about $10^{14}M_{\odot}$ with near perfect alignment between the source, lens and observer (Paczynski, 1987). Galaxies in the clusters are also anticipated to produce irregularities in the arc image. In fact, as reported by Petrosian (1989), the arc in the cluster of galaxies 2244-02 exhibits knots or breaks near the center of the image. This distorsion of the image can be caused by two reasons. First, it can be an irregularity in the lensed galaxy (for example, Hammer, 1989). Second, it is a distorsion produced by mini–lenses of galaxies in the cluster.

The mini–lens effect, the effect of the random distribution of galaxies in a cluster on the imaging, has been investigated by Grossman and Narayan (1988) using a Monte Carlo Method. Wetake an alternative approach, i.e., the Markov Method. This approach enables us to calculate a probability distribution of the deflection of light due to a random distribution of galaxies. This is originally formulated by Chandrasekhar (1943) for treating problems of scattering of test particles in stellar dynamics. We have applied this successfully for the gravitational microlensing of distant sources (Deguchi and Watson, 1987, 1988).

2. Modeling:

Probability distribution functions of the deflection angle can be calculated by the standard procedure in the Markov process. The deflector plane containing galaxies is taken to be perpendicular to the line between the center of the source and the observer. In this calculation, we assume that the mass of galaxies are the same. We adopt the King's model for the distribution of galaxies. The average deflection of the light must be subtracted from the deflection angle. In contrast to the microlensing, the deflection of light by galaxies does not increase infinitely at the perfect alignment. This leads a Gaussian probability distribution function for the deflection angle.

For investigating the irregularity of the arc image, we consider the spatial correlation of the intensity fluctuation at two separate points $\mathbf{r_a}$ and $\mathbf{r_b}$:

$$< (l(\mathbf{r_a})- < l(\mathbf{r_a}) >).(l(\mathbf{r_b})- < l(\mathbf{r_b}) >) > \qquad (1)$$

where $l(\mathbf{r})$ is the intensity of the image at the position $\mathbf{r}$ and $<>$ is the ensemble average. For this purpose, we have to calculate the probability distribution function of deflecton angles $\mathbf{f_a}$ and $\mathbf{f_b}$ at two position $\mathbf{r_a}$ and $\mathbf{r_b}$. We can obtain the Gaussian distribution function of the deflection angle $\mathbf{f_a}$ and $\mathbf{f_b}$ in the first order approximation as

$$P(\mathbf{f_a}, \mathbf{f_b}) = cexp(-xAx^t) \tag{2}$$

where vectors $x = (\mathbf{f_a}, \mathbf{f_b})$ and x^t is the transformed vector of f. Here A is the 4×4 matrix and c is the normalization constant. The matrix A can be calculated numerically at given two points $\mathbf{r_a}$ and $\mathbf{r_b}$. Then the spatial intensity correlation (1) can be calculated from equation (2).

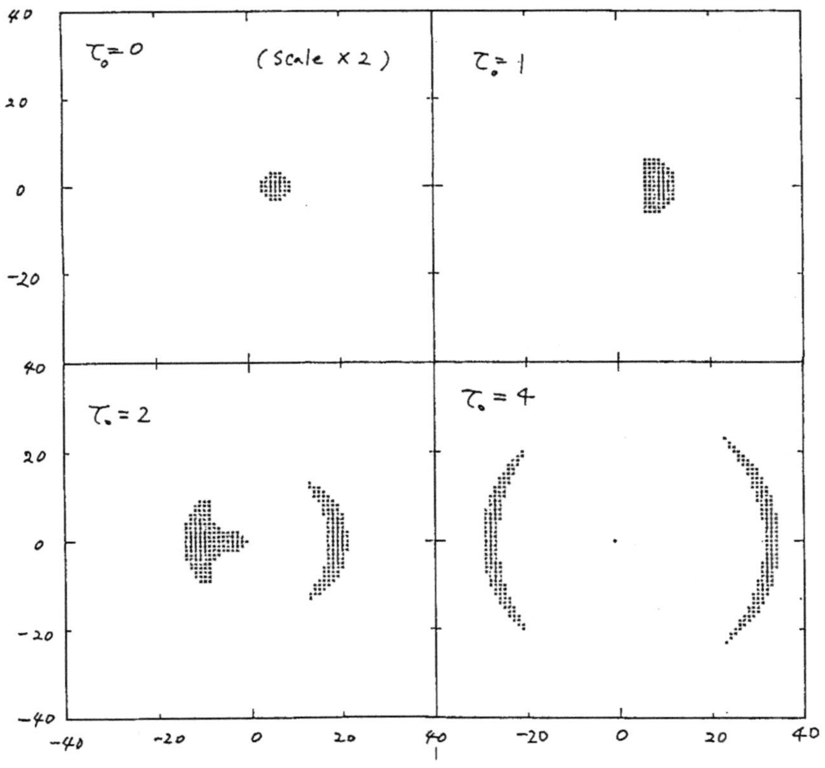

Figure 1. Average arc images in the different optical depth. The cluster is assumed to be circularly symmetric. The optical depth at the cluster center τ_0 is varied.

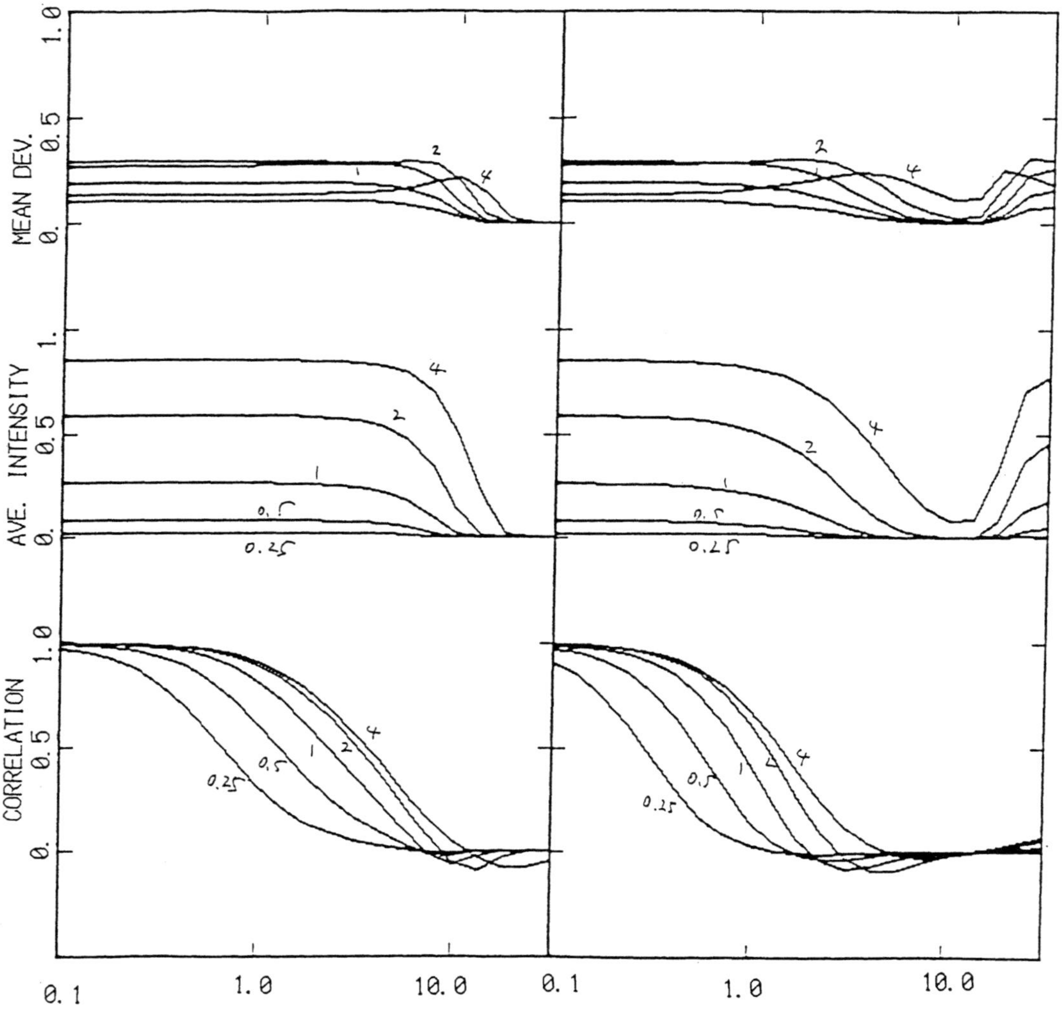

Figure 2. Mean intensity deviation, average intensity, and intensity correlation of fluctuation. The origin of the x axis is the center of the arc (the brightest point in the arc), and x axis is taken along the arc (left panel) and across the arc (right panel). Note that the correlation is normalized by the standard deviation at each points.

Figure 1 shows the average image of the circularly symmetric lens with various optical depths. The light source is assumed to be Gaussian brightness distribution. The arc feature (and second and third images) appear in the case of $\tau_0 = 2$. Figure 2 shows the mean intensity deviation, mean intensity, and the spatial intensity correlation for the model of $\tau_0 = 2$. The origin of the x axis in figure 2 is at the brightest point in arc ($x = 19, y = 0$ in the lower left panel in figure 2. The point $\mathbf{r_a}$ is fixed at the brightest point and $\mathbf{r_b}$ is moved along the arc in the left panel and across the arc (toward the center) in the right panel. Of course, the correlation of the intensity fluctuation remains at the larger distance when $\mathbf{r_b}$ is along the arc. However, it should be noted that the correlation decreases faster than the average intensity does.

3. Conclusions:

It is difficult to discriminate the mini–lens effect to the irregularities in the original source at present. The calculation in this paper show that at least the observed knots in the arc are consistent with mini–lens model. The observation of colors of the arc with high–spatial resolution will greatly help to solve the problem.

References:

Chandrasekhar, S., 1943: Rev. Mod. Phys. **15**, 1

Deguchi, S., and Watson, W.D., 1987: Phys. Rev. Letters **59**, 2814

Deguchi, S., and Watson, W.D., 1988: Ap. J. **335**, 67

Grossman, S.A., and Narayan, R., 1988: Ap. J. (Letters) **324**, L37

Hammer, F., 1989, in this proceeding

Paczynski, B., 1987: Nature **325**, 572

Petrosian, V., 1989, in this proceeding

Soucail, G., Fort, B., Mellier, Y., and Picat, J.P., 1987: Astr. Ap. **172**, L14

OPTICAL-INFRARED STUDIES OF ARCS IN ABELL 370

Alfonso Aragón Salamanca & Richard S. Ellis
Physics Department, Durham University, U.K.

SUMMARY. *We demonstrate the potential of optical-infrared photometry in deciphering the nature of high redshift galaxies responsible for the arcs seen in foreground rich clusters. Using the arcs in Abell 370 as an example, we find that photometry of various portions of the giant arc is consistent with a common spectral energy distribution equivalent to that of a background spiral galaxy as determined spectroscopically by Soucail* et al. *However, the faint arc A5 is* not *detected in a deep 2μm exposure, suggesting that the lensed z>1 galaxy is at least as blue as the most extreme star-forming galaxies observed today.*

1. Introduction

The giant arc in the rich cluster Abell 370 (z = 0.37) is the most impressive example of several now known. The pioneering spectroscopic techniques of Soucail *et al.* (1988) led to a convincing discovery of an emission line throughout the arc supporting the hypothesis of a gravitationally-lensed background galaxy at z = 0.72. A fainter arc in the same cluster (A5) is now claimed to be due to a lensed galaxy with z ~ 1.3 (Soucail, this meeting). Arcs arising from such distant sources promise to be valuable tools in studies of otherwise inaccessibly faint objects. Since most arcs will be too faint for spectroscopic studies, we examine the potential of optical-infrared photometry in studying the stellar populations of serendipitously-discovered background galaxies.

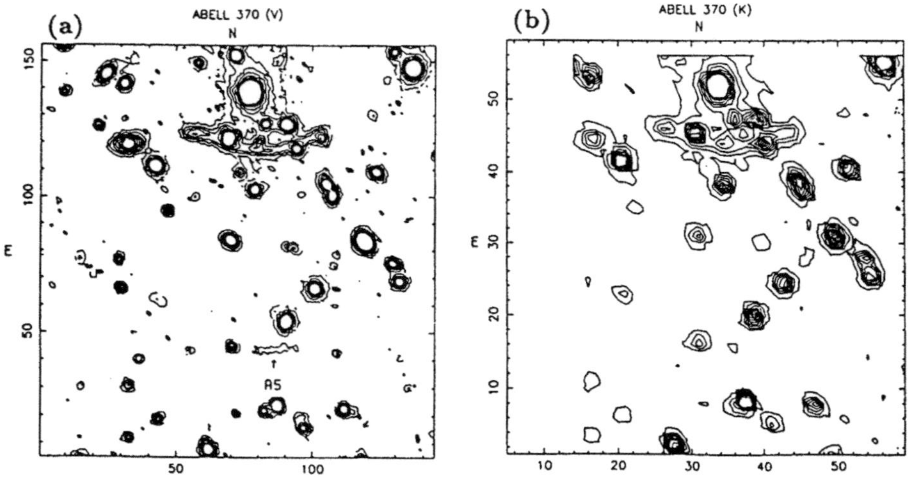

Fig. 1. (a) 4 hour V-band image of Abell 370 obtained with the 1.5m Danish telescope at 0.47 arcsec pixel^{-1}. The lowest contour has μ_V=26^m.0 arcsec^{-2}. **(b)** 1.7 hour K-band image obtained with the InSb 62x58 IRCAM array on UKIRT at 1.2 arcsec pixel^{-1}. The lowest contour has μ_K=21^m.7 arcsec^{-2}.

Fort *et al.* (1988) claim to detect a large number of possible *arclets* in A370 mostly with blue B-R colours. Tyson (this meeting) confirms this in other clusters and proposes such a background population could be that seen in deep galaxy counts, and inferred by Cowie & Lilly (1989) to be the prime source of heavy elements in the Universe. A major question therefore is whether the blue arcs found are indicative of normal spirals or extreme star-forming galaxies.

Optical CCD images of A370 were taken at the AAT through 4 intermediate-band filters centred at 418, 502, 685 and 862nm plus broad-band U (see MacLaren *et al.* (1988)). Deep V-band images were also obtained at the 1.5m Danish telescope at La Silla as part of a search for distant supernovae (Nørgaard-Nielsen *et al.* 1989). A summed 4 hour exposure in V shows both the giant and A5 arcs used in this analysis (Figure 1(a)). Infrared K ($\equiv 2\mu$m) images were taken using the InSb IRCAM array on the 3.8m UKIRT at a coarser spatial resolution (Figure 1(b)); the giant arc is clearly visible but A5 is not detected.

2. Spectral Energy Distribution of the Giant Arc

Since the giant arc is heavily contaminated by cluster members, to test the lensing hypothesis we determined SEDs for four uncontaminated portions. Are the four portions consistent with a single SED and can the SED can be understood in terms of a typical field spiral at z=0.72 (the spectroscopic redshift of Soucail *et al.*) ? Although the arc is prominent in U (no contaminating galaxies can be seen), this is largely because the cluster galaxies are themselves very red; the arc is not *intrinsically* very blue. The 4 portions share the same SED to within photometric uncertainties and can readily be understood in terms of a Sb or Sc galaxy at the spectroscopic redshift z = 0.72 (Figure 2(b). The K measurement is somewhat too high for a galaxy at the cluster redshift, z = 0.37 (Figure 2(a)). Thus our broad-band photometry provides reasonable independent support for the lensing of a normal background spiral.

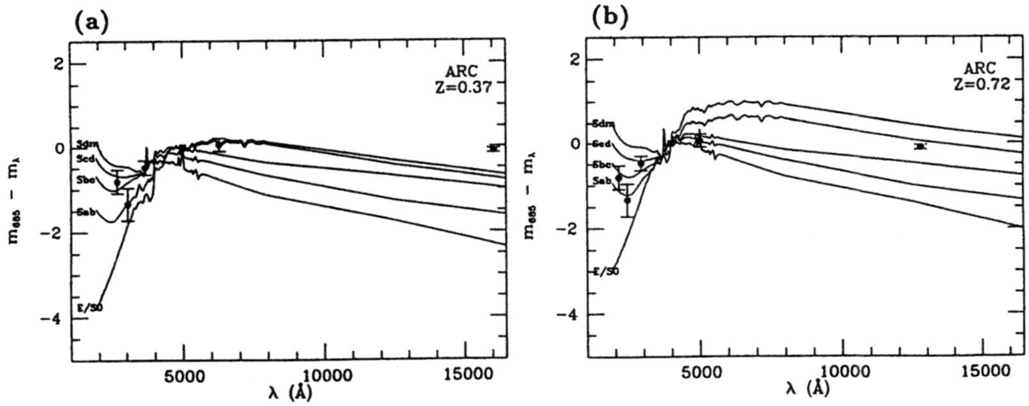

Fig. 2. Mean rest-frame SED of the giant arc in Abell 370 compared with present-day predictions of different morphological types, assuming the arc has **(a)** z=0.37 – the cluster redshift, and **(b)** z=0.72 – the spectroscopic redshift of Soucail *et al.*. Error bars represent the scatter in colour within 4 portions uncontaminated by cluster galaxies.

MacLaren *et al.* (1988) found many spectroscopically-confirmed cluster E/S0s showed an excess of ultraviolet light (UVX) which they interpreted in terms of recent star formation. The spatial distribution of such objects was interesting; few were found in the cluster core and of 4 galaxies lying on/close to the arc, 3 are UVX. Could this arise because of widespread contamination of UV light from arcs similar to the giant arc, or is the congregation of UVX galaxies near the giant arc simply a coincidence? There is little evidence for elongations in the UV images of the UVX objects, and also little correlation between Fort *et al.*'s elongated blue structures and the MacLaren *et al.* UVX E/S0s.

3. The Nature of the Arc A5

Precise redshifts are not necessarily required to make interesting statements about the galaxy population behind the cluster. The arc A5 is a good example; it is considerably fainter than the giant arc, much bluer and now a tentative redshift z=1.3 is available (Soucail, this conference).

In a 1.7 hour integration at 2μm with IRCAM, we can place interesting constraints on the nature of the galaxy producing the arc A5. At a limiting contour $\mu_K=21^m.7$ arcsec^{-2} we see no source across several pixels. In integrated optical-infrared colour A5 must thus have V-K$\leq$3.7 which, at the proposed redshift, is an upper limit bluer than all but the most spectacular star-forming galaxies known (for example, the Sm/Irr NGC4449 placed at z=1.3 would have V-K=3.3). If subsequent 2μm imagery fails to detect A5 then might it not be at the high redshift proposed? Actually even at z=0.37, the V-K limit is only just consistent with a normal Scd. More likely the object is one example of a population of intense star-forming galaxies at z >1. Depending on the magnification, the absolute magnitude must be at least as faint as $M_B \sim -19^m.3$ (H_0=100). The connection between such a population and that inferred in faint field galaxy surveys is discussed in some detail by Tyson and Ellis in these proceedings.

4. Conclusions

Arcs offer an important new way to study the population of serendipitously-found high redshift galaxies. Our study clearly shows the importance of securing accurate optical and infrared photometry and not just spectra. The role of the infrared for high redshift galaxies is particularly important and demonstrated for 2 arcs in Abell 370. The giant arc can be naturally explained as a z=0.72 intermediate-class spiral, whereas the A5 arc is potentially a most interesting object at least as blue as the most intense star-forming galaxies known today.

References

Cowie, L. & Lilly, S.J., 1989. in *The Universe of Galaxies*, ed. Kron, R. (in press).

Fort, B., Prieur, J.L., Mathez, G., Mellier, Y., Soucail, G., 1988. *Astron. Astrophys.*, **200**, L17.

MacLaren I., Ellis, R.S. & Couch, W.J., 1988. *Mon. Not. R. astr. Soc.*, **230**, 249.

Nørgaard-Nielsen, H.U., Hansen, L., Jørgensen, H.E., Aragón Salamanca, A., Ellis, R.S. & Couch, W.J., 1989. *Nature*, **339**, 523.

Soucail, G., Mellier, Y., Fort, G., Mathez, G. & Cailloux, M., 1988. *Astron. Astrophys.*, **191**, L19.

SPECTROSCOPY OF ARCS IN Cl2244–02 AND A370 (A5)*

G. Soucail, Y. Mellier, B. Fort, G. Mathez, M. Cailloux

Observatoire Midi-Pyrénées, Toulouse (France)

We give a redshift determination of the giant arc in Cl 2244–02 and a possible identification of spectral features in the "mini-arc" A5 in A370.

1. The giant arc in Cl 2244–02: spectral evidence for a galaxy at z= 2.238

1.1. Photometric and spectroscopic data

Photometry comes from B,V and R CCD frames obtained by Hammer *et al.* (1989) at the Prime Focus of the CFHT. In summary, the integrated magnitudes of the arc are $B = 21.40 \pm 0.2$, $V = 20.87 \pm 0.2$ and $R = 20.63 \pm 0.2$, while the color index B–R is 0.77 ± 0.2. The surface brightness ($\mu_B = 25.2$) corresponds approximatively to 6 % of the sky brightness.

The spectroscopic data come from 4 observing runs and are presented on Figure 1:

- The first spectrum (1) comes from ESO, with the UV300 grism (6 Å/pixel, spectral range 3600–6500 Å) mounted on the focal reducer EFOSC and a special curved slit punched with the PUMA system. 3 frames were co-added, with a total integration time of 16,000 sec.
- The second spectrum (2) comes from ESO with the B300 grism (6.5 Å/pixel, spectral range 3800–7200 Å), a long slit and a total integration time of 20,000 sec.
- The third spectrum (3) comes from CFHT, with the focal reducer, the low-dispersion grism (8 Å/pixel, spectral range 4500–7500 Å) and a curved slit. The total integration time is 27,000 sec.

Such a large amount of data was necessary to obtain a large enough S/N ratio for that low surface brightness object.

1.2. Analysis and results

These 3 spectra (without flux calibration) are superimposed in Figure 1, in order to cross-correlate the most significant absorption and emission features. The main features detected are listed in Table 1, as well as their identification with lines redshifted at z=2.238. To confirm this result, Figure 2 presents a flux calibrated spectrum (with the B300 grism), superposed with a synthetic spectrum from Guiderdoni and Rocca-Volmerange (1987), redshifted at z=2.238. The correspondence between the two spectra is amazing!

This results appears as a new evidence of the *gravitational origin of the giant arcs*. This arc is the second one, after the arc in A370, for which a redshift is measured. We took advantage of the *gravitational magnification* (factor about 15 for Cl2244–02) to get the spectrum of a VERY distant galaxy, which is most probably a FIELD galaxy. Without amplification, this galaxy should have a magnitude $B \simeq 24.5$, and should be unobservable spectroscopically.

---

* Based on observations made with the 3.6m telescope at the European Southern Observatory (La Silla, Chile)

2. A5: The brightest of the "mini–arcs" in A370

Photometry was published in Fort *et al.* (1988) from data obtained at the Prime Focus of the CFHT. The integrated magnitude of the arc is B $\simeq$ 22.4, and the color index is B–R $\simeq$ 0.9. The surface brightness ($\mu_B = 24.9$) corresponds to 7 % of the sky brightness.

The spectroscopic data in Figure 3 correspond to 6 hours of integration with the B300 grism at ESO. The spectrum is flux calibrated (in arbitrary units) and some features are indicated. A tentative redshift from these structures is presented as well, taking into account the identification of the significant absorption line at $\lambda = 6450$ Å as the λ MgII 2800 line, redshifted at **z = 1.305.** Some other less significant lines are marginally detected, but the continuum correlation is less convincing.

There exists a strong interest to determine the redshift of this object as it should be the second well identified arc inside the same cluster. Two different redshifts should add very strong constraints on the modelling of the gravitational potential of the cluster. But we have to wait for more data (ESO run in Sept. 89) in order to confirm this preliminary determination.

REFERENCES :

Fort, B., Prieur, J.L., Mathez, G., Mellier, Y., Soucail, G., 1988, *Astron. Astrophys.* **200**, L5

Guiderdoni, B., Rocca-Volmerange, B., 1987, *Astron. Astrophys.* **186**, 1

Hammer, F., Le Fèvre, O., Jones, J., Rigaut, F., Soucail, G., 1989, *Astron. Astrophys.* **208**, L7

Figure 1

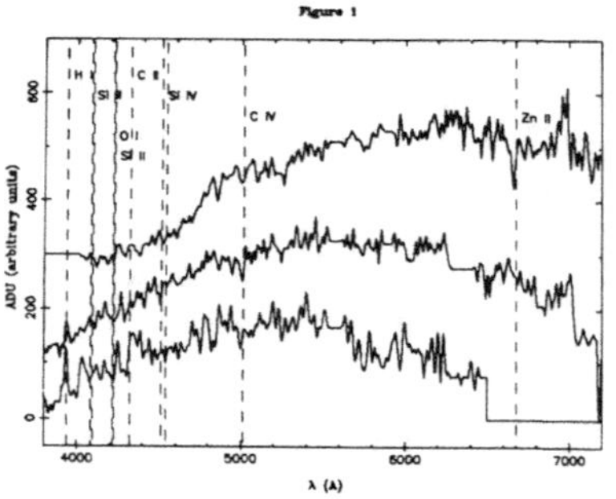

Figure 2

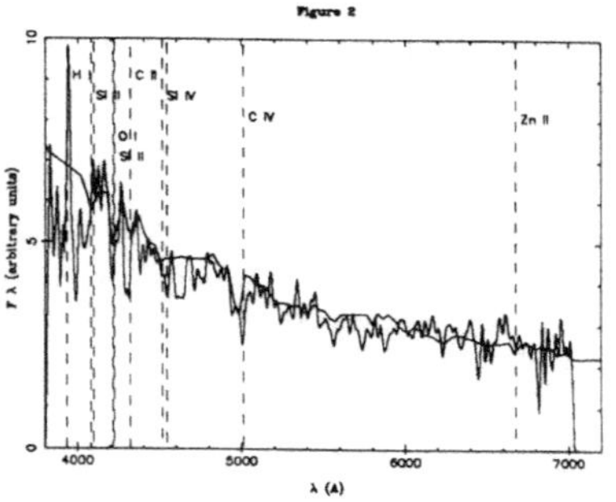

TABLE I

Spectral lines identified in the spectrum at a redshift of z=2.238

Line	λ_0	$\lambda_0(1+z)$
Ly α	1215.7	3936
Si II	1260.4, 1264.8	4081, 4095
O I, Si II	1302.2, 1304.4	4216, 4223
C II	1334.5	4321
Si IV	1393.8, 1402.8	4512, 4542
C IV	1548.2	5013
Zn II	2062.0	6677

Figure 3

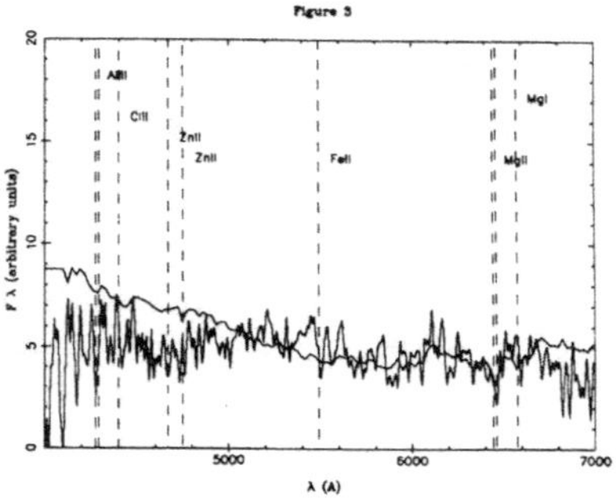

FIRST RESULTS ON THE SPECTROSCOPY OF THE ARC-LIKE OBJECT IN ABELL 2390 *

R. Pello[1], J.F. Le Borgne[2], G. Mathez[2], Y. Mellier[2], B. Sanahuja[1], and G. Soucail[2]

[1] Departament de Fisica de la Atmosfera, Astronomia i Astrofisica. U. de Barcelona (Spain)
[2] Observatoire Midi-Pyrenees, Toulouse (France)

HISTORY:

During an observing run for the "Arcs and Gravitationally Distorted Images Survey" with the 2.5m INT in July 1988, RP and JFLB pointed out that the rich cluster A2390 (z=0.23) contains a strange object which, in spite of its straight shape, could be a so called arc. The existence of this object was confirmed by GS in August 1988 at CFHT. Spectra of this arc and of the cluster galaxies were obtained during 3 runs: CFH 3.6m telescope (September 1988, GM and YM), ESO 3.6 m telescope (October 1988, YM and GS), WHT (4.2m) (August - September 1989, RP, JFLB and BS).

In this poster we focus to preliminary results concerning the WHT spectroscopy of the arc and suggest a redshift of 0.91 for this structure.

IMAGING AND PHOTOMETRY OF THE ARC

The first images obtained with the INT, and later with the CFHT, show that this structure has a rectangular shape 15" long and 1.3" arcsec wide (see Mellier,1989 and this workshop). The arc is resolved on the CFHT pictures. This "straight" arc is divided by 2 breaks into 3 regions of 3.0, 5.6 and 6.4 arcsec long, respectively. The total magnitude of the arc is B=21.9 ± 0.2. The surface brightness is variable, but the color indices seem to be constant all along the structure. Their values are (± 0.2 mag.): B-g=1.0, g-r=0.6, B-R=2.0. The arc is bluer than the galaxies in the cluster the mean g-r color of which is about 1.2.

SPECTROSCOPY

The CFH and ESO spectra have been obtained using both multi-object spectroscopy (PUMA) and a long slit; they range between 4500 and 6800 Å for CFHT and between 4000 and 7000 Å for ESO. The spectra of the cluster galaxies lead to a redshift of 0.23 for the cluster. The galaxy near the arc also belongs to the cluster (z=0.230), but none of these arc spectra did show any strong feature allowing to determine the redshift. The WHT spectra were obtained using the FOS in the long slit mode: they range roughly between 5000 and 9500 Å (red mode: first order selected by red filters). The dispersion is about 9 Å per pixel. The CCD used is a coated low noise GEC (10 e^- rms). The exposures on the arc were of 1 hour, taken with the slit aligned on it. The total exposure time was about 15 hours, with a seeing ranging between 0.8" and 1.5". The present results were obtained by selecting the 7 exposures with the highest S/N, best seeing and when the contamination by the light of the nearby galaxy is minimum.

RESULTS AND CONCLUSION

The spectrum of the arc shows a strong emission line at 7130 Å, that exists all along the arc (see figure). The general shape of the continuum seems to be the same in all the 3 regions and

---

* *Based on observations made with the Isaac Newton and William Herschel telescopes operated on the island of La Palma by the Royal Greenwich Observatory in the spanish Observatorio del Roque de los Muchachos of the Instituto de Astrofisica de las Canarias, the 3.6m telescope at the European Southern Observatory (La Silla, Chile) and the Canada - France - Hawaii 3.6m telescope*

is in good agreement with the color indices. Region 2 clearly shows the contamination from the nearby galaxy, that slightly affects the shape of the continuum. Hα, Hβ and Hγ at a redshift of 0.23 can be identified in this region, overlapping the spectrum of the arc.

If we identify the emission line as [OII] (3727 Å), the redshift is 0.913. The identification of [OII] takes into account the shape of the continuum: very flat spectrum, break at $\sim$ 2900 Å. When we compare the spectrum of region 3 (which is the less contaminated by the nearby galaxy) with a synthetic spectrum of a Sd galaxy redshifted at z=0.91 (Rocca-Volmerange and Guiderdoni, 1988; dotted lines in the figure), they are in good agreement. The shape of the continuum is compatible with a break at 7650 Å (4000 Å break at 0.91). However, the S/N is very low for $\lambda > 7500$ Å. Other lines may be identified with this redshift, as the Mg II (2795 Å) line. A more detailed study will follow after the reduction of all the data. In any case, spectroscopy confirms that this structure is due to a galaxy gravitionally lensed by the cluster A2390. This the third arc-candidate which is now confirmed (Soucail et al., this workshop).

ACKNOWLEDGEMENTS

We would like to thank B. Fort and M. Cailloux for useful comments and help during the reduction of the data. The observers at WHT appreciated very much the interest and the help of M. Breare before and during the FOS run.

REFERENCE

Rocca-Volmerange, B., and Guiderdoni, B.: 1988, Astron. Astrophys. Suppl. Ser., **75**, 93

Mellier, Y.: 1989,Proceedings of the STSCI on "Clusters of galaxies". Baltimore, USA, M. Fitchett ed.

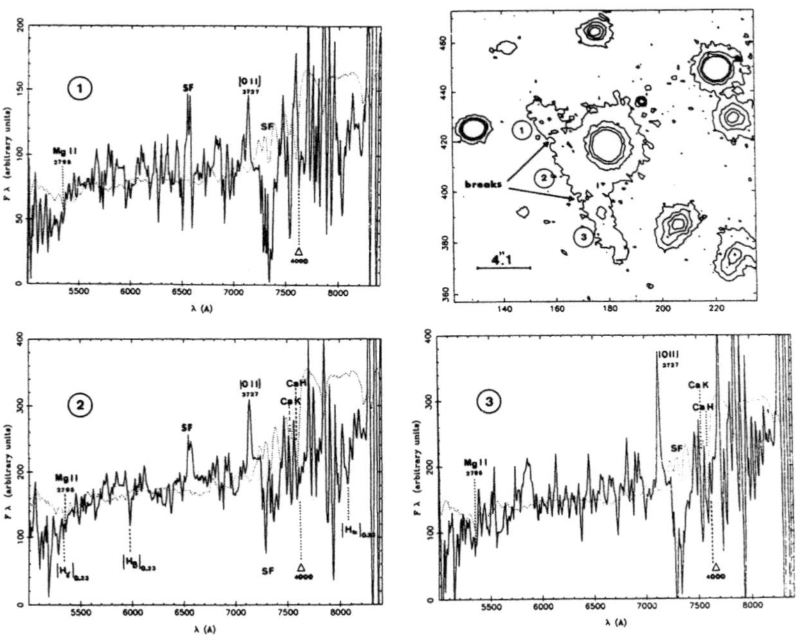

DEEP PHOTOMETRY OF Cl.2244-02 IN U COLOUR WITH AN ELECTRONOGRAPHIC CAMERA.

G. Wlérick[1], C. Vanderriest[1], F. Hammer[1], G. Lelièvre [1], D. Horville[1], L. Renard[1], J. Arnaud[2], C. Gouiffes[3]

(1) Observatoire de Paris, U.R.A. 176 and 335 5 place Janssen, 92195 Meudon (FRANCE).
(2) Canada-France-Hawaii Telescope corp. P.O. Box 1597 Kamuela 96743 Hawaii (U.S.A.)
(3) European Southern Observatory Casilla 19001 Santiago 19 (CHILE)

Introduction:

The giant arc detected in the cluster of galaxies 2244-02 (Lynds and Petrosian, 1986) is believed to be the distorted gravitational image of a distant background galaxy. More exactly, it consists of 2 adjacent, highly amplified images with opposite parities. Gravitational lens models then predict the existence of 3 other faint images, whose precise magnifications and locations depend of the detail of the mass distribution in the cluster. Photometric data in B,V,R colours obtained at the CFH telescope with a CCD revealed 2 good candidates ("objects" S and T, see Hammer et al., 1989) with colour indices similar to those of the arc itself. The last missing image could have escaped detection because it is very faint and/or hidden by a galaxy of the lens cluster. Observations in U colour would help to confirm this identification of the faint images. It could also throw some light on the nature of the "patches" seen in the arc and interpreted as the bulge and arms of the source galaxy, supposed to be spiral (Hammer and Rigaut, in press).

Observations:

Petrosian (1988) states that no positive detection of the arc have been obtained in U with CCDs. We present here the data obtained at the CFH telescope in june 1987 with a wide field electronographic camera (Wlérick et al, 1983). Two long exposures (105 mn and 115 mn) were made on Kodak Electron Microscope film, digitized with a PDS microphotometer, processed in Meudon with the STII and EVE softwares and coadded. On the resulting picture (Fig. 1), the resolution is 1.1"(FWHM). The arc displays the same knotty structure than in other colours and we detect also S and T as well as the brightest galaxies of the cluster.

The photometric calibration was done with the help of shorter exposures, by transfer of a nearby photometric sequence in 3C 445 established at ESO. Table 1 summarizes our measurements. The U-B colour indices of the galaxies are generally compatible with ellipticals at $z \simeq 0.33$ (Pence, 1976; Guiderdoni and Rocca-Volmerange, 1988), galaxy O being a possible exception, and the colours of S and T are in fair agreement with that of the arc. Good measurement of the sky level is critical; changing its value by 0.1% changes the magnitudes by about 0.07 for the arc and as much as 0.15 for the faint galaxies. The slight displacement ($\simeq$ 1") of image T from the positions observed by Hammer et al. and by Petrosian could also result from a poor subtraction of the background combined with low S/N.

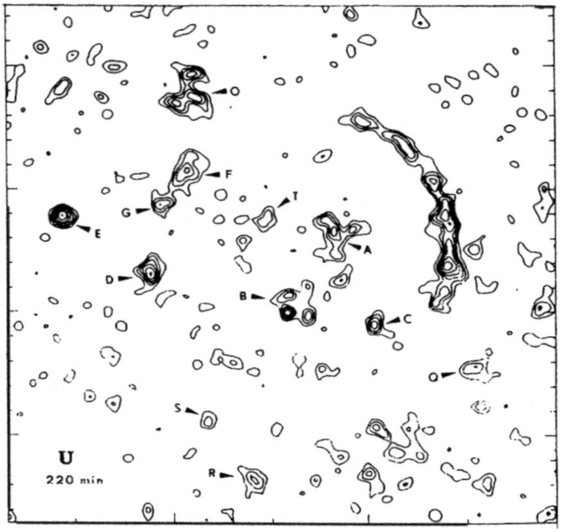

Fig. 1: Isophotes on the processed U picture (sum of 2 exposures). Objects (or images) have been labeled according to Hammer et al. (1989).

We tried to measure individually the patches P_1 to P_6 (fig. 2) and found significant colour differences. A possible interpretation of the UV excess in the bulge of the source galaxy (patches P_2 and P_5) is that it contain an active nucleus. Maybe the (Lyα ?) emission observed in the spectrum of the arc (Soucail et al., this workshop) is limited to this central region. Pictures through an appropriate interference filter would be interesting. Note also that the sum of the fluxes for P_1 to P_6 does not correspond to the total flux of the arc in the paper of Hammer et al., while it does for our data in U; reanalysis of the B and U data within exactly the same areas is presently under way.

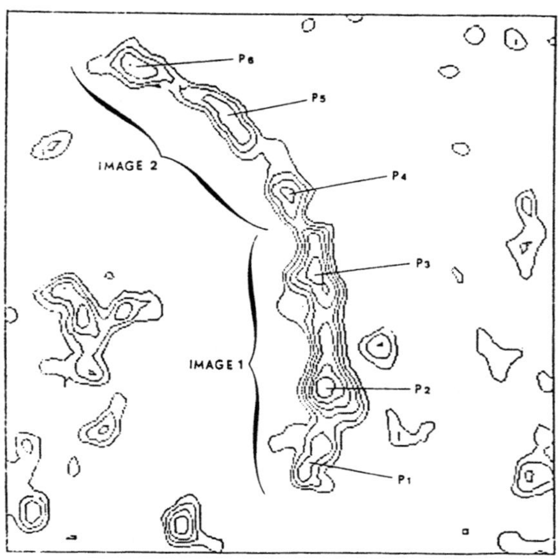

Fig.2: Close-up on the arc; identification of the condensations P_1 to P_6.

Table1: Photometric measurements:

	"object"	area in arcsec.^{2}	U (this work)	B (Hammer et al. 1989)	$U - B$
	Arc	58.4	20.85 ± 0.1	21.51	−0.66
image 1	P_1	10.	22.98 ± 0.2	23.4	−0.42
	P_2	10.	22.43 ± 0.2	23.2	−0.77
	P_3	9.2	22.65 ± 0.2	23.6	+0.05
image 2	P_4	6.7	23.34 ± 0.2	23.35	−0.01
	P_5	12.5	22.81 ± 0.2	23.4	−0.59
	P_6	10.	23.06 ± 0.2	23.35	−0.29
	S	8.4	23.57 ± 0.3	24.75	−1.18
	T	6.5	23.48 ± 0.3	24.25	−0.77
	O	23.	21.86	22.31	−0.45
	Q	13.	22.85	22.31	0.54
	R	17.	22.83	22.27	0.56

Conclusion and suggestions:

These observations agree with the current model interpreting S and T as faint additional images of a distant spiral galaxy whose 2 brightest images make up the arc.

We would like to suggest further observations, feasible with the present instrumentation and site quality of the C.F.H. telescope, that would help to improve the model.

- Deeper U pictures would enable us to measure the faint images with minimal contamination from the cluster galaxies [the contrast is increased by more than 1 mag. with respect to B colour]. It would be interesting to detect the fifth image or, at least, have a better upper limit on its flux. In the model of Hammer et al., a slight elongation of image S should be seen, which is not observed on the present (poor S/N) pictures. Improving S/N would obviously help to adapt the model. Such deep U data could be obtained either by CCD with good UV coating, or by electronography: here, the limit magnitude was about $U_{lim} \simeq 24$, but values of $U_{lim} \simeq 25$ have already been reached by this technique (Wlérick et al., to be published).

- On pictures with the highest possible spatial resolution (0.5" - 0.6"), the arc should probably be resolved in the radial direction and the "patches" better seen tangentially. This would put better constraints on the model and lead possibly to a good reconstruction of the shape of the source galaxy. The configuration is more favourable than for A370, because the arc in Cl.2244 apparently does not suffer from local distortion by individual galaxies.

References:

- Guiderdoni B., Rocca-Volmerange B., 1988, Astron. Astrophys. Suppl., **74**, 185.
- Hammer F., Le Fèvre O., Jones J., Rigaut F., Soucail G., 1989, Astron. Astrophys., **208**, L.7.
- Lynds R., Petrosian V., 1986, Bull. A.A.S., **10**, 1014.
- Pence W., 1976, Astrophys. J., **203**, 39.
- Petrosian V., 1988, in "Gravitational lenses", Lecture Notes in Physics, **330**, 109.
- Wlérick G. et al., 1983, in "Instrumentation in Astronomy V", p.143.

V

Concluding Remarks

CONCLUDING REMARKS

R. D. Blandford
Theoretical Astrophysics, Caltech,
Pasadena, CA 91125, U. S. A.

We have gathered in the home of Pierre Fermat, originator of the principle that is so helpful in understanding gravitational lenses, (and also, according to local tradition, the inventor of the calculus) at a particularly opportune moment in the development of this subject. There have been many exciting observational developments over the past year. Several more multiple-imaged quasars have been found in systematic surveys. Microlensing, exhaustively simulated by theorists and used as a panacea for any statistical discomfort we might have felt, may actually have been detected in a location where it is hard to avoid. Two radio rings have been announced and more are confidently expected. The natural arcs, co-discovered by our gracious hosts, (and a far more appropriate celebration of French culture than the artificial ones recently proposed to commemorate the construction of the Eiffel Tower) are now known to be a common property of high redshift clusters. 1989 is also an anniversary as it is 70 years since the first relevant observational (Dyson, Eddington and Davidson 1920) and theoretical (Lodge 1919) papers on this subject and ten years since the discovery of the first *bona fide* gravitational lens (Walsh, Carswell and Weymann 1979).

This workshop has provided an opportunity to take stock of a decade's progress and contrast it with the hopes and expectations of the astronomical community at the time when 0957+561 appeared. It also gives us a chance to ignore the strong warning implicit in this comparison and to look ahead to the next decade. Looking back, gravitational lensing has had many prophets, many of them false. Even Einstein misunderstood some key issues. By popular acclaim, our most reliable prophet has been Sjur Refsdal who has, over a 25 year career in a series of individual and collaborative papers, laid much of the foundation for interpreting the observations. It is therefore fitting that the organizing committee chose to make this workshop a celebration of his contributions to the subject. I hope that some of the contributions to the proceedings will be as durable as his have proved to be.

Quasars

Gravitational lensing of quasars is rare and so most telescope time has been devoted to finding more examples of this phenomenon. It is therefore very important to devise observational strategies to optimise the yield. As was discussed by Surdej, Swings and Magain, one way to accomplish this is to restrict attention to high luminosity quasars.

That this should help a bit was clear as soon as the importance of "amplification bias" was emphasized by Turner, Ostriker and Gott(1984). However, at a limiting magnitude of 18.5^m, this should only increase the lens frequency by, at most, a factor three, if we use conventional quasar counts. As far as I can tell, this survey seems to have done better than this, and I wonder why? Is it possible that the faint quasar counts are much greater than reported and that some of the new examples are highly magnified?

In a similar vein, Webster, Magain and Kovner analysed reports of an over-density of galaxies around cosmologically distant quasars. Overdensities of a factor $\lesssim 2$ are reasonable and to be expected, but there are too few faint quasars to account for the factor 4.4 overdensity that was originally reported by Webster and her colleagues. Conservatives can take comfort in reports that the discrepancy is diminishing and, in particular, that the galaxy overdensity around X-ray selected quasars, previously reported by Stocke *et al.*(1987) is no longer believed to be significant (Rix and Hogan 1989, preprint). We all know that gravitational lensing is the only possible explanation for a true over-density of galaxies in front of quasars. (I trust yesterday's lecture on the fate of the Albigensians will concentrate the minds of any doubters.)

Burke reviewed radio surveys, emphasizing that when you discover a multiply-imaged quasar which is also a strong radio source, then you eventually get far more assistance in modelling. Unfortunately, as Heflin described, this has still not been enough to understand 2016+112. (However, correcting a misunderstanding about the position angle sign convention in one of Narasimha's models may yet allow a solution to this outstanding problem. Despite this, I suspect that, as we see two galaxies, two lens combine non-linearly.)

Many people have remarked that we have once again become too casual in making lists of gravitational lenses. For several reasons, I believe that observers should adopt some sort of code of conduct and always be prepared to describe lens candidates as either "probable" or "possible" and to be prepared to promote or demote from the latter category. Of course, assignment to these categories is a subjective matter, but I propose that, in the light of our experience with 1146+111 and PKS1145-071, spectroscopic similarity or statistical improbability alone never be sufficient for a lens candidate to be regarded as probable. I think that there should always be some additional circumstantial evidence such as detection of a lensing galaxy or cluster in the right location and of sufficient mass to create the image splitting. Alternatively, an image configuration characteristic of lensing like a cloverleaf or similar VLBI structure in at least two components, should be observed. A third possible criterion, discussed by Israel Kovner, is the observation of "arclet" images of background galaxies. Of the 14 multiple lens source candidates listed by Surdej, I believe that only 7 satisfy these more stringent criteria.

Nowhere are these considerations more important than in the dark matter lenses, notably 1635+267 and 2345+007. For the former, the case for lensing may have been strengthened by Turner *et al.*'s(1988) possible detection of a galaxy continuuum underlying the brighter component. (This may be a common occurrence, Magain.) However, as it does not seem credible that such a galaxy could provide the image splitting of 4" unassisted, I believe that it is equally reasonable to say that the two images are intrinsically different and therefore not lensed. Similar arguments apply to

2345+007 where the fainter image has been reported to comprise at least two subcomponents, (Nieto *et al.*1988). Here the image splitting is over 7", again too large for a single normal galaxy. It is precisely because unambiguous detection of isolated dark matter has such far-reaching cosmological implications that extreme scepticism is necessary. I argue that, until we have stronger evidence that these are true multiple images, we are still no more justified in describing these as probable gravitational lenses than we would be in assuming that two papers on lensing attributed to Schneider (or Burke) are written by the same author. Besides, as Kovner and Meylan reminded us, the discovery of close quasar pairs may be more interesting than finding another gravitational lens, as it allows us to study how quasars interact with each other, their host galaxies and the surrounding medium.

One topic that was not much discussed is the relationship between gravitational lenses and BL Lacs. It has long been clear that some BL Lacs, notably AO0235+164, have intervening galaxies which ought to produce multiple macroimages and don't. It is important to understand why. By contrast, Ostriker and collaborators have proposed that the nominally low redshift BL Lacs are in fact high redshift OVV quasars microlensed by low redshift elliptical galaxies. This still seems to me to be quite unlikely because BL Lacs have quite distinct properties like large variable polarisation, centimetre excess and so on which cannot easily be attributed to lensing. Nevertheless, there are some high redshift OVV quasars including AO0235+164 that are reported to be surrounded by nebulosity and these must be good microlensing candidates. This idea deserves closer scrutiny.

Arcs

To most of us, the highlights of this workshop have been the new observations of arcs. Fort listed six large arcs and five additional examples of "arclets". Four of the arcs have redshifts (Soucail, Pello), all so far consistent with the lens interpretation. This is pleasantly reassuring for those of us who have not seriously questioned the conventional interpretation of redshifts. However, I really wonder if it will be possible to understand the cluster mass distributions well enough to deduce a Friedmann-Robertson-Walker density parameter on account of the notorious insensitivity of the ratio of the lens-source angular diameter distance to the observer-source distance. It may however set useful limits on Eddington-Lemaître cosmologies.

Looking at this data from the point of view of cluster dynamics, it seems much easier to assume a simple cosmology and use the arcs to trace the mass distribution. Much modelling has been carried out and it seems that the disagreements are mostly at a level that can only be resolved by future observations. On the basis of their simulations, Tyson and colleagues have concluded that the cluster mass does appear to trace the galaxy light. However, the conclusion must be tentative at the moment.

An optical feature that presents a challenge to the modellers, is the oxymoronic "straight arc" discovered by Pello, (Mellier, Soucail). This is because arcs, formed when the source lies close to a cusp, are generically curved. One way in which the lens origin can be verified is if it appears that the feature has three colinear parts with the center part a mirror image of the outer two images. However, failure to detect this pattern will not rule out a lens origin, because it is possible to form singly imaged arcs at cusps.

Tyson presented strong evidence that there is a dense, blue background galaxy population with $1 \lesssim z \lesssim 3$. This came as a surprise to Ellis who has carried out redshift surveys to brighter limiting magnitude and does not observe such strong color evolution. To further confuse matters, Grossman reported that he could isolate the background galaxy population using redness in the filter combination $R-I$. As far as I can tell there are no *direct* contradictions so far in the observations and if this continues then we must surely have learnt something of great importance about galaxy formation. It must be of relevance that the co-moving density of Tyson's population exceeds that of local bright galaxies by nearly an order of magnitude. We may either be seeing dwarf galaxies in early stages of rapid star formation, or we be detecting proto-galactic fragments prior to their agglomeration to make galaxies (*e.g.*Barron and White 1987).

The question of what constitutes a "probable" case of a gravitational lens arc has also come up and I would propose either the measurement of suitable lens and arc redshifts or successful modelling of detailed image structure are necessary. Apparently eight of Fort's list of eleven arcs and both radio rings pass this test.

Cluster arcs do not seem to be rare although we may have to wait a few years before Fort can quantify this. Before the discovery of arcs, most extragalactic astronomers seemed to believe that the vast majority of rich clusters would have sub-critical surface densities and were sceptical that they would form multiple images. It is now very important to decide if we are just looking at the high surface density tail of a slowly evolving distribution, or if we have been wrong all along about the core radii of local clusters? Note that, as Fort has emphasized, it is the comparatively low redshift clusters like A1699, A545 and A2218 that have yielded so many of the observed arclets and which must have very high central surface densities.

Rings

The discovery of two radio rings by Hewett, Langston and collaborators was particuarly happy. As Kochanek explained, the optics is sufficiently simple to justify quite elaborate inversion procedures that have predictive power and now await more detailed optical scrutiny to see if the lens is as required by theory. I am personally relieved to learn from Tyson that the light seen in 1131+0456 and which appeared to be elongated along a position angle perpendicular to the predicted direction, probably derives from the source rather than the lens.

Tyson and Fort have reported the discovery of optical rings around cluster galaxies. Although these are very naturally interpreted in terms of the magnification of an Einstein ring by a cluster mass distribution to a size larger than that of the lens galaxy, the interpretation can only be designated as "possible" because we know of at least two other mechanisms for producing rings, namely those responsible for shell galaxies and for polar ring galaxies. Nevertheless, assuming that Tyson's features really are Einstein rings, magnified by the surrounding cluster, then they promise to be very powerful probes of dark matter. In particular, the presence of complete rings can be used to place a limit on the ratio of shear to convergence and so on both the overall mass distribution and the masses of the underluminous objects. (They must not be much heavier than galaxies.)

Theory

To a theorist, the biggest surprise in the development of this subject has been how fascinating the physics has become and how slow many of us have been to appreciate the richness of classical geometrical optics, a subject with which astronomers should be intimately familiar. Of course, many of the tools have been there all along as Nityananda described in his pedagogically sweet overview of the many complementary approaches to the problems of gravitational lens optics. As he clearly demonstrated, there is no preferred formalism for handling lenses. Ray-tracing, wavefront evolution, scalar, vector and curvature formalisms, catastrophe optics, and optical scalar equations can all be called upon to illuminate specific aspects of the problem. That our language for describing gravitational lenses increasingly resembles that once used to design large refracting telescopes is one way in which Zwicky's (1937) dream has come true.

These general principles were applied to more specific instances of lensing by subsequent speakers, notably Kovner, Nottale, Schneider, Wambganss and Witt and it is particularly gratifying to see how they have been combined with increasingly sophisticated numerical algorithms and incorporated into specific modeling procedures by Deguchi, Grossman, Hammer, Kochanek, Mellier, Narasimha, and Petrosian. Despite all this ingenuity, it seems that many theoretical investigations remain computer-limited.

I ought, in fairness, to propose a code of conduct for theorists and have two suggestions. Firstly, there now seems little point in publishing models of specific lenses just to demonstrate that a model is possible. We now have enough experience of the art of modeling to require serious discussion of the volume of solution space permitted by the observations and suggestions as to how this volume may be shrunk by future observations. Secondly, it is all too easy to carry out numerical simulations of gravitational lensing. What is harder but of more lasting value is to extract the crucial physical principles that govern these simulations. Perhaps, theorists should strive to understand the optics so well, that simulations eventually become redundant.

Future Prospects

One way that gravitational lenses have not fulfilled their promise is that they have not yet been very useful to other astronomers. We thought they would measure the Hubble constant. Instead, even in the best case of 0957+561, we have learnt that taking the data is hard, painstaking work, (Schild, Vanderreist), and yet still

may not satisfy your immediate colleagues (Falco). Modelling the potential well is uncertain because we have no guarantee that mass traces the light–a large sheet of matter is indistinguishable from a change in the Hubble constant, microlensing can change the intensity ratio (Schneider, Vanderriest) and there remains sensitivity to the assumed cosmological mass distribution (Nottale). Applications for telecope time often include claims like "photons accelerated in quasars will map out the potential wells of galaxies and clusters in much the same way as beams of relativistic electrons are used to measure the depths and effective ranges of nuclear potential wells". Despite this rhetoric, we have not yet learnt anything reliable about galaxy or cluster potential wells from gravitational lenses. Another "unfulfilled campaign promise", made first by Zwicky, is that gravitational lenses will act as giant telescopes to "super-resolve" the cores of quasars. This hasn't happened either.

My reason for bringing this up is not to discourage but to encourage. This is because the analogy with radio pulsars might be helpful. Although pulsars were discovered in 1967, it took a good decade before they started to fulfil their promise as astronomical tools. Most pulsar research in the first decade was devoted to trying to figure out how they worked. It is only in recent years that the focus has turned to using them as testers of relativity, accurate clocks, sources for scintillation and magnetoionic experiments, probes of globular cluster dynamics and fossil records of stellar evolution. Perhaps the same will turn out to be true for lenses. We may be on the threshold of finding the few good lenses that are sufficiently isolated and overspecified as to allow non-controversial determinations of H_0 (still of course subject to the above caveats). The are other optimistic signs. The great redundancy in models of cluster rings and arcs is starting to give direct quantitative measurements of surface densitites and velocity dispersions that are more accurate than direct determinations. If the tantalising measurements of image variability in 2237+0305 (Corrigan, Yee) hold up, then the source is unreasonably well-suited to testing the theoretical predictions of accretion disk models, especially their spectra (Grieger). (I must mention that current models of quasars associate the 1-2μ source with the outer parts of accretion disks (Sanders *et al.*1989) and these are much larger than the visual continuum source sizes. Yee's tentative suggestion that he might be seeing this, may one day turn into a serious test of quasar models.)

Future prospects in detecting more lens candidates seem good. Swings, Burke and Webster seem to understand their selection criteria sufficiently well that they are able to predict future lens yields. I believe that this is justified, but would caution against making this research too programmatic because there may well be totally new classes of lensing phenomena out there waiting to be discovered. After all multiple-imaged quasars and arcs were both found by surprise. Perhaps cosmic strings really do exist and will be found in a similar manner.

We look forward to changes in the status of possible lenses like 1635+267, 2345+007 and 1120+019 particularly if lensing galaxies can be confirmed on one of the images and the magnification ratios understood. Surdej's equally tantalising claim (reported by Magain) that many high luminosity quasars are superposed on intervening galaxies will surely be studied by others. Much hope hangs on HST for providing superior resolution of existing lenses and perhaps for studying galaxy-quasar superpositions. The notion that lensing biases the statistics of quasar absorption lines (suggested originally by Nottale and discussed further here by Thomas and Angenin) seems bound to be true at the level at which these studies are beginning to be carried out.

Despite all this, it is the arcs that hold most promise. Large samples of over a hundred rich clusters at optimal redshifts $0.3 \lesssim z \lesssim 1$ have been defined and redshifts are being taken using multi-slit and fibre optic spectroscopy. Extremely deep images will undoubtedly be taken using giant CCD's and current indications are that many will be super critical. ROSAT X-ray observations will complement these studies (Kellogg) and should be useful, in conjunction with lens models, for understanding the thermal evolution of the intergalactic gas.

From the theorists, we now have a good qualitative understanding of lensing and quantitative formalisms adequate for modelling individual macrolenses and discussing statistical distributions (Kovner, Nemiroff). To my mind, the current challenges are to derive convenient ways to handle compound lenses as may be required in 2016+112, to interpret more physically the caustic networks created by the modest stellar optical depths in the centers of galaxies paying particular attention to correlations in stars and to make the formalisms for handling inhomogeneous cosmology more user-friendly and capable of handling highly irregular structures. (Schneider reported on some developments in the last two of these.)

However, for audacity and vision I must finish as I began with Sjur Refsdal. That we be able to use Cassini or some similar spacecraft to measure spatial as well as temporal gradients in microlensed quasar flux is an imaginative proposal. Success in this endeavour would tell us much about the proper motions of distant objects. No one who has attended this meeting would want to say "It won't happen".

Acknowledgements

I am sure I speak for many when I pay tribute to our gracious hosts, especially Bernard Fort who initiated the planning of the meeting, Genvieve Soucail who carried the burden of the local organisation and Yannick Mellier who will be responsible for editing the proceedings. That this workshop seems to have finished far too soon is a measure of their success. Personally I acknowledge invaluable discussion with participants too numerous to mention, helpful comments and the first draft of this discussion by Bernard Fort, Chris Kochanek, Israel Kovner, Peter Schneider and Jean Surdej and support under NSF grant AST86-15325.

References

Barron, E. J. and White, S. D. M., 1987. *Astrophys. J.*, **322**, 585

Dyson, F. W., Eddington, A. S. and Davidson, C., 1920. *Phil Trans. Roy. Soc.*, **220A**, 291

Lodge, O. J., 1919. *Nature*, **104**, 354

Nieto, J. L., Roques, S., Llebaria, A., Vanderiest, Ch., Lelievre, G., di Serego, Alighieri, S., Macchetto, F. D. and Perryman, M. A. C., 1988. *Astrophys. J.*, **325**, 644

Sanders, D. B., Phinney, E. S., Neugebauer, G., Soifer, B. T. and Matthews, K., 1989. *Astrophys. J.*, , (in press)

Stocke, J. T., Schneider, P., Morris, S. L., Gioia, I. M., Maccacaro, T. and Schild, R. E., 1987. *Astrophys. J. Lett.*, **315**, L11

Turner, E. L., Hillenbrand, L. A., Schneider, D. P., Hewitt, J. N. and Burke, B. F., 1988. *Astronom. J.*, **96**, 1682

Turner, E. L., Ostriker, J. P. and Gott, J. R., 1984. *Astrophys. J.*, **284**, 1

Walsh, D., Carswell, R. F. and Weymann, R. J., 1979. *Nature*, **279**, 381

Zwicky, F., 1937. *Phys. Rev. Lett.*, **51**, 290,697

Annexes

Gravitational lens candidatos

Source name (QSO/AGN)	Number of images	Image brightnesses/ Abs. Mag.	separation/ time delay	zs	Lens	Lens brightness	zl	Ml(MO)/ (rad.)/ Mic.Lens	Detection (O/R)	Reference
1. Presently accepted cases of multiply lensed source images										
0957+561 A-B (QSO)	2	B=17.5 (A) 17.7 (B) M=-29.2	6.1"(A-B) (1.14 yrs)	1.41	G/C	R=18.5	0.36	8.7(11) (3.1") ML	O/R	Walsh et al., 1979
1115+080 A1/2-C (QSO)	4 (5?)	B=17.2 (A1) 17.2 (A2) 18.7 (B) 18.2 (C) M=-30.1	0.5"(A1-2) 1.8"(A-B) 2.3"(A-C) (dys,mnths)	1.72	G	R=19.8	?	? ML	O	Weymann et al., 1980
2016+112 A-C' (QSO)	3	i=22.9 (A) 23.2 (B) 24 (C') M=-25.6	3.4"(A-B) 3.8"(A-C') (0.5-1 yrs)	3.27	2Gs	i=21.9	1.01 ?	ML?	R/O	Lawrence et al., 1984
2237+030 A-D (QSO)	4	r=17.6 (A) 17.8 (B) 18.1 (C) 18.4 (D) M=-29.2	1.8"(A/B) 1.4"(A/C) 1.7"(A/D) (dys)	1.69	G	r=14.5	0.04	1.0(10) (0.9") ML	O	Huchra et al., 1985
0142-100 A-B (QSO)	2	B=17.0 (A) 19.1 (B) M=-30.2	2.2"(A-B) (7 weeks)	2.72	G	R=19	0.49	2.4(11) (2.2")	O	Surdej et al., 1987
3C324 A-B (AGN)	>2	R=22.7 (A) 23.3 (B)	2" (A-B)	1.206	G		0.84	1.0(12)	R/O	Le Fèvre et al., 1987
1413+117 A-D (QSO)	4	18.3 (A) 18.5 (B) 18.6 (C) 18.7 (D) M=-29.8	0.8"(A/B) 0.9"(A/C) 1.1"(A/D) (1 mnth)	2.55	G?	?	1.4? 1.7?	5.0(11) ML?	O/R	Magain et al., 1988
0414+053 A-D (AGN?)	4	R=23 (A/D)	3" (A/D)	1.2?			1.0?		R/O	Hewitt et al., 1989
2. Additional suspected cases of multiply lensed source images										
2345+007 A-B1/2 (QSO)	3	B=19.5 (A) 22.0 (B1) 22.0 (B2) M=-27.3	7.3"(A-B) 0.4"(B1-2)	2.15	?	J>25.5		ML?	O	Weedman et al., 1982
1635+267 A-B (QSO)	2	19.2 (A) 20.8 (B) M=-28.1	3.8"(A-B)	1.96	G?	R=20	0.57?	1.0(12)?	O	Djorgovski and Spinrad, 1984
0023+171 AB-C (AGN)	2	r=22.8 (AB) 23.4 (C)	4.8"(AB-C)	0.95					R/O	Hewitt et al., 1987b
1429-008 A-B (QSO)	2	R=17.7 (A) 20.8 (B)	5.1"(A-B)	2.08	?		1.6?		O	Hewett et al., 1989
1042+178 A-D (AGN)	4	21 (A/D)	1.6"(A-B)	0.92	?		0.5?		R/O	Hewitt et al., 1987a
1120+019 A-B (QSO)	2	V=16.2 (A) 20.8 (B) M=-29.1	6.5"(A-B)	1.46	G?/C?	R=22.5	0.6?		O	Meylan and Djorgovski, 1989
3. Radio rings										
1131+045 A-B	2+ ring	radio lobe	2.1" (ring diam.)	?	?	R=22			R/O?	Hewitt et al., 1988
1654+134	ring	radio lobe	2.1" (ring diam.)	1.74	G	r=18.7	0.25	9.5(10) (2.1")	R/O?	Langston et al., 1989
4. Other proposed cases of multiply lensed and/or magnified source images										
3C194, 3C225A										Le Fèvre and Hammer 1988
3C238, 3C241, 3C305.1										Le Fèvre, Hammer and Jones, 1988b
3C13, 3C256										Le Fèvre et al., 1988a
3C208.1										Le Fèvre and Hammer, 1989
5. Multiply lensed QSO images with large separations (?)										
M82 A-C (QSO)	3	V=20.5 (A) 20 (B) 20.5 (C)	216"(A/C)	2.04			?			Burbidge et al., 1980
6. Twin galaxies imaged by string loops (?)										
0249-186 A-D (AGN)	4X2		2.5"(A-A') 2.6"(B-B') 2.0"(C-C') 2.6"(D-D')	0.43 0.43 0.20 0.4	string?				O	Cowie and Hu, 1987
7. Lensed gamma-ray bursters (?)										
B1900+14 A-C (AGN?)	3		? (hrs, dys)		G?			1.0(10)		Paczynski, 1986

Explanations: G: Galaxy   ML: Micro-lensing   O: optical detection
C: Cluster   rad.: observed angular radius for the estimated mass of the lens   R: radio detection

See Surdej (1989, these proceedings) for the exact references.

Identification fields and optical spectra of gravitational lens candidates

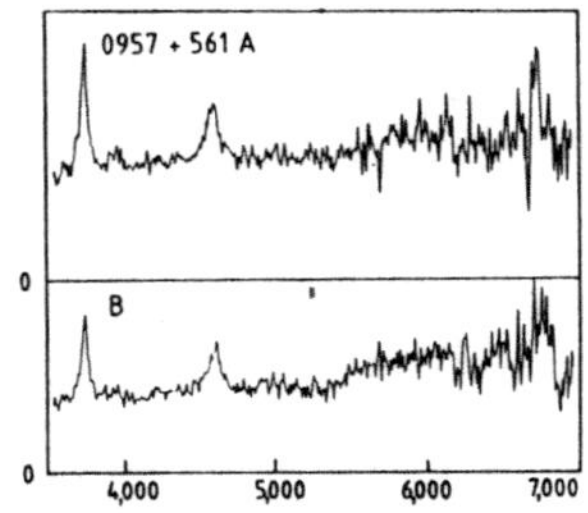

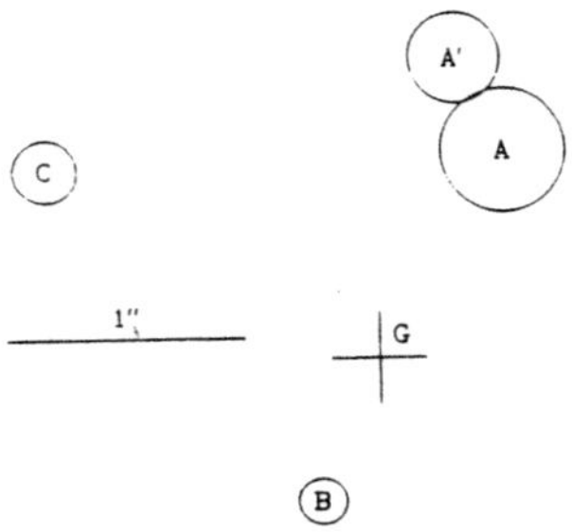

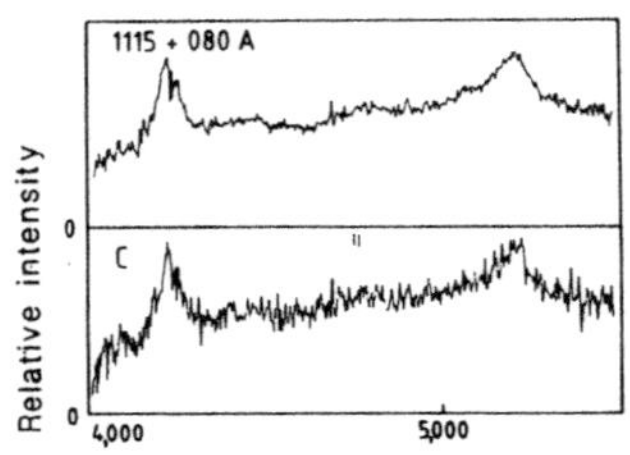

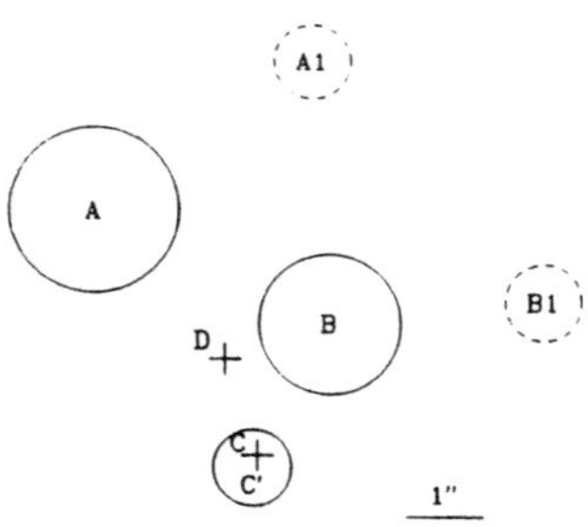

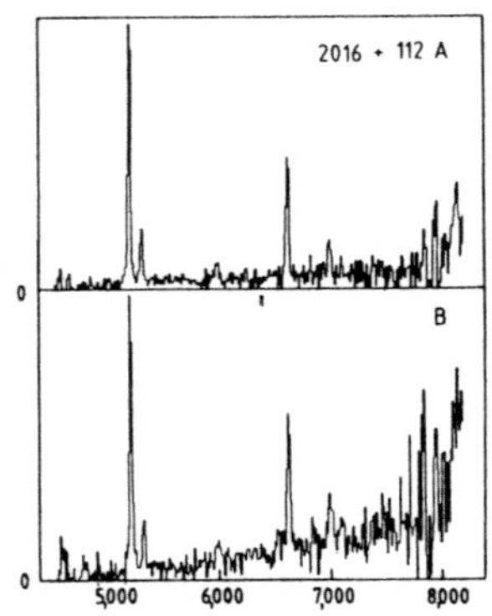

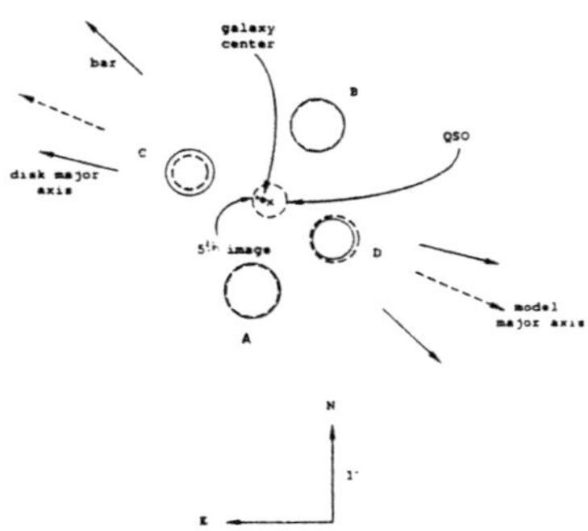

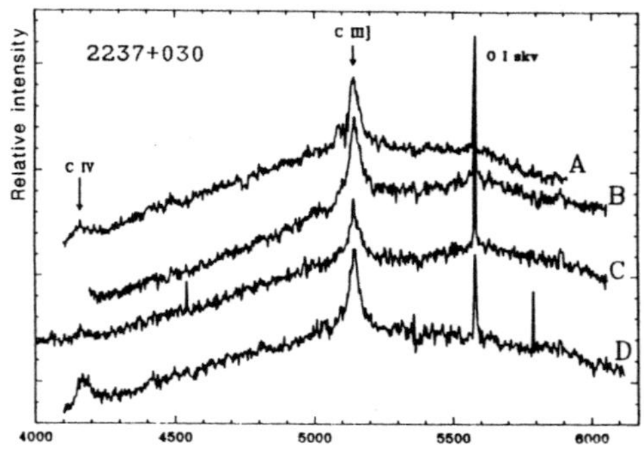

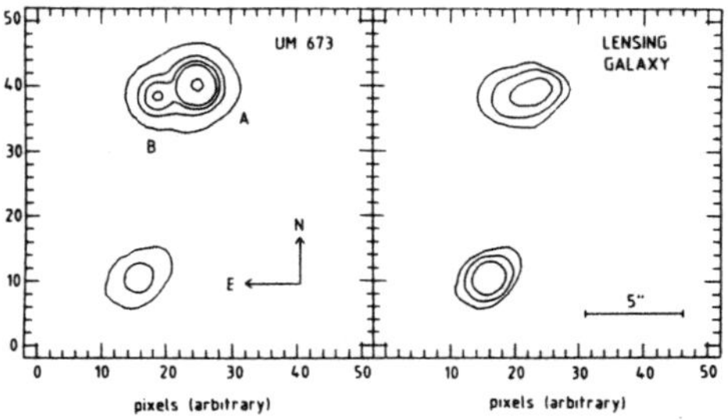
UM 673
LENSING GALAXY
A
B
N
E
5"
pixels (arbitrary)
pixels (arbitrary)

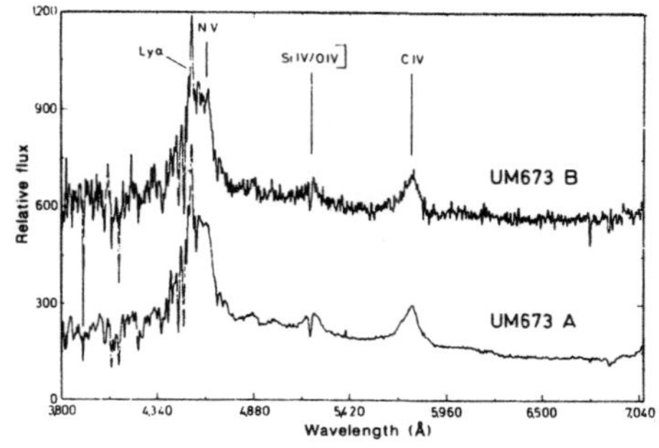
Lyα
N V
Si IV/O IV
C IV
UM673 B
UM673 A
Relative flux
Wavelength (Å)

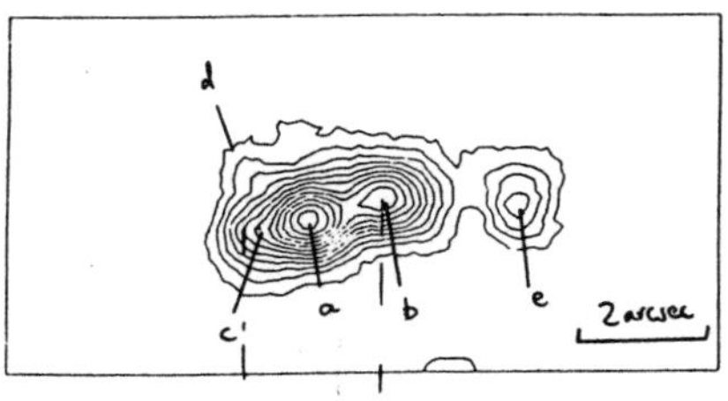
d
c'
a
b
e
2 arcsec

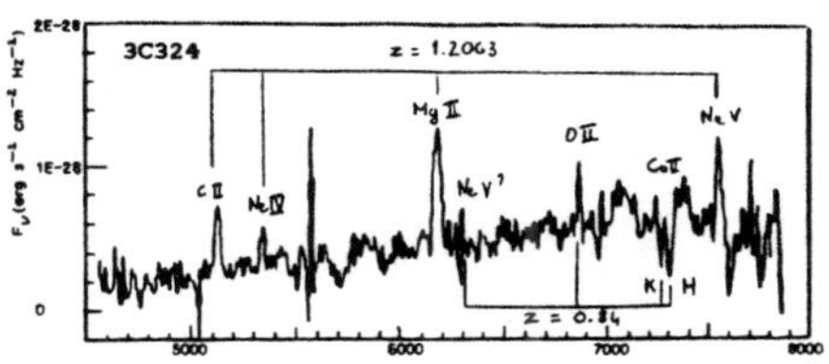
3C324
z = 1.2063
C II
Ne IV
Mg II
O II
Ne V
K
H
z = 0.84

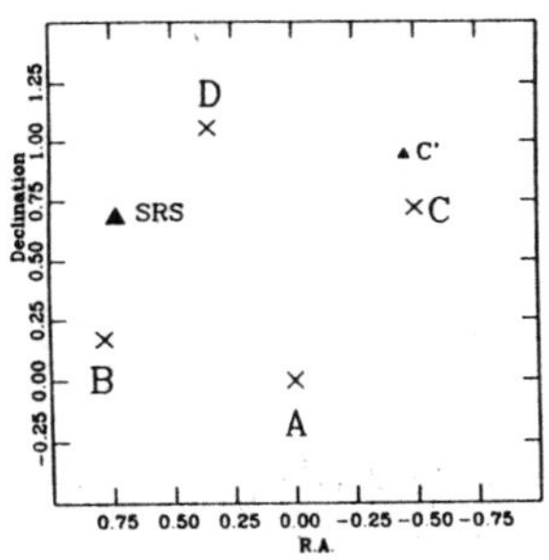
D
C'
C
SRS
B
A
Declination
R.A.

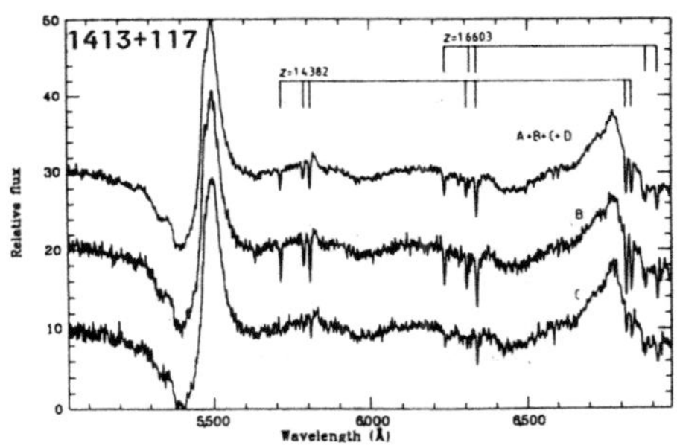
1413+117
z=1.6603
z=1.4382
A+B+C+D
B
C
Relative flux
Wavelength (Å)

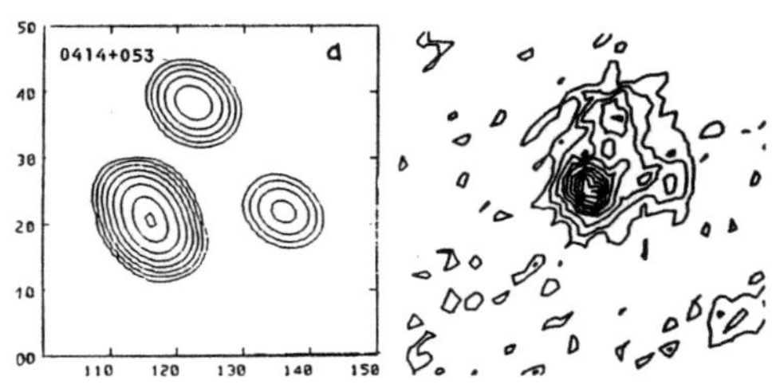
0414+053
a

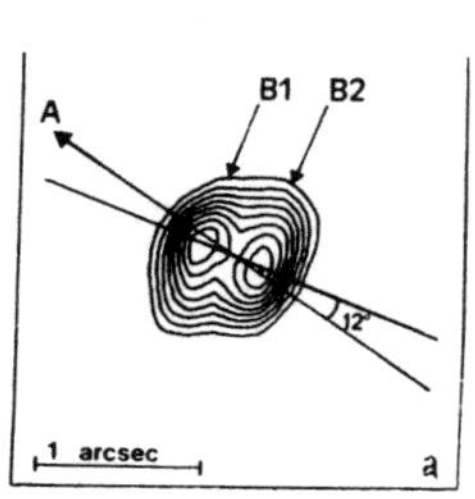
A
B1
B2
12°
1 arcsec
a

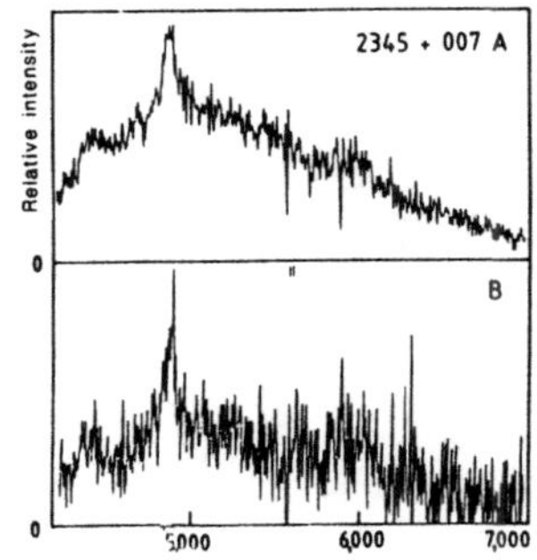
2345 + 007 A
B
Relative intensity

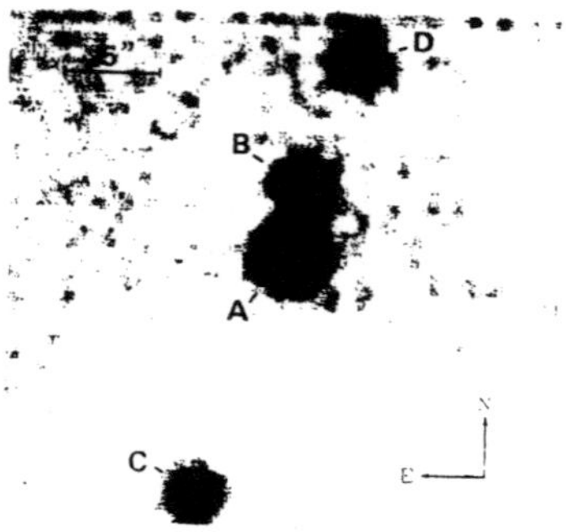
D
B
A
C
N
E

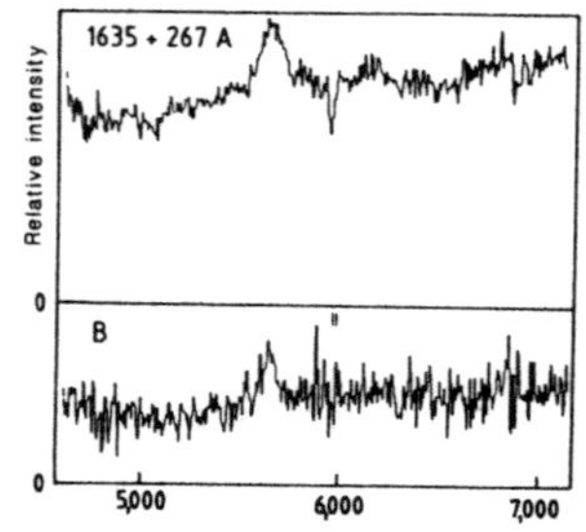
1635 + 267 A
Relative intensity
B
0
5,000
6,000
7,000

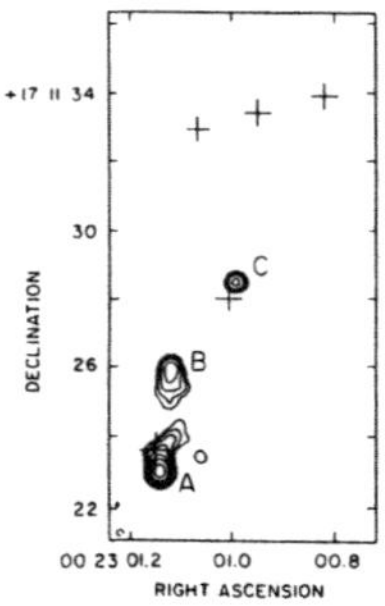
+17 11 34
30
26
22
DECLINATION
C
B
A
00 23 01.2
01.0
00.8
RIGHT ASCENSION

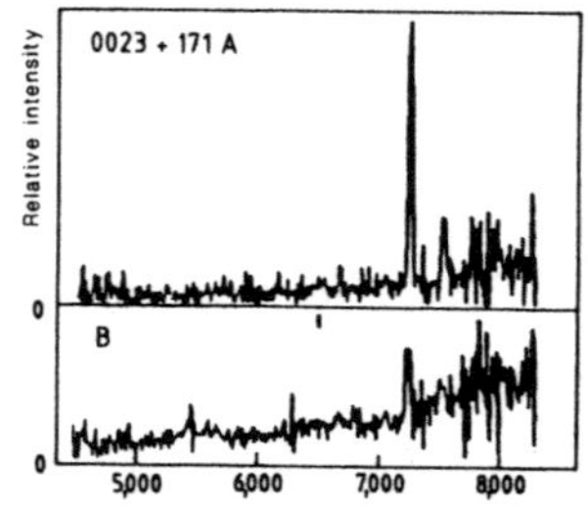
0023 + 171 A
Relative intensity
B
0
5,000
6,000
7,000
8,000

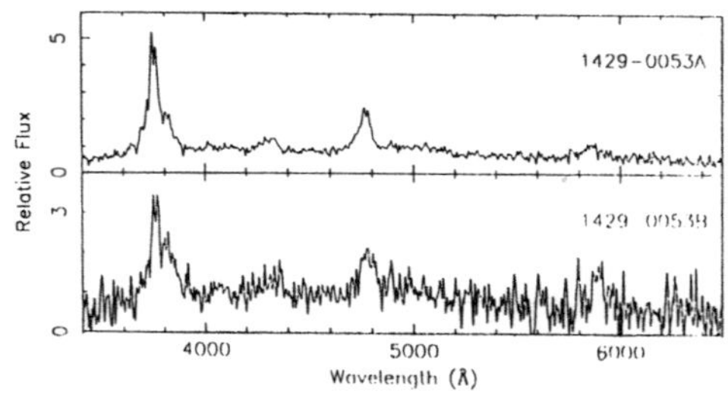
1429−0053A
Relative Flux
5
0
3
4000
5000
6000
Wavelength (Å)

1042+178

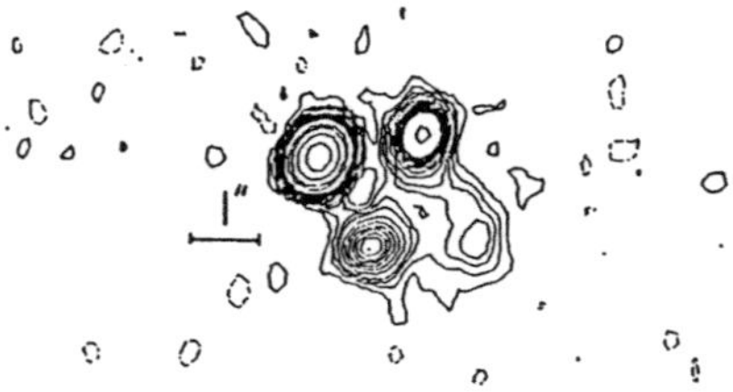
1"

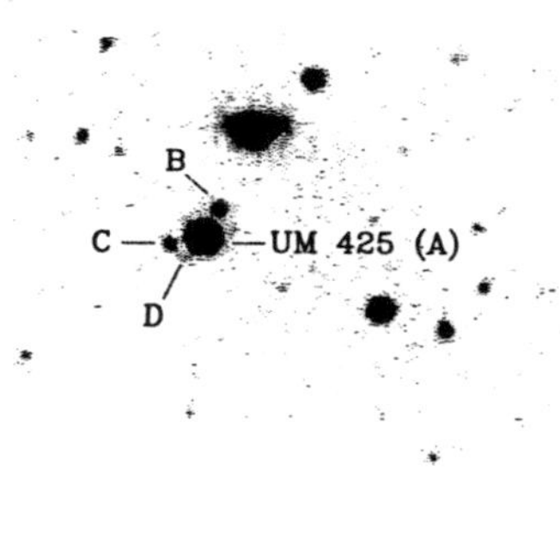
B
C
UM 425 (A)
D

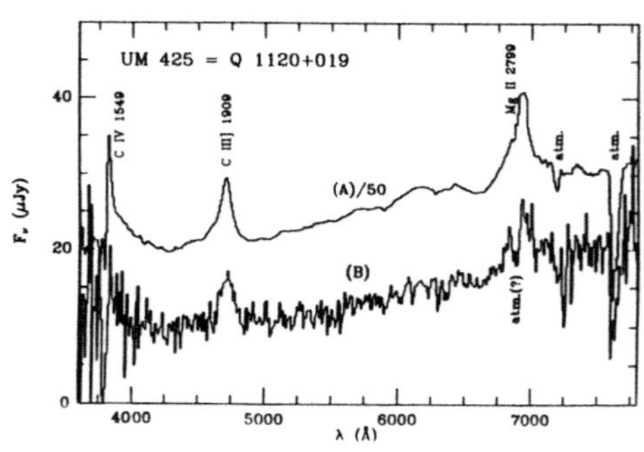
UM 425 = Q 1120+019
C IV 1549
C III] 1909
Mg II 2799
(A)/50
(B)
atm.
atm.(?)
F_ν (μJy)
40
20
0
4000
5000
6000
7000
λ (Å)

Gravitational lens candidates : arcs in clusters

Cluster	Z	C	σ km/s	M/R Virial	M/R Lens	Lx	Comments & references
A 370	0.374	40	1350/1700	80/100	90 ± 30	9.710^{44}	Z large arc = 0.724 (Soucail et al 88) mini arcs (Fort et al 88) Lynds & Petrosian 88
Cl 2244	0.328	–	–	–	70 ± 20	–	Lynds & Petrosian 88 Hammer et al 89 Z Large arc=2.23 Soucail et al 89
A 2218	0.171	218	1400	90	–	6.510^{44}	Mini arc Pello 87 ring around a galaxy
A 2390	0.23	72	1600	100 ± 30	–	9.910^{44}	Mellier et al 89 Z arc=0.92 Pello et al 89
A 963	0.206	134	1100	60	60	–	2 Arc around the CD Lavery & Henry 88
Cl 0500-24	0.321	200	1300	80 ± 20	100 ± 30		Giraud, Schneider et al 89
A 1689	0.181	228	1800 km/s				50/70 Arclets Tyson 89
3C 295	0.461		2800 km/s 3×1000 km/s			9.210^{44}	50/70 distorded images 10/15 Arclets Tyson 89
A 545	0.154	234					one large arc. Tyson 89
A 1525	0.181	186					one possible Arclet. Giraud 88 to be confirmed
Cl 0024+16	0.391	–	1300	–	–	–	Koo 1988

Lecture Notes in Mathematics

Vol. 1236: Stochastic Partial Differential Equations and Applications. Proceedings, 1985. Edited by G. Da Prato and L. Tubaro. V, 257 pages. 1987.

Vol. 1237: Rational Approximation and its Applications in Mathematics and Physics. Proceedings, 1985. Edited by J. Gilewicz, M. Pindor and W. Siemaszko. XII, 350 pages. 1987.

Vol. 1250: Stochastic Processes – Mathematics and Physics II. Proceedings 1985. Edited by S. Albeverio, Ph. Blanchard and L. Streit. VI, 359 pages. 1987.

Vol. 1251: Differential Geometric Methods in Mathematical Physics. Proceedings, 1985. Edited by P.L. García and A. Pérez-Rendón. VII, 300 pages. 1987.

Vol. 1255: Differential Geometry and Differential Equations. Proceedings, 1985. Edited by C. Gu, M. Berger and R.L. Bryant. XII, 243 pages. 1987.

Vol. 1256: Pseudo-Differential Operators. Proceedings, 1986. Edited by H.O. Cordes, B. Gramsch and H. Widom. X, 479 pages. 1987.

Vol. 1258: J. Weidmann, Spectral Theory of Ordinary Differential Operators. VI, 303 pages. 1987.

Vol. 1260: N.H. Pavel, Nonlinear Evolution Operators and Semigroups. VI, 285 pages. 1987.

Vol. 1263: V.L. Hansen (Ed.), Differential Geometry. Proceedings, 1985. XI, 288 pages. 1987.

Vol. 1265: W. Van Assche, Asymptotics for Orthogonal Polynomials. VI, 201 pages. 1987.

Vol. 1267: J. Lindenstrauss, V.D. Milman (Eds.), Geometrical Aspects of Functional Analysis. Seminar. VII, 212 pages. 1987.

Vol. 1269: M. Shiota, Nash Manifolds. VI, 223 pages. 1987.

Vol. 1270: C. Carasso, P.-A. Raviart, D. Serre (Eds.), Nonlinear Hyperbolic Problems. Proceedings, 1986. XV, 341 pages. 1987.

Vol. 1272: M.S. Livšic, L.L. Waksman, Commuting Nonselfadjoint Operators in Hilbert Space. III, 115 pages. 1987.

Vol. 1273: G.-M. Greuel, G. Trautmann (Eds.), Singularities, Representation of Algebras, and Vector Bundles. Proceedings, 1985. XIV, 383 pages. 1987.

Vol. 1275: C.A. Berenstein (Ed.), Complex Analysis I. Proceedings, 1985–86. XV, 331 pages. 1987.

Vol. 1276: C.A. Berenstein (Ed.), Complex Analysis II. Proceedings, 1985–86. IX, 320 pages. 1987.

Vol. 1277: C.A. Berenstein (Ed.), Complex Analysis III. Proceedings, 1985–86. X, 350 pages. 1987.

Vol. 1283: S. Mardešić, J. Segal (Eds.), Geometric Topology and Shape Theory. Proceedings, 1986. V, 261 pages. 1987.

Vol. 1285: I.W. Knowles, Y. Saitō (Eds.), Differential Equations and Mathematical Physics. Proceedings, 1986. XVI, 499 pages. 1987.

Vol. 1287: E.B. Saff (Ed.), Approximation Theory, Tampa. Proceedings, 1985–1986. V, 228 pages. 1987.

Vol. 1288: Yu. L. Rodin, Generalized Analytic Functions on Riemann Surfaces. V, 128 pages, 1987.

Vol. 1294: M. Queffélec, Substitution Dynamical Systems – Spectral Analysis. XIII, 240 pages. 1987.

Vol. 1299: S. Watanabe, Yu.V. Prokhorov (Eds.), Probability Theory and Mathematical Statistics. Proceedings, 1986. VIII, 589 pages. 1988.

Vol. 1300: G.B. Seligman, Constructions of Lie Algebras and their Modules. VI, 190 pages. 1988.

Vol. 1302: M. Cwikel, J. Peetre, Y. Sagher, H. Wallin (Eds.), Function Spaces and Applications. Proceedings, 1986. VI, 445 pages. 1988.

Vol. 1303: L. Accardi, W. von Waldenfels (Eds.), Quantum Probability and Applications III. Proceedings, 1987. VI, 373 pages. 1988.

Lecture Notes in Physics

Vol. 336: M. Müller, Consistent Classical Supergravity Theories. VI, 125 pages. 1989.

Vol. 337: A.P. Maclin, T.L. Gill, W.W. Zachary (Eds.), Magnetic Phenomena. Proceedings, 1988. VI, 142 pages. 1989.

Vol. 338: V. Privman, N.M. Švrakić, Directed Models of Polymers, Interfaces, and Clusters: Scaling and Finite-Size Properties. VI, 120 pages. 1989.

Vol. 339: F. Ehlotzky (Ed.), Fundamentals of Laser Interactions II. Proceedings, 1989. XI, 317 pages. 1989.

Vol. 340: M. Peshkin, A. Tonomura, The Aharonov-Bohm Effect. VI, 152 pages. 1989.

Vol. 341: E.F. Milone (Ed.), Infrared Extinction and Standardization. Proceedings, 1988. III, 79 pages. 1989.

Vol. 342: M. Balabane, P. Lochak, C. Sulem (Eds.), Integrable Systems and Applications. Proceedings, 1988. VII, 342 pages. 1989.

Vol. 343: M. Month, S. Turner (Eds.), Frontiers of Particle Beams; Observation, Diagnosis and Correction. Proceedings, 1988. IX, 509 pages. 1989.

Vol. 344: M. Rascle, D. Serre, M. Slemrod (Eds.), PDEs and Continuum Models of Phase Transitions. Proceedings, 1988. VII, 229 pages. 1989.

Vol. 345: H. Holden, A. Jensen (Eds.), Schrödinger Operators. Proceedings, 1988. V, 458 pages. 1989.

Vol. 346: D. Lüst, S. Theisen, Lectures on String Theory. VII, 346 pages. 1989.

Vol. 347: Ph. Blanchard, J.-P. Dias, J. Stubbe (Eds.), New Methods and Results in Non-linear Field Equations. Proceedings. 1987. VII, 133 pages. 1989.

Vol. 348: A. Bohm, M. Gadella, Dirac Kets, Gamow Vectors and Gel'fand Triplets. VII, 119 pages. 1989.

Vol. 349: Y.A. Kubyshin, J.M. Mourão, G. Rudolph, I.P. Volobujev, Dimensional Reduction of Gauge Theories, Spontaneous Compactification and Model Building. X, 80 pages. 1989.

Vol. 350: G. Tenorio-Tagle, M. Moles, J. Melnick (Eds.), Structure and Dynamics of the Interstellar Medium. Proceedings, 1989. XXI, 537 pages. 1989.

Vol. 351: A. Dervieux, B. Larrouturou (Eds.), Numerical Combustion. Proceedings, 1989. VII, 481 pages. 1989.

Vol. 352: K.B. Wolf (Ed.), Lie Methods in Optics II. Proceedings, 1988. XII, 197 pages. 1989.

Vol. 353: M. Barthes, J. Léon (Eds.), Nonlinear Coherent Structures. Proceedings, 1989. X, 280 pages. 1990.

Vol. 354: J. De Coninck, F. Dunlop (Eds.), Wetting Phenomena. Proceedings, 1988. VII, 112 pages. 1990.

Vol. 355: R. Lima, L. Streit, R. Vilela Mendes (Eds.), Dynamics and Stochastic Processes. Proceedings, 1988. V, 298 pages. 1990.

Vol. 356: R. Wehrse (Ed.), Accuracy of Element Abundances from Stellar Atmospheres. Proceedings, 1988. V, 109 pages. 1990.

Vol. 357: J.L. Lumley (Ed.), Whither Turbulence? Turbulence at the Crossroads. Proceedings, 1989. V, 535 pages. 1990.

Vol. 358: J. Kovalevsky, Astrométrie moderne. VI, 294 pages. 1990.

Vol. 359: K. Kirchgässner (Ed.), Problems Involving Change of Type. Proceedings, 1988. XII, 207 pages. 1990.

Vol. 360: Y. Mellier, B. Fort, G. Soucail (Eds.), Gravitational Lensing. Proceedings 1989. XV, 315 pages. 1990.

GPSR Compliance

The European Union's (EU) General Product Safety Regulation (GPSR) is a set of rules that requires consumer products to be safe and our obligations to ensure this.

If you have any concerns about our products, you can contact us on ProductSafety@springernature.com

In case Publisher is established outside the EU, the EU authorized representative is:

Springer Nature Customer Service Center GmbH
Europaplatz 3
69115 Heidelberg, Germany

Batch number: 10620854

Printed by Printforce, the Netherlands